NMR Spectroscopy in Food Analysis

RSC Food Analysis Monographs

Series Editor:
P.S. Belton, *School of Chemical Sciences, University of East Anglia, Norwich, UK*

Titles in this Series:
 1: Dietary Fibre Analysis
 2: Chromatography and Capillary Electrophoresis in Food Analysis
 3: Quality in the Food Analysis Laboratory
 4: Mass Spectrometry of Natural Substances in Food
 5: The Maillard Reaction
 6: Extraction of Organic Analytes from Foods: A Manual of Methods
 7: Trace Element Analysis of Food and Diet
 8: Analyses for Hormonal Substances in Food Producing Animals
 9: Mass Spectrometry and Nutrition Research
10: NMR Spectroscopy in Food Analysis

How to obtain future titles on publication:
A standing order plan is available for this series. A standing order will bring delivery of each new volume immediately on publication.

For further information please contact:
Book Sales Department, Royal Society of Chemistry, Thomas Graham House, Science Park, Milton Road, Cambridge, CB4 0WF, UK
Telephone: +44 (0)1223 420066, Fax: +44 (0)1223 420247,
Email: booksales@rsc.org
Visit our website at http://www.rsc.org/Shop/Books/

NMR Spectroscopy in Food Analysis

Apostolos Spyros and Photis Dais
Chemistry Department, University of Crete, Heraklion, Crete, Greece
Email: *aspyros@chemistry.uoc.gr; dais@chemistry.uoc.gr*

RSC Publishing

RSC Food Analysis Monographs No. 10

ISBN: 978-1-84973-175-1
ISSN: 1757-7098

A catalogue record for this book is available from the British Library

Published by The Royal Society of Chemistry,
Thomas Graham House, Science Park, Milton Road,
Cambridge CB4 0WF, UK

Registered Charity Number 207890

For further information see our web site at www.rsc.org

Printed in the United Kingdom by CPI Group (UK) Ltd, Croydon, CR0 4YY, UK

Preface

During the last few decades, NMR spectroscopy has found increased use in the analysis of food materials at an academic, industrial and government regulation level. The first NMR experiments aimed at studying food were probably the product of the curiosity of NMR spectroscopists (it was easier to find food in close proximity to NMR spectroscopists than NMR spectrometers close to food scientists!). Presently, the full exploitation of NMR spectroscopy in food analysis requires the involvement of food scientists with a deep understanding of food materials, in designing and implementing NMR experiments directed to specific analytical problems. The need for a book that could serve as an introductory tool in NMR spectroscopy for food scientists is thus clear and present. We hope we will succeed, at least partially, in covering this void with our present contribution.

The organization of the material is such that in the first few chapters, theoretical aspects and instrumentation are discussed, followed by sample preparation and various experimental aspects. After briefly describing chemometrics, several chapters are devoted to NMR applications in specific food categories. In each chapter, relatively simple NMR methodologies are discussed first, followed by more advanced techniques. The presentation of the chapter material is organized so that the reader can obtain easily an "NMR overview" of a specific food category. Ample use of figures (especially NMR spectra/images) has been made, in an effort to make the material more appealing and easier to digest. The coverage of NMR applications in different food categories is quite extensive, although not complete, as this would extend beyond the scope of any introductory book.

It is our hope that after reading this book food scientists, especially those not very familiar with NMR spectroscopy, will possess the ability to comprehend

RSC Food Analysis Monographs No. 10
NMR Spectroscopy in Food Analysis
By Apostolos Spyros and Photis Dais
© Apostolos Spyros and Photis Dais 2013
Published by the Royal Society of Chemistry, www.rsc.org

NMR related applications in the ever growing "NMR in food" literature, to acquire the basic skills necessary to design an NMR experimental approach, and to select the NMR methodology suitable for addressing the food-related problem at hand.

A. Spyros, Ph. Dais

Contents

RSC Food Analysis Monographs No. 10
NMR Spectroscopy in Food Analysis
By Apostolos Spyros and Photis Dais
© Apostolos Spyros and Photis Dais 2013
Published by the Royal Society of Chemistry, www.rsc.org

CHAPTER 1
Introduction

Nuclear magnetic resonance (NMR) spectroscopy is an effective analytical technique, which has been used systematically in food analysis and authentication in recent years. Its origin is traced back to 1946 when two groups of scientists at Harvard University (Purcell, Torrey and Pound) and at Standford University (Bloch, Hansen, and Packard), working independently, observed proton resonance signals from paraffin wax and water, respectively. For their discovery, Purcell and Bloch were jointly awarded the Nobel Prize in Physics in 1952.

The first application of NMR in food science dates back to 1957 when low-resolution NMR measured moisture in foods. Consistent and widespread application of NMR in food science started in the 1980's mainly due to deficiencies in instrumentation and the complexity of food matrices. Since then, an explosive publication of research and review articles dealing with NMR applications in food science has appeared in scientific journals and several books. Figure 1.1 shows diagrammatically the explosion of publications in food science after 1988.

Also, numerous oral and written communications have been presented in domestic and international conferences. In particular, an *International Conference on Application of Magnetic Resonance in Food Science* is held in Europe every two years. This conference started in 1992 and gave the opportunity for scientists worldwide to present new applications of NMR in food science and technology. It is worth mentioning that NMR methods have been approved as official methods by the European Union (*e.g.* detection of wine fraud). There are several reasons for this development: (a) the increasing sophistication and the user-friendly NMR instrumentation; (b) the increasing need of the food industry to understand and innovate its products and processes; and (c) the necessity for the development of new and more effective analytical techniques for the quality control and authentication of foods and thereby the reinforcement of pertinent legislation.

RSC Food Analysis Monographs No. 10
NMR Spectroscopy in Food Analysis
By Apostolos Spyros and Photis Dais
© Apostolos Spyros and Photis Dais 2013
Published by the Royal Society of Chemistry, www.rsc.org

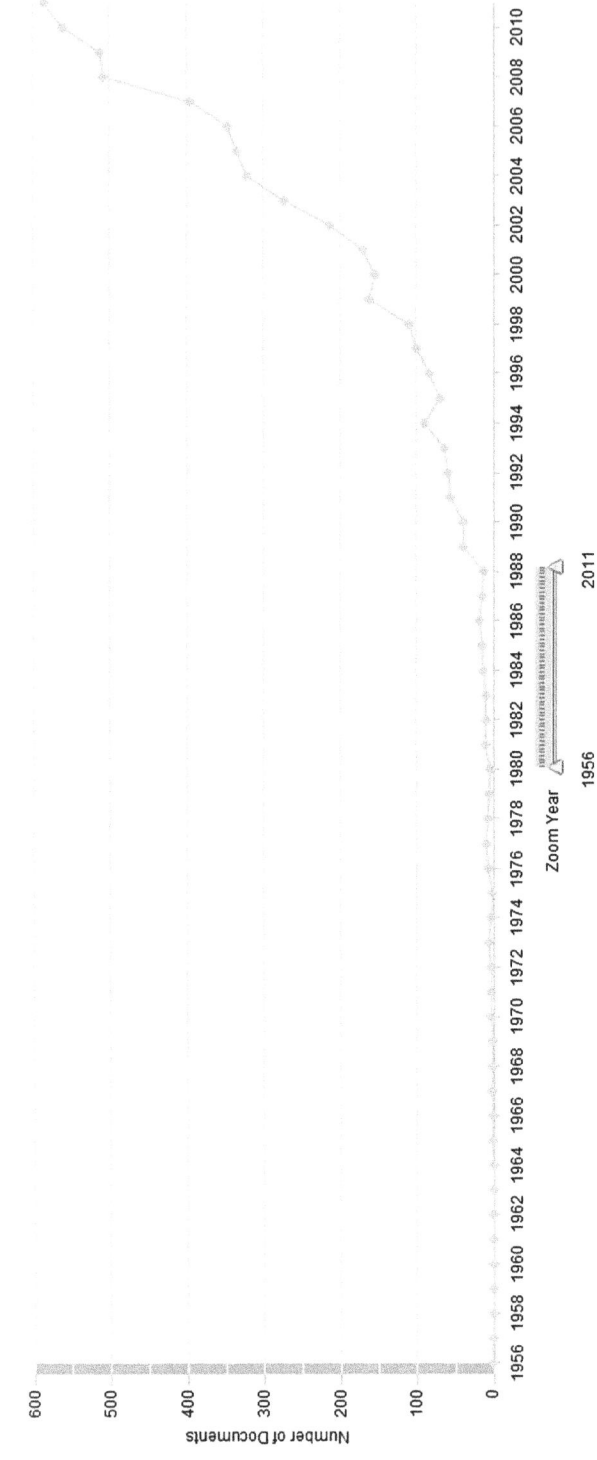

Figure 1.1 The graph presents the number of publications per year that include the words 'food' and 'NMR' or 'MRI' (either as acronyms or in full text) in their title, keywords or abstract, as indexed in Scopus. (Retrieved on May 25, 2012 from www.scopus.com.)

Foods are very complex and highly heterogeneous systems comprising a large number of chemical compounds, the composition of which varies considerably under certain circumstances (*e.g.* agronomical or slaughter practices, industrial processes, storage, maturation, *etc.*). To this direction, one-dimensional (1D) liquid or solid-state high-resolution NMR spectroscopy can provide in a single experiment a wealth of structural and quantitative information in the form of the NMR parameters, namely chemical shifts, coupling constants and signal intensities. For the same sample the researcher can choose different nuclei, such as ^{1}H, ^{13}C, ^{31}P, ^{19}F—to mention the most popular nuclei—that allow the study of food samples under different perspectives and to extract the maximum information about its natural or industrial condition. These experiments need no separation of the various food components and no serious sample pre-treatment. Moreover, NMR spectroscopy is sensitive to dynamics, which allows differentiation between molecules or groups of molecules with different mobility through spin–lattice and/or spin–spin relaxation measurements.

In cases where the complexity of the food sample is so severe, causing extensive signal overlap in 1D spectra, the arsenal of NMR spectroscopy provides a large number of analytical techniques starting from the homonuclear and heteronuclear multi-dimensional NMR to its hyphenation with effective separation techniques, such as liquid chromatography (LC-NMR). In particular, two-dimensional (2D) NMR techniques, such as COSY, TOCSY, NOESY, HSQC, *etc.*, based on the inherent 'communication' of nuclei with each other (through spin–spin and dipolar coupling), spread out the spectroscopic information in two dimensions unravelling hidden nuclear connectivities and facilitating the structural characterisation of the molecules in the food sample. Although NMR spectroscopy is not a destructive analytical technique and recovery of the analyte can be easily achieved after experimentation, industrial needs may require the examination of food products under different processing conditions by non-invasive means. Magnetic resonance imaging (MRI), used extensively in medicine, has been exploited in recent years in food analysis. The ability of MRI to show spatial resolution within the food product and the judicial application of MRI techniques allows the monitoring of the fate of certain molecules (*e.g.* water) and reveals various molecular interactions and changes in tissue structure that occur during food processing or storage (*e.g.* food freezing and thawing).

The combination of NMR spectroscopy with multivariate statistical methods provided an alternative possibility of analysing and maximising the information recovery from complex NMR spectral data of foods. This methodology, usually called metabonomics, does not necessarily require the identification of the individual signals in the spectrum as in quantitative NMR, but seeks to find subtle spectral features that can identify unequivocally the presence of metabolites or useful biomarkers. Pattern recognition techniques (supervised or unsupervised) can be used to map the NMR spectra of a large number of samples, and locate spectral fingerprints that reflect either metabolic changes or used to distinguish sample classes.

The disadvantage of the early days of NMR spectroscopy related to the low sensitivity and high cost of the analysis does not hold true for the NMR instrumentation of the present day. These drawbacks have been largely compensated by the development of modern hardware comprising strong magnetic fields up to 23.5 T and cryogenic probes that allow easy detection of food components at the level of μg and even ng. Moreover, the progress in sophisticated software and innovations in automation allow the screening of a large number of samples (overnight run), reducing the experimental time to a few minutes even for the less sensitive nuclei (*e.g.* ^{13}C, ^{15}N).

In concluding this introductory chapter, we could add that it is not only the unique information that NMR provides, but also the versatility of methods, instruments and probes that make it an important tool for qualitative and quantitative analysis.

This book has been organised as follows: *Chapter 1* is the book's introduction. *Chapter 2* gives an account of the theory underlying the physical phenomenon of NMR and grouping the most useful NMR techniques to better understand the core principles, which appear in subsequent chapters. Since NMR is a well-documented spectroscopic technique and it is well described in several introductory and advanced books, this chapter will be kept to a minimum. *Chapter 3* describes the NMR instrumentation in an attempt to familiarise the reader with the hardware and software components of modern high-resolution and solid-state NMR spectrometers, and their functions and automation. Relevant information about NMR spectrometers and the implementation of its components may help the reader to choose the right spectrometer and accessories for their needs. Also, this chapter includes useful information about the hardware systems and experimental designs to perform sophisticated experiments, such as (HP)LC-NMR, time-domain NMR, and high-throughput and on line NMR. Appropriate guidance for obtaining pure samples from various food matrices that are suitable for NMR experiments will be presented in *Chapter 4*, whereas the experimental conditions described in *Chapter 5* may help the NMR user to choose the right input values for the critical parameters in the experimental setup in order to obtain the maximum possible information from the NMR experiment, and to perform quantitative analyses with high accuracy and precision. *Chapter 6* presents a few aspects of the supervised and unsupervised pattern recognition statistical methods employed for data exploration, classification of food samples, and the build-up of calibration–prediction models giving special attention to NMR metabonomics. The applications of NMR spectroscopy and its specialties to different food systems are discussed in *Chapters 7–11*. A detailed presentation of the available NMR methodologies and techniques for each food category is provided, whereas practical guidance and tips for performing concrete experiments is afforded. Every chapter starts with a short abstract and ends with relevant bibliographic coverage.

CHAPTER 2
Theoretical Aspects

2.1 Nuclear Spins and Energy States

An atomic nucleus is a collection of protons and neutrons (nucleons) that possess a quantum mechanical property called spin, which is characterised by spin angular momentum. Spin angular momentum is an intrinsically quantum mechanical property that does not have a classical analog. All subatomic particles are spin $\frac{1}{2}$ particles. The nucleus itself has a total spin angular momentum formed by the coupling of the individual spin angular momenta of its constituent protons and neutrons. The total nuclear spin angular momentum quantum number I may therefore take values: 0, 1/2, 1, 3/2, 3, 5/2, *etc*. A nucleus with non-zero quantum numbers I behaves as a small magnet or magnetic dipole with a magnetic moment μ. The magnetic moment is an intrinsic property of the nucleus, and it is associated with the angular momentum of the nucleus. As a vector, nuclear magnetic moment has two properties: magnitude and direction. The magnitude of μ is quantised and given by eqn (2.1):

$$\mu = \gamma \hbar \sqrt{I(I+1)} \tag{2.1}$$

where $\hbar = h/2\pi$ is Plank's constant and γ is the magnetogyric ratio, an inherent property of the nucleus. This parameter is unique for each nucleus. When $I = 0$, then μ is zero, and the nucleus does not have magnetic properties. Nuclei with $\mu = 0$, such as ^{12}C, ^{16}O, ^{32}S cannot be studied by NMR. The smallest magnetic moments belong to protons and is called the nuclear magneton, μ_N; its value is calculated from $\mu_N = e\hbar/2m_p c$ (e is the electric charge, m_p is the mass of proton, and c is the velocity of light) to 5.0505×10^{-27} J T^{-1}. Under the influence of an external magnetic field of intensity (strength) B_0 fixed along the z-axis of a static Cartesian coordinate system, the magnetic moment assumes discrete orientations (the nuclear Zeeman effect) governed by the magnetic quantum number m_I. The allowed values of m_I are $-I, -I+1, -I+2, \ldots, I-1, I$, giving rise to

RSC Food Analysis Monographs No. 10
NMR Spectroscopy in Food Analysis
By Apostolos Spyros and Photis Dais
© Apostolos Spyros and Photis Dais 2013
Published by the Royal Society of Chemistry, www.rsc.org

$2I+1$ possible orientations. Each orientation defines an energy level or state, with energy:

$$E = -\mu_z B_0 = -m_I \gamma \hbar B_0 \tag{2.2}$$

with $m_I = +\frac{1}{2}$ or $-\frac{1}{2}$ for nuclei with $I=\frac{1}{2}$. $\mu_z(=m_I\gamma\hbar)$ is the projection of the magnetic moment along the z-axis. The orientation of μ_z with respect to B_0 defines the nuclear energy states. For $I=\frac{1}{2}$, μ_z parallel to B_0 defines the energy state with the lower energy, whereas its anti-parallel orientation identifies the energy state with the higher energy. It is traditional to label the low and high-energy states with the Greek letters α and β, respectively. The α state with $m_I = +\frac{1}{2}$ is often described as 'spin up', and the β state with $m_I = -\frac{1}{2}$ as 'spin down' (Figure 2.1).

The energy space (eqn (2.3)) between the two states is:

$$\Delta E = E_\beta - E_\alpha = \left(\frac{1}{2}\gamma\hbar B_0\right) - \left(-\frac{1}{2}\gamma\hbar B_0\right) = \gamma\hbar B_0 \tag{2.3}$$

where, E_β and E_α are the energies of the upper and lower energy state, respectively. One important conclusion derived from eqn (2.3) is that the energy gap is variable and increases with increasing magnetic field strength B_0 (Figure 2.1). Transitions between energy states are induced by the magnetic component of the electromagnetic irradiation emitting in the radiofrequency (RF) region and having the same energy with the energy gap between the states. Excitation occurs provided that the NMR selection rule ($\Delta m_I = \pm 1$) is satisfied. The radiation frequency inducing transitions for a given type of nucleus is given by eqn (2.4).

$$\nu_0 = (\gamma/2\pi)B_0 \quad \text{or} \quad \omega_0 = \gamma B_0 \quad (\omega_0 = 2\pi\nu_0) \tag{2.4}$$

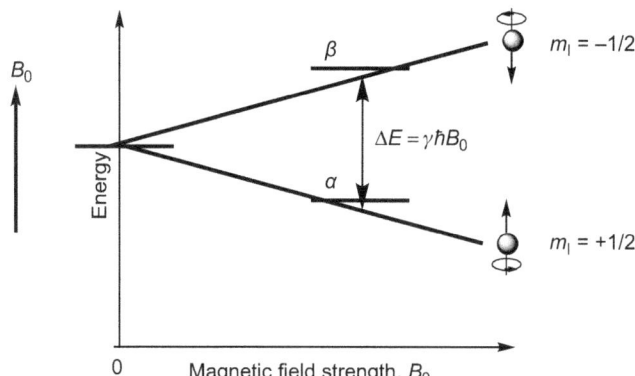

Figure 2.1   The orientation of the magnetic moment of a nuclear spin with $I=\frac{1}{2}$. Each orientation defines an energy state. The energy gap ΔE depends on the static magnetic field strength B_0 and the gyromagnetic ratio of the nucleus.

This equation indicates that at a given magnetic field strength each type of nucleus has its own resonance frequency, inasmuch the magnetogyric ratio is unique for each nucleus. Consequently, different types of nuclei cannot be detected in the same NMR experiment. Table 2.1 summarises the magnetic properties of selected nuclei.

Each food sample contains a huge number of molecules (1 mole contains 6.023×10^{23} molecules) and each molecule may comprise a large number of nuclei that occupy the available energy states. At room temperature, under the influence of the magnetic field B_0, states of lower and higher energy are both occupied by nuclei (Figure 2.2).

This occurs because the energy difference between the two states is more than 100 times smaller than the energy of the thermal motion ($k_B T$) of the molecules in solution at room temperature. The equilibrium population of the nuclear spins in the energy states is governed by the Boltzmann distribution law with a slight excess of nuclei in the state with the lower energy. According to Boltzmann law, the relative populations of the lower (N_α) and higher (N_β) energy levels at room temperature (eqn (2.5)) are given by:

$$\frac{N_\alpha}{N_\beta} = \exp\left(\frac{\Delta E}{k_B T}\right) = \exp\left(\frac{\gamma \hbar B_0}{k_B T}\right) \tag{2.5}$$

At $T = 303$ K and magnetic field strength of 7.05 T, only 25 nuclei out of 1 million are in excess in the lower energy state. Nevertheless, this tiny excess of nuclei at the low energy state is the reason of the occurrence of the NMR phenomenon and at the same time the main cause of the low sensitivity characterising the NMR spectroscopy. The NMR sensitivity is lower compared to other branches of spectroscopy (*e.g.* UV-vis, IR, Raman). However, the low sensitivity of NMR is in part compensated by the wealth of information that can be extracted from typical NMR spectra. The sensitivity for nuclei with $I = \frac{1}{2}$ is measured by the signal-to-noise ratio, S/N, which depends, among others, on the magnetic field strength B_0, the gyromagnetic ratio, the natural abundance of the nucleus, the temperature of the experiment, and the sample concentration (the population of nuclei at the lower state). According to eqn (2.5), the sensitivity is higher for nuclei with higher gyromagnetic ratio and natural abundance. For a given nucleus, the sensitivity is enhanced by the increasing magnetic field strength B_0 (eqn (2.5)) or by decreasing the temperature. The sensitivity is also augmented by manipulating other factors including the instrumental setup and data processing (see chapters 3 and 5).

In principle, all nuclear spins with magnetic properties (acquiring magnetic moment) can be studied by NMR. The most popular nuclei used in food analysis are ^{1}H, ^{13}C, and to a lesser extent ^{15}N, ^{19}F, and ^{31}P. The nuclear charge of these nuclei with $I = \frac{1}{2}$ is symmetric with a relatively low magnetic moment and affords signals with narrow line widths. This property facilitates greatly the interpretation of complex NMR spectra. The ^{1}H nucleus, characterised by high gyromagnetic ratio and natural abundance (99.99%), is the most preferred nucleus used in food analysis. ^{1}H NMR spectroscopy provides valuable

Table 2.1 Magnetic properties of selected nuclei.

Isotopes	Spin quantum number (I)	Gyromagnetic ratio (γ) (10^7 rad T^{-1})	Magnetic moment[a] (μ/μN)	Theoretical NMR frequency for free atom[b] (MHz)	Natural abundance (%)	Relative Receptivity[c]	Quadrupole moment (Q) (fm^2)[d]
^{1}H	$\frac{1}{2}$	26.7522	2.79284	100.000	99.9885	1.000	
^{2}H	1	4.1066	0.85744	15.350	0.0115	0.0096	0.286
^{10}B	3	2.8747	1.80064	10.746	19.9	0.0199	8.459
^{11}B	3/2	8.5847	2.68886	32.090	80.1	0.1650	4.059
^{13}C	$\frac{1}{2}$	6.7283	0.70241	25.150	1.07	0.0159	
^{14}N	1	1.9338	0.40376	7.229	99.636	0.00101	2.044
^{15}N	$\frac{1}{2}$	−2.7126	−0.28319	10.140	0.364	0.00104	
^{17}O	5/2	−3.6281	−1.89379	13.561	0.038	0.02910	−2.558
^{19}F	$\frac{1}{2}$	25.179	2.62687	94.057	100	0.8320	
^{29}Si	$\frac{1}{2}$	−5.3190	−0.55529	19.883	4.685	0.00786	
^{31}P	$\frac{1}{2}$	10.8394	1.13160	40.5178	100	0.06650	
^{33}S	3/2	2.0557	0.64382	7.6842	0.75	0.00227	−6.78
^{35}Cl	3/2	2.6242	0.82187	9.8093	75.76	0.00472	−8.165
^{37}Cl	3/2	2.1844	0.68412	8.1652	24.24	0.00272	−6.435
^{79}Br	3/2	6.7256	2.10640	25.1404	50.69	0.07940	30.5
^{81}Br	3/2	7.2498	2.2706	27.100	49.31	0.09950	25.4
^{127}I	5/2	5.3896	2.81133	20.1462	100	0.09540	−71

[a]The relative magnetic moment is expressed in units of nuclear magneton, μ_N.
[b]It is calculated from γ and scaled to ^{1}H = 100 MHz.
[c]Relative to equal number of protons; it is proportional to $\gamma/(I+1)$.
[d]1 $fm^2 = 10^{-30}$ m^2.

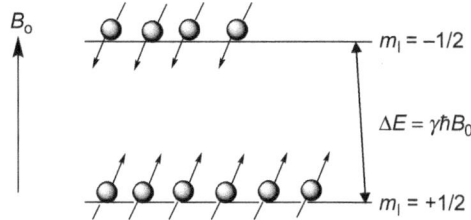

Figure 2.2 Occupation of the energy states by a collection of nuclear spins with $I = \frac{1}{2}$. At equilibrium, there is always a tiny excess of nuclear spins on the lower energy state determined by the Boltzmann distribution law.

information about minor components of foods (*e.g.* phenolic compounds, sterols, terpenes) in a reasonable experimental time. On the other hand, ^{1}H NMR spectra of most foodstuffs are quite complex. Strong signal overlaps and scalar couplings within a relatively narrow range (15 ppm) of resonance frequencies make their interpretation a difficult task. Fortunately, the use of modern 2D NMR spectroscopy has largely mitigated this problem.

The ^{13}C nucleus is the second commonly used nucleus in the NMR analysis of foods. ^{13}C NMR spectroscopy provides additional information to that already obtained by ^{1}H NMR, and can be considered as a complementary technique, which could facilitate the interpretation of complex ^{1}H NMR spectra. The much wider range (250 ppm) of ^{13}C resonance frequencies and the fact that ^{13}C NMR experiments are conducted with proton decoupling results in simplified spectra even for complex food matrices. Moreover, the phenomenon of NOE that is developed during proton decoupling increases the signal intensities of the protonated carbons by a factor of *ca.* 2 (see below). The main disadvantage of ^{13}C nucleus relative to ^{1}H nucleus is its lower sensitivity due to its lower magnetogyric ratio (by a factor of *ca.* 4) relative to proton, and its low natural abundance (*ca.* 1.1%). Consequently, recording a ^{13}C spectrum needs much more analytical time to show signals of minor components. This problem becomes even worse for quantitative analysis, where the long relaxation time of ^{13}C nucleus should be taken into consideration in setting up the quantitative NMR experiment (see section 5.2).

Some of the problems showed by ^{1}H and ^{13}C NMR are removed by using the ^{31}P nucleus, when possible. The large range of chemical shifts (*ca.* 1000 ppm) reported for the ^{31}P nucleus ensures a good separation of signals obtained under proton decoupling, whereas its 100% natural abundance (the phosphorus atom has only one isotope), and its high relative sensitivity which is only *ca.* 15 times less than that of proton, make ^{31}P NMR experiments a reliable analytical tool to determine amounts of the order of μmol, or lower, depending on the available instrumentation. Limitation of this method may be the fact that ^{31}P nucleus is not widespread in foods as proton and carbon. As we shall see in section 7.1.2, a novel ^{31}P NMR methodology developed recently has extended the application of this nucleus to components bearing no phosphorus.

Nuclei with $I \geq 1$ are associated with an electric quadrupole moments due to the non-spherical distribution of the nuclear charge, which has the symmetry of an ellipsoid (prolate or oblate). Due to the rapid quadrupolar relaxation, the NMR signals of quadrupolar nuclei are usually much wider than those of spin $\frac{1}{2}$. Amongst these, ^2H has the lowest quadrupole moment, and hence shows relatively narrow signals in ^2H NMR spectra. It has been used as a valuable analytical method to reveal geographic or biosynthetic origin of foods and to detect fraud, particularly for juices, wines and beverages.

2.2 A Collection of Spins—the Vector Model

Under the influence of the field B_0, a collection of identical nuclei with spin $\frac{1}{2}$ will all precess with the frequency ω_0, the so called Larmor frequency, but with random phases relative to each other, so that on the average the overall (macroscopic) magnetisation vector M_0 (the vector sum of all individual magnetic moments) remains along the positive z-axis (Figure 2.3(a)).

This is because the population in the ground state ($m_I = +\frac{1}{2}$) is very slightly greater than the higher energy state ($m_I = -\frac{1}{2}$), as mentioned previously. Since the vectors precess randomly, no transverse magnetisation is observed on the (xy) plane. The system is now irradiated by an electromagnetic radiation in the RF region of the spectrum, which is produced by the transmitter of the NMR spectrometer. The magnetic component of the radiation, B_1, is orientated along

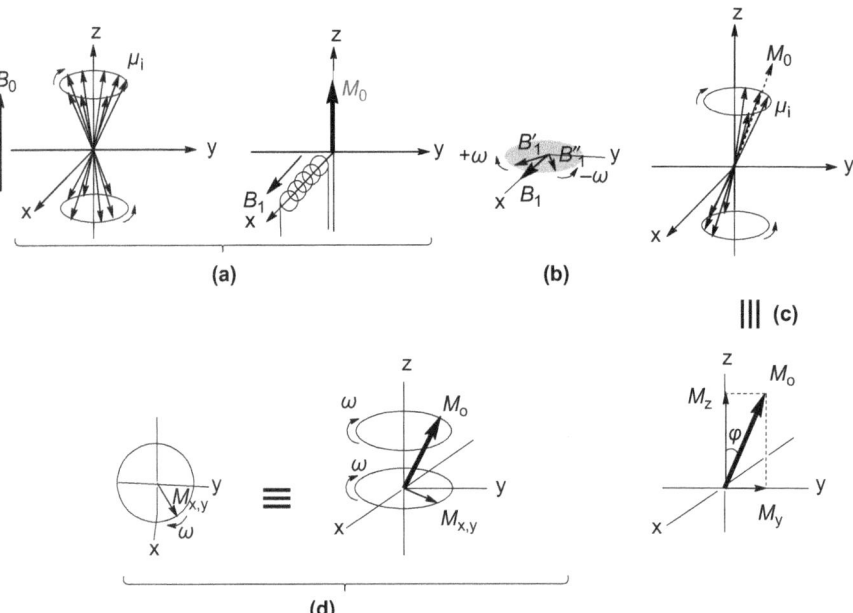

Figure 2.3   (a) Graphical description in the laboratory frame of the precession of a collection of $I = \frac{1}{2}$ nuclei around B_0. M_0 is the net nuclear magnetisation at equilibrium, *i.e.* the vector sum of the individual nuclear magnetic moments. (b) The orientation of the linearly polarised field B_1. (c) Deflection of M_0 from its equilibrium position during irradiation by B_1. (d) Precession of M_0 and the transverse component M_{xy} on the (xy) plane.

the x-axis, perpendicular to the applied magnetic field B_0. As a matter of fact, the oscillating field B_1 is linearly polarised and results from the vector sum of two circularly polarised magnetic fields B_1' and B_1'' of the same magnitude that are counter-rotating on the (xy) plane (Figure 2.3(b)). One of these fields, say B_1', rotates clockwise with frequency $+\omega$ (B_1'' rotates anti-clockwise with frequency $-\omega$) in the same direction as the nuclear magnetic moments. When the frequency $+\omega$ of B_1' becomes exactly equal to the Larmor frequency, ω_0 of the nuclear spins (resonance condition), some of the spins flip from the ground to the excited state, but most importantly the vectors of magnetic moments of the individual nuclei start precessing in phase, *i.e.* they become phase coherent as shown in Figure 2.3(c). The individual magnetisation vectors no longer cancel each other. Instead, the overall macroscopic vector 'tips' away by some angle φ with respect to the z-axis (Figure 2.3(c)). The magnetisation vector M_0 continues to precess with frequency ω_0 on a cone around the z-axis along with its projection on the (xy) plane. At time zero, the y-component is given by $M_0 \sin \varphi$ (Figure 2.3(c)); at time t the vector has rotated by an angle $\omega_0 t$ (Figure 2.3(d)). The components of magnetisation along the y and x axes are given in eqn (2.6):

$$M_y = M_0 \sin \varphi \cos \omega_0 t \qquad M_x = M_0 \sin \varphi \sin \omega_0 t \qquad (2.6)$$

Shutting down the field B_1, the nuclear spins start losing the absorbed energy and the magnetisation is relaxing back to its equilibrium position. In the context of the vector model, when the field B_1 turns off, the nuclear magnetic moments lose their phase coherence (Figure 2.3(c)) and once again adopt a regular distribution over the precessing cone (Figure 2.3(a)). There are two processes, whereby nuclei can lose energy. The first involves transfer of energy from the nucleus to surrounding molecules as heat by a process known as spin-lattice or longitudinal relaxation, and the second involves energy exchange with non-excited nuclei of a different type (with a different Larmor frequency) referred to as spin–spin or transverse relaxation. The rates of both relaxation mechanisms are exponential and are defined by time constants T_1 and T_2, respectively. The two relaxation processes operate simultaneously and for a distant observer the magnetisation vector appears to return into the Boltzmann equilibrium by performing a spiral pathway (Figure 2.4). Relaxation processes will be discussed in more details below.

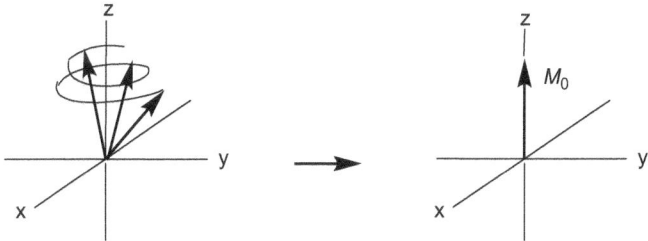

Figure 2.4 Return of M_0 to its Boltzmann equilibrium state during relaxation in the laboratory frame.

2.3   NMR Parameters

NMR has long been recognised as an effective analytical tool for qualitative and quantitative analysis. This is mainly because valuable structural information and quantitative results of mixtures is readily obtained from parameters measured in NMR spectra. These include the chemical shift (δ), the scalar spin–spin coupling constant (J), and the signal intensity. In view of the fact that there are many excellent references that discuss extensively these parameters, only a brief overview will be given here. It should be noted that additional NMR parameters could be measured by performing specific NMR experiments. NOE and the spin–lattice (T_1) and spin–spin (T_2) relaxation times are extensively used for structural and conformational analysis of small and large molecules, and the study of the molecular dynamics.

2.3.1   Chemical Shift

The electron cloud of the bonding electrons surrounds an atomic nucleus. The structure of the electron clouds depends on the atoms forming the bond and its relative position within the molecule. If the atom is subjected to an externally applied magnetic field B_0, a rotational motion is induced in the electron cloud. This movement creates a small local magnetic field B_{loc} that may oppose the external field B_0, so that the nucleus experiences an effective field, B_{eff}, which is slightly reduced relative to B_0. According to the Lenz rule: $B_{loc} = \sigma B_0$, thus:

$$B_{eff} = B_0 - B_{loc} = B_0 - \sigma B_0 = B_0(1 - \sigma) \tag{2.7}$$

The parameter σ is called the screening or shielding constant and is dependent on the density and distribution of the electron cloud surrounding the nucleus. It takes values ranging from 10^{-6} for the lighter nuclei to 10^{-3} for the heavier ones. The electron density in turn depends on the chemical environment within the molecule yielding differences in the resonance frequency of the nucleus relative to the Larmor frequency ω_0; this is expressed by eqn (2.8) that was derived upon combining eqn (2.4) and (2.7).

$$\omega_{eff} = \omega_0(1 - \sigma)B_0 \tag{2.8}$$

The frequency difference owing to electronic environments yields the chemical shift of the nucleus. Practically, the chemical shift designates the position of a certain signal in the NMR spectrum with respect to the signal of an internal reference that defines the beginning of the so-called δ scale, expressed as:

$$\delta = \frac{\omega_{eff} - \omega_{ref}}{\omega_{ref}} \times 10^6 = \frac{(\sigma_{ref} - \sigma_{eff})}{1 - \sigma_{ref}} \times 10^6 = (\sigma_{ref} - \sigma_{eff}) \times 10^6 \quad \text{(ppm)} \tag{2.9}$$

where ω_{ref} is the resonance frequency of the protons of the reference. The factor 10^6 is included to obtain convenient values; ppm (parts per million) can be added to δ values. The difference $(\sigma_{ref} - \sigma_{eff})$ is obtained by taking into account that $\sigma_{ref} \ll 1$.

Nuclei that are shielded resonate at lower frequencies or higher magnetic fields, whereas those nuclei that are de-shielded resonate at higher frequencies or lower magnetic fields. According to eqn (2.9), shielded nuclei show smaller chemical shifts with respect to the internal reference than de-shielded nuclei, which appear at larger chemical shifts. For organic solvents, tetramethylsilane (TMS) is the internal reference of choice.

One of the most important determinants of the chemical shift of a nucleus is its proximity to an electronegative atom or group of atoms (functional group) that perturbs the electron density around the nucleus. For instance, the presence of electron-withdrawing substituents one to three bonds away from a proton nucleus decreases the electron density around it (inductive effect through sigma bonds) inducing a strong diamagnetic de-shielding. This effect shifts the proton resonance in the far left side (higher frequencies or larger chemical shifts) relative to TMS. The opposite effect is observed with an electron-donating substituent; the proton resonance is shifted in the right side (lower frequencies or smaller chemical shifts) of the NMR spectrum, close to TMS.

Moreover, the spatial anisotropy of the local magnetic fields induced by the circulation of the bonding electrons of single and multiple bonds nearby the proton in question play a decisive role in the chemical shift of this nucleus. ^{1}H nuclei shows different δ values in groups, such as $-C{\equiv}H$, $>C{=}H$, $HC{=}O$, or $>CH_2$. The anisotropic, local magnetic fields create two cones spanning from the centre of the bond: protons inside the cone are de-shielded, and shielded outside the cone. Protons in groups $>C{=}H$, $HC{=}O$ are inside the de-shielding cone and show relative large chemical shifts, whereas protons $>CH_2$ lie outside the de-shielding cone of the single bond and appears at smaller chemical shifts. The reverse field anisotropy is apparent for the triple bond, and protons within the two cones spanning from the centre of the bond are shielded (Figure 2.5(a)).

That is why the acetylenic protons show the smallest chemical shift of all groups. One of the most pronounced effects of local magnetic fields is observed in aromatic rings (Figure 2.5(b)). The circulation of *p*-electrons below and above the aromatic ring (*e.g.* benzene), the so-called ring current, generates an anisotropic local magnetic field with shielding and de-shielding regions (Figure 2.5(c)). ^{1}H nuclei above or below the plane of the aromatic ring are shielded and show smaller δ values than protons situated near this plane. This is why the protons of benzene at the periphery of the ring appear with considerable shift (δ 7.26) from TMS. Chemical shifts of the functional groups of different nuclei including protons and carbons are given in the literature in a tabular or graphical form.

2.3.2 Spin-Multiplicity and Coupling Constant

Scalar spin–spin coupling is a property of the nuclear spins that is manifested in the NMR spectrum by the signal splitting in several components. The quantum

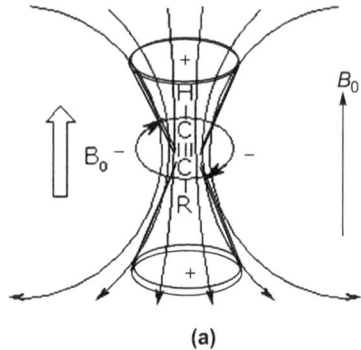

(a)

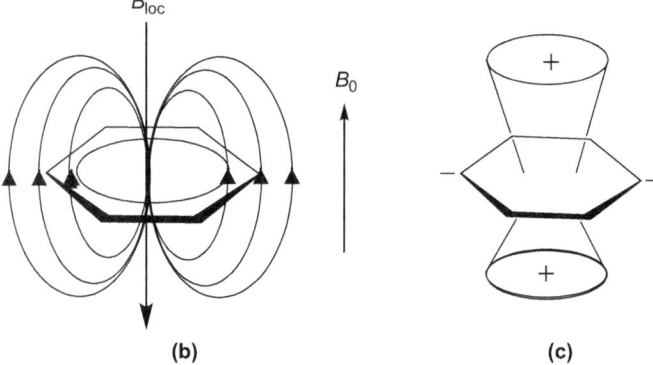

(b)   (c)

Figure 2.5   Anisotropic magnetic fields of (a) a triple bond and (b) a benzene ring.
(c) Schematic representation of the magnetic anisotropic effect of the ring
current; ($+$) and ($-$) are the shielding and deshielding zones, respectively.

mechanical description of the scalar coupling is associated with transitions
between energy levels or spin states. For a coupled two-spin $\frac{1}{2}$ system, there are
four energy levels associated with the orientations of the nuclear spins up and
down (with respect to B_0). In terms of the labels α and β, the four spin states for
a homonuclear spin system AX are arranged according to the diagram in
Figure 2.6(a).

The quantum mechanical treatment of the two spin system AX, assuming
weak coupling, results in the complete set of transitions that involves the energy
of the states, the allowed transitions according to the selection rule, the frequency for each allowed transitions and the signal intensities. If ω_A and ω_X are
the Larmor frequencies of the two spins A and X, respectively, the corresponding spectrum of the spin system AX appears on the right of Figure 2.6(a).
The spectrum consists of two doublets each split by J_{AX} centred at the Larmor
frequencies ω_A and ω_X of spins A and X. Contrary to the chemical shift, the
coupling constant J_{AX} (measured in Hz) is field independent. To obey the

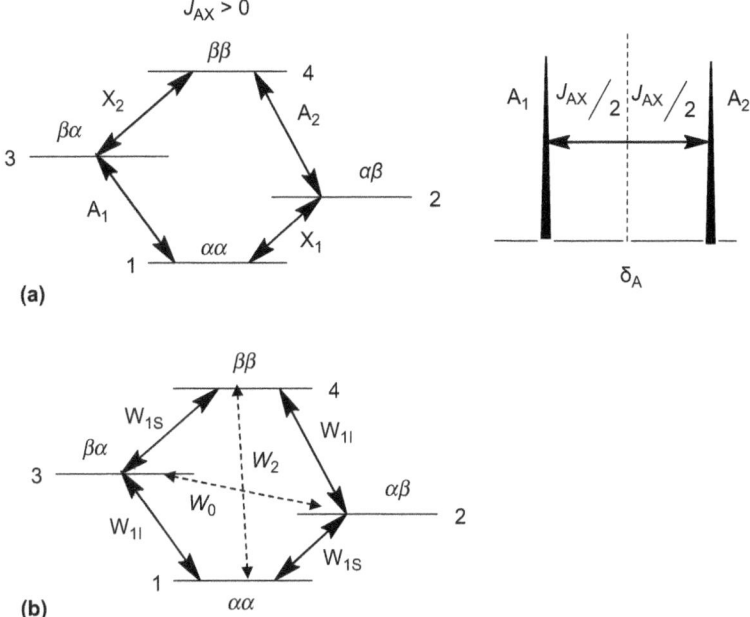

Figure 2.6 (a) Energy level diagram and spectrum of two weakly coupled nuclei AX. (b) Solomon diagram describing the NOE phenomenon for a two-spin system IS.

selection rule, each allowed transition corresponds to one of the spins changing the magnetic quantum number m_I by one from one spin state to the other, while the state of the other spin remains fixed. For example, transition $1 \rightarrow 3$ involves a spin A going from α to β while spin X remains in the α state. It is said that spin A is active and spin X is passive in the α state. This transition is allowed because it obeys the selection rule shown in eqn (2.10).

$$M_T = \sum_n m_I = \pm 1, n = 1, 2 \tag{2.10}$$

For instance, the $\alpha\alpha \rightarrow \alpha\beta$ transition is allowed, because

$$M_T = (\alpha a - a\beta) = (1/2) + (1/2) - (1/2) - (-1/2) = +1 \tag{2.10a}$$

The allowed transitions in Figure 2.6(a) are referred to as single-quantum transitions, because the energy of one quantum is sufficient to induce each of these transitions. There are two more transitions in the two-spin systems that are considered forbidden by the selection rule. The first forbidden transition is between states 1 and 4 in which both spins flip ($\alpha\alpha \rightarrow \beta\beta$); according to eqn (2.10), $M_T = -2$. This is called a double-quantum transition. The second forbidden transition is between states 2 and 3, where both spins flip ($\alpha\beta \rightarrow \beta\alpha$); here $M_T = 0$, and this transition is called the zero-quantum transition.

The above treatment corresponds to weakly coupled spin systems, where the chemical shift is much larger than the coupling constant, *i.e.* $(\omega_A - \omega_X)/J_{IS} \gg 1$, and involves relatively simple calculations even for systems with more than two spins. For strongly coupled systems, where the chemical shift is comparable or smaller than the coupling constant $(\Delta\omega/J \leq 1)$, the computation of the energies of spin states and frequencies of the allowed transitions is more demanding and for more than two spins requires the use of appropriate software packages to solve the quantum mechanical problem.

The spectra of the weakly coupled spins follow the rules of the so-called first-order spectra, whereas the strongly coupled spins give second-order spectra. In the first-order spectra of nuclei with $I = \frac{1}{2}$ (^{1}H, ^{13}C, ^{15}N, ^{19}F, and ^{31}P), multiplicity (the number of components of each signal), which is another valuable parameter for structure elucidation, follows the $N+1$ rule (N is the number of the chemically equivalent neighbouring nuclei to the proton in question), and the Pascal triangle for the intensity pattern. For instance, when $N = 1, 2, 3$, the signals are split in to doublets, triplets, and quartets, respectively, with relative intensities $1:1$, $1:2:1$, $1:3:3:1$ in that order. For chemically non-equivalent nuclei, the splitting pattern follows the composite rule $(N+1)(N'+1)\ldots$, where N and N' are the numbers of the adjacent non-equivalent nuclei. For protons with $N = 1$ and $N' = 1$, a doublet of doublets is observed in the spectrum.

The coupling can range from one-bond to five-bonds; of these, one-bond, two-bonds (known as germinal coupling), and three-bonds (known as vicinal coupling) coupling are of structural value. The three-bond coupling is the most informative of all because of its relation to the dihedral angle formed by the three bonds (H–C–C–H). This relationship has been described in semiquantitative terms by the Karplus equation or Karplus curve.

Long-range coupling of four or more bonds usually occurs in aromatic systems and between protons in specific unsaturated segments (allylic and homoallylic coupling) of organic molecules. Long-range coupling can also be observed between protons in saturated systems, in which the intervening bonds have the zigzag configuration. Coupling occurs between heteronuclear and homonuclear spin pairs. Finally, both coupled nuclei need to be NMR active, *i.e.* ^{12}C does not cause splitting.

2.3.3   Signal Intensity

Signal intensity is defined as the area encompassed by the NMR signal, and is measured by digital integration. Also, it can be expressed as signal height provided that the signal widths at half height of all resonances in the spectrum are comparable, something which rarely occurs. Signal intensity is directly proportional to the number of equivalent nuclei in the sample and is commonly used for the spin-multiplicity analysis, but frequently for quantitative analysis. As we shall see later, quantification of food components requires signals acquisition under non-saturating condition by applying time intervals between pulses long enough for the spins to maintain their Boltzmann equilibrium.

2.4 Nuclear Relaxation

Spontaneous nuclear relaxation is almost negligible. This is due to the very small energy gap between the spin states. Nuclear relaxation is stimulated by the surrounding environment (the lattice) of the nucleus and/or by exchanging energy with neighbouring non-excited nuclei. These two relaxation processes that lead the spin populations of the nuclear states back to their equilibrium distribution are called spin–lattice or longitudinal relaxation and the spin–spin or transverse relaxation with time constants T_1 and T_2, respectively. The spin–lattice relaxation process is realised by a variety of relaxation mechanisms that allow the excited nuclear spins to release their energy to the lattice. Kinetically, T_1 relaxation is a first order process described by the differential equation (eqn 2.11a):

$$\frac{dM_z(t)}{dt} = -\frac{M_z(t) - M_0}{T_1}$$

(2.11a)

with solution:

$$M_z(t) = M_0(1 - e^{-t/T_1})$$

(2.11b)

Eqn (2.11b) denotes that the return of the magnetisation vector to its equilibrium value along the z-axis is exponential. Note that T_1 relaxation is generally strongly dependent on the NMR frequency and so vary considerably with magnetic field strength B_0. There are several methods for measuring T_1 although two experimental methods, the progressive saturation Fourier transform (PSFT) and the inversion recovery Fourier transform (IRFT) are the most popular.[1,2] The first method is faster and is used for measuring T_1s longer than 5 s, whereas the second pulse sequence is more accurate. However, the fast version of IRFT (FIRFT) does not involve long waiting times reducing thus significantly the duration of the experiment.[2] Factors affecting the accuracy in T_1 measurements have been discussed thoroughly by Craik and Levy.[2]

First order reaction follows, also, the decay of magnetisation along the y-axis.

$$\frac{dM_y(t)}{dt} = -\frac{M_y(t)}{T_2}$$

(2.12a)

The solution of eqn (2.12a) describes an exponential decay of the transverse magnetisation, *i.e.* eqn (2.12b):

$$M_y(t) = M_0 e^{-t/T_1}$$

(2.12b)

In reality however, the spin–pin relaxation decay (the FID as we shall see later) is affected by such factors as magnetic field inhomogeneity (*i.e.* microscopic regions within the sample with slightly different magnitudes of B_0), unresolved coupling, temperature gradients, and other factors. Because of these effects, the decay constant of spin–spin relaxation is called T_2^* (T_2 star) rather than T_2. T_2^* is an instrumentally dependant parameter and it determines the line width of an NMR resonance observed in the spectrum. T_2, on the other hand, is the natural relaxation parameter independent of the field inhomogeneity, J coupling and other influences. T_2 is always greater than or equal to T_2^*. The spin-echo sequence (see below) has been used to measure T_2 by applying a train of 180_x° pulses interrupted by time intervals t_1 and acquiring the successive spin-echoes (Carr–Purcell–Meiboom–Gill sequence).[3]

Small amounts of paramagnetic substances in a liquid sample speed up relaxation very much. The magnetic moment of the lone electron is about six times higher than that of protons, generating thus much stronger local magnetic fields that induce relaxation. In this respect, sample degassing is required to remove the dissolved paramagnetic oxygen for the measurement of T_1 and T_2 of liquid samples.

2.5 NOE

NOE arises from dipole–dipole coupling, *i.e.* though space interaction of the magnetic moment of a nucleus with the magnetic field generated by a neighbouring nucleus. As we shall see below the dipolar interaction is related to the distance that connects the dipoles. To understand the nature of NOE, we have to look at a two-spin system I and S with spin $\frac{1}{2}$. Since NOE involves polarisation, *i.e.* population differences between the α and β states, the energy level diagram in Figure 2.6(b) will be used. In this Figure, the single-, double-, and zero-quantum transitions are labelled as W_{1I} or W_{1S}, W_2, and W_0, respectively. Ws represent transition probabilities or rates at which certain transitions can take place. W_{1I} or W_{1S} correspond merely to spin–lattice relaxation of the two spins I and S. W_2, and W_0 cannot be observed because they have much lower probability to occur (remember that these transitions are forbidden), but they are allowed for relaxation. If the population of the α and β states of one spin *e.g.* spin I, is changed by saturation, creating equal population in both states, then T_1 relaxation will force the population of spin I back to equilibrium Boltzmann distribution. Spin I will just relax through W_1 mechanism without effecting spin S. However, the other two mechanisms, W_2, and W_0 will have an effect on spin S. While the population of spin I goes back from saturation to equilibrium, W_0 mechanism will cause the neighbouring (so far unperturbed) spin S to deviate from its Boltzmann equilibrium towards a decrease in population difference between states α and β. This will result in a decrease in signal intensity for S, *i.e.* a 'negative NOE effect'. On the other hand, W_2 mechanism will cause the population difference of the undisturbed spin S to increase, corresponding to an increase in signal intensity, a 'positive NOE effect'. The magnitude of NOE can be calculated according to the Solomon's equation shown in eqn (2.13):

$$\text{NOE} = 1 + \frac{\gamma_I}{\gamma_S} \frac{(W_2 - W_0)}{W_0 + 2W_{1S} + W_2} \tag{2.13}$$

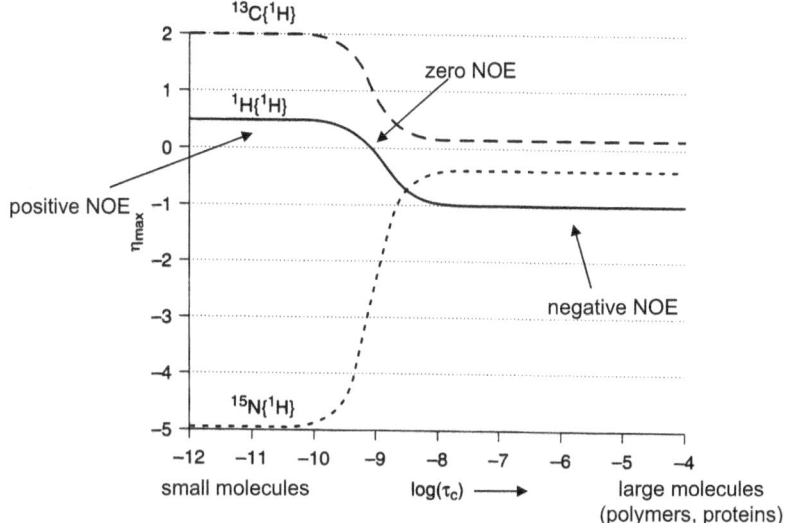

Figure 2.7 Variation of the maximum steady state NOE as a function of the molecular correlation time. Positive NOE is observed for fast molecular
motions at the extreme narrowing limit, while negative NOE is attained
for slow motions. Also negative NOE is displayed for nuclei with
negative gyromagnetic ratios (*e.g.* ^{15}N).

Both magnetogyric ratios are the same for the same type of spins, *e.g.* protons.
The magnitude of NOE depends strongly on the difference $(W_2 - W_0)$. When
$W_2 < W_0$, then NOE is negative, and it is positive when $W_2 > W_0$. The relative
magnitude of W_2 and W_0 depends on the molecular motion. A microscopic
measure of the rotational motion of a molecule is the so-called molecular correlation time, τ_c (the time it takes a molecule to rotate one radian, $360°/2\pi$); τ_c is
related to the transition probabilities Ws and the resonance frequencies ω_I and
ω_S of nuclei I and S, respectively, with rather complex equations. The magnitude
of τ_c depends primarily on molecular size, and increases with it. Rapidly moving
small sized molecules have a short correlation time, while slowly rotating heavier
molecules of approximately the same shape have a long correlation time. The
magnitude of τ_c ranges from 10^{-9} s for large molecules (*e.g.* polymers) to 10^{-11}–
10^{-12} s for small molecules. Moreover, fast motion favours the W_2 mechanism,
and hence a positive NOE is expected, while slow motion advances the W_0
mechanism and a negative NOE. Figure 2.7 shows the NOE dependence on the
molecular correlation time for two protons undergone dipolar interaction.

For very short correlation time satisfying the condition $\omega_0\tau_c \ll 1$, the so-called
'extreme narrowing limit', the NOE reaches its maximum value of $+0.5$ (eqn
2.14). As the correlation time increases, the NOE decreases, passes through zero
and then increases in negative values (Figure 2.7). (eqn 2.14)

$$\text{NOE(max)} = \frac{\gamma_S}{2\gamma_I} \qquad (2.14)$$

The maximum NOE is 0.5 and 1.987 for ^{1}H and ^{13}C nuclei, respectively. Positive NOE values are exploited to increase the signal intensity in proton decoupled NMR spectra of the less sensitive nucleus ^{13}C.

This analysis indicates that the intensity of NMR signal depends strongly on molecular dynamics. Moreover, for large molecules (*e.g.* polymers,) tumbling outside the extreme narrowing condition ($\omega_0\tau_c > 1$), the magnitude of NOE is dependent on the applied magnetic field B_0. In fact, NOE decreases with increasing magnetic field, and reaches faster its zero value. Therefore, the NMR sensitivity of large molecules may be not benefited by NOE using NMR spectrometers equipped with large magnets. Negative signal intensities (inverted signals) can be observed for nuclei whose gyromagnetic ratio is negative (*e.g.* ^{29}Si, ^{15}N, ^{17}O) even if the molecule is tumbling within the extreme narrowing limit (Figure 2.7). If less than the full negative NOE is generated, the resonance may disappears altogether, as the negative NOE cancels the natural signal. To avoid this annoying situation, it is customary to run the spectra of nuclei with negative gyromagnetic ratios in the absence of NOE. Methodologies to suppress NOEs are the inverse gated decoupling (see section 5.2.1), polarisation transfer methods (see section 2.8.3) and the less commonly use of paramagnetic relaxation reagents (see section 5.2.1).

Another parameter that affects NOE is the inter-nuclear distance, r. The probability rates W are proportional to $1/r^6$; NOE is only sensitive to neighbouring nuclei no more than $5-6$ Å away. Steady-state NOE has proven to be an additional effective means for spectral assignment, especially for its ability to connect hidden signals in complex spectra. In addition, the dependence of NOE to inter-proton distances has been exploited heavily to determine the three-dimensional structure of small and large molecules through its 2D variant, NOE spectroscopy (NOESY).

2.6   Pulsed Fourier Transform NMR

2.6.1   Rotational Frame of Reference (RFR)

The RFR is a theoretical device that simplifies the description of the spin system using the vector model. The RFR is a Cartesian coordinate system, whose z'-axis coincides with the z-axis of the stationary laboratory frame, while its $(x'y')$ plane rotates with an arbitrary frequency, ω_R (Figure 2.8(a)).

The behaviour of the magnetisation vector M_0 and its transverse component M_{xy} in the laboratory frame and RFR after excitation are shown in Figure 2.8(b) and Figure 2.8(c), respectively. In the static laboratory frame both the magnetisation vector M_0 and its transverse component rotate with the Larmor frequency ω_0. Also, the static frame shows field components B_1' and B_1'' rotating in opposite directions (section 2.2). The whole picture changes in the RFR, which rotates with the Larmor frequency (equal to the frequency $+\omega_0$ of the field component B_1'). The B_1' component appears to be static and directed along the x'-axis of the RFR, whereas at the same time vector M_0 and its transverse component M_{xy} appears to be stationary (Figure 2.8(c)). The RFR has

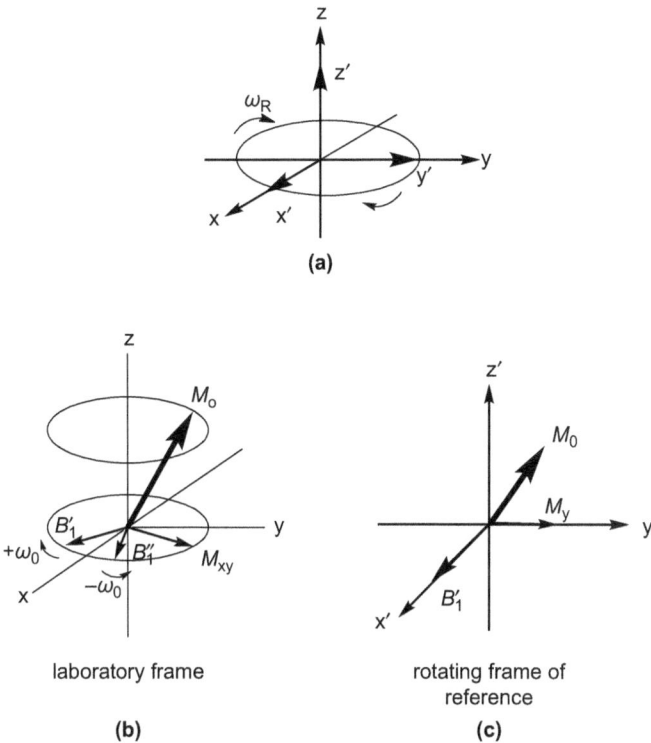

Figure 2.8 (a) The fixed laboratory frame and the rotating frame of reference with frequency ω_R. (b) Precession of the magnetization vector M_0 and its transverse component M_{xy} with the Larmor frequency ω_0 in the laboratory frame; the rotating field components B′ and B″ are also shown. (c) Motions of both vectors are frozen in the rotating frame of reference, when $\omega_R = \omega_0$.

succeeded to remove the time dependence of the applied magnetic field B_1. Since the precessional motion of the magnetisation vector was induced by the static magnetic field B_0, this is also no longer present in the RFR. The condition that the RFR has the same frequency with the precessional frequency of the nucleus (Figure 2.8(c)) is called 'on resonance'. The 'off resonance' condition occurs when the vector rotates with a frequency ω greater or lower than the frequency ω_0 of the RFR (Figure 2.9(a)). This difference frequency $\Omega = \omega_0 - \omega$, which is positive when $\omega_0 > \omega$ and negative when $\omega_0 < \omega$, is called the offset frequency. Also, RFR is useful to describe the motion of a coupled nucleus with coupling constant J. The situation is analogous to what happened with chemical shift. In this case, since there are two new energy levels for the spin, we get two counter-rotating vectors. Provided that the nucleus is on-resonance, its evolution will depend on the magnitude of J, not ω_0: the vector splits into two counter-rotating components with frequencies $+J/2$ and $-J/2$ (Figure 2.9(b)). For the sake of brevity, unprimed axes will be used hereafter for the RFR.

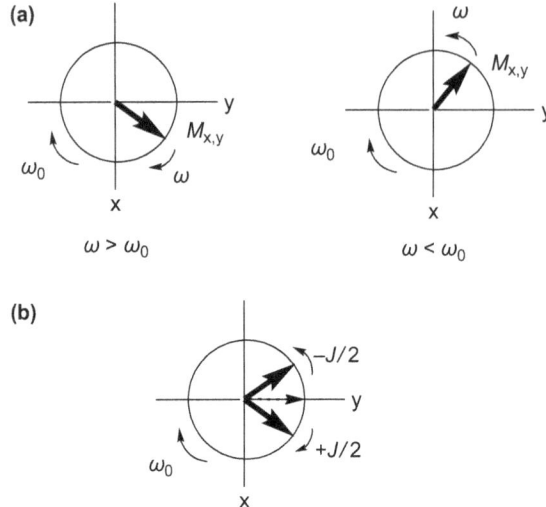

Figure 2.9   (a) Off-resonance conditions with vectors rotating faster ($\omega > \omega_0$) or slower ($\omega < \omega_0$) than the rotating frame of reference with frequencies $\Omega = \omega - \omega_0$ and $\Omega = -(\omega - \omega_0)$, respectively. (b) Coupling in the rotating frame. Vectors evolve in opposite directions according to their coupling constant J.

2.6.2   The Single-Pulse Experiment

NMR spectrometers during the first two decades of their appearance used to acquire NMR spectra by a technique known as continuous wave (CW) spectroscopy. According to this method, the magnetic field B_0 was kept constant, while sweeping the frequency ω_1 of the electromagnetic radiation over the frequency range of interest (frequency sweep). Alternatively, NMR spectra could be recorded using a fixed radiofrequency source ω_0 and varying the magnetic field B_1 (field sweep). Both CW methods generated a signal whenever the Larmor frequency of a nucleus was in resonance with the frequency of the radiating source. The final outcome of this technique was the 1D NMR spectrum, *i.e.* a plot of the signal frequencies against their intensities. Despite the revolution in physical and analytical chemistry brought by NMR, CW spectroscopy was inefficient in time and sensitivity. In time, because the sweep rate was slow (usually $1\,\mathrm{Hz\,s^{-1}}$, which means $1000\,\mathrm{s}$ or $0.28\,\mathrm{h}$ for a spectral width of $1000\,\mathrm{Hz}$), while a wide range of scanned frequencies were vacant of signals. As the NMR signal is intrinsically weak, this would lead to a poor signal-to-noise ratio (S/N). On the other hand, the procedure of signal averaging, *i.e.* adding spectra from repeated frequency or field sweeps, to increase signal sensitivity lengthens tremendously the duration of the experiment.

Modern NMR spectroscopy developed during the 1970s employed pulses to excite at once all nuclei within a particular frequency range. The RF coil of the transmitter of the NMR spectrometer generates NMR pulses. The output

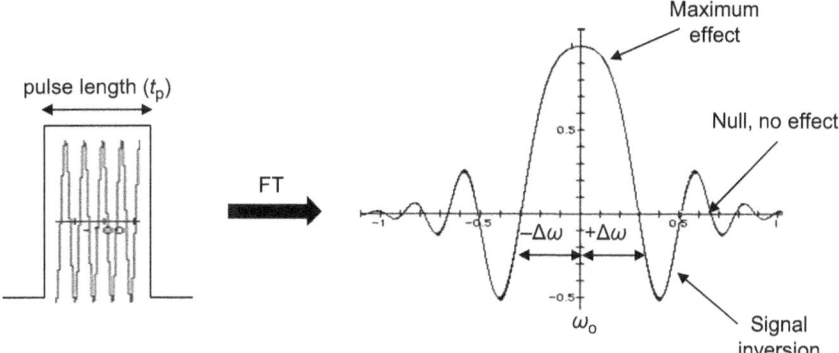

Figure 2.10   Rectangular pulse and its FT. Apart from the principal excitation bandwidth, the frequency profile involves side-lobes that induce inverted or no signals at all.

power of the transmitter in pulsed NMR must rise very quickly from zero to its highest value (*ca.* 100 W) and then falls back to zero at the end of the pulse in a very short time. A pulse is characterised by its amplitude or power, which is a function of the magnetic field B_1, and its duration or pulse length, or pulse width, t_p. A train of pulses with frequency ω_0 and small t_p, produces side bands within a frequency range of $\pm 1/t_p$ (Figure 2.10).

All nuclei having resonance frequencies within the range $\Delta\omega = + 1/t_p$ or $-1/t_p$ are excited simultaneously in a period of time much shorter than that consumed in CW spectroscopy for the same frequency range. Short pulses (also called hard pulses), typically of a few microseconds, cover a wider range of frequencies, whereas longer pulses (also called soft pulses) of a few milliseconds can excite selectively a narrow range in the spectrum.

The effect of a pulse on a spin system is most easily described visually by adopting the vector model, and the RFR. Applying an RF pulse on a spin system tips the total or bulk or macroscopic magnetisation M_0 from its equilibrium position at an angle φ (see Figure 2.3(c)). The deflection or flip angle φ depends on B_1 and the pulse length t_p according to eqn (2.15):

$$\varphi = (\gamma/2\pi)B_1 t_p. \tag{2.15}$$

Different flip angles can be achieved by varying B_1 or t_p, or both. Two most common angles are $\varphi = 90°$, deflecting clockwise the bulk magnetisation vector M_0, in the (xy) along the y-axis of the RFR, and $\varphi = 180°$, which inverts the spin population of the energy levels and aligns M_0 with the negative z'-axis (see below).

The single pulse experiment is shown schematically in Figure 2.11.

The preparation period, T, is required to establish the Boltzmann equilibrium among the nuclear spins in the sample before the onset of the experiment. The spin system is then disturbed by a 90° RF pulse, and after the pulse,

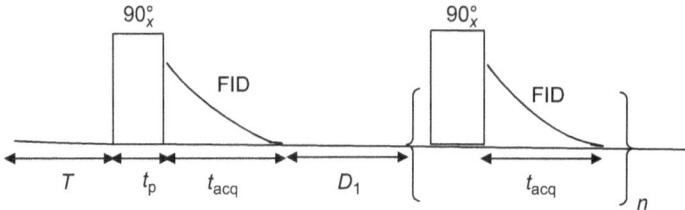

Figure 2.11   The single pulse experiment.

the signal in the time domain, the so-called free induction decay (FID), is detected during the acquisition time, t_{acq}. A time period D_1, known as relaxation delay, is inserted just after the acquisition period. The length of D_1 is adequate to allow full relaxation of the excited spins before the application of the next pulse. The relaxation delay is an important factor while designing the experiment for quantitative analysis (section 5.2). The single pulse experiment is repeated n times to increase the S/N of the experiment. The repetition of the experiment in a reasonable time period for obtaining a sufficient S/N ratio is a prerequisite for insensitive nuclei, such as ^{13}C and ^{15}N. Since data acquisition is so fast, requiring about 1 s to sample the FID, the accumulation of several hundred FIDs before carrying out Fourier transform is a matter of minutes.

   When pulse is turned off the transverse magnetisation M_y along the y-axis starts decaying, while the longitudinal component M_z is increasing along the z-axis, inasmuch the excited nuclei are relaxing back to their equilibrium position, At the same time the various components of M_y lose their phase coherence due to spin – spin coupling and the effect of field homogeneity and precess with different Larmor frequencies on the (xy) plane. As mentioned in section 2.4, the driving force of the M_y decay is the spin – spin relaxation process with time constant T_2^*, and that of the M_z built up along the z-axis is the spin – lattice relaxation process with time constant T_1. The built up of the equilibrium magnetisation along the z-axis, and the decay of the transverse components on the (xy) plane are expressed by the following expressions (eqn (2.16)) following the deflection of M_0 by a 90° along the y'-axis.

$$M_z(t) = M_0 e^{t/T_1} \quad M_y = M_0 \cos \omega_0 t e^{t/T_2^*} \quad M_x = M_0 \sin \omega_0 t e^{t/T_2^*} \qquad (2.16)$$

   The decaying transverse magnetisation induces an oscillating electric current on the receiver coil of the NMR spectrometer according to the Faraday law. The amplified intensity of this current follows the decay of M_y and can be seen on the screen of the NMR spectrometer. This time domain signal, $F(t)$, is the free induction decay mentioned previously. All the information concerning the NMR spectrum is contained in the FID and as we shall see later a large part of the spectrometer operation is devoted to sample, amplifying, recording and analysing this signal. Fourier transform of FID affords the signal in the frequency domain $F(\omega)$, the NMR spectrum. For one or two types of nuclei in the

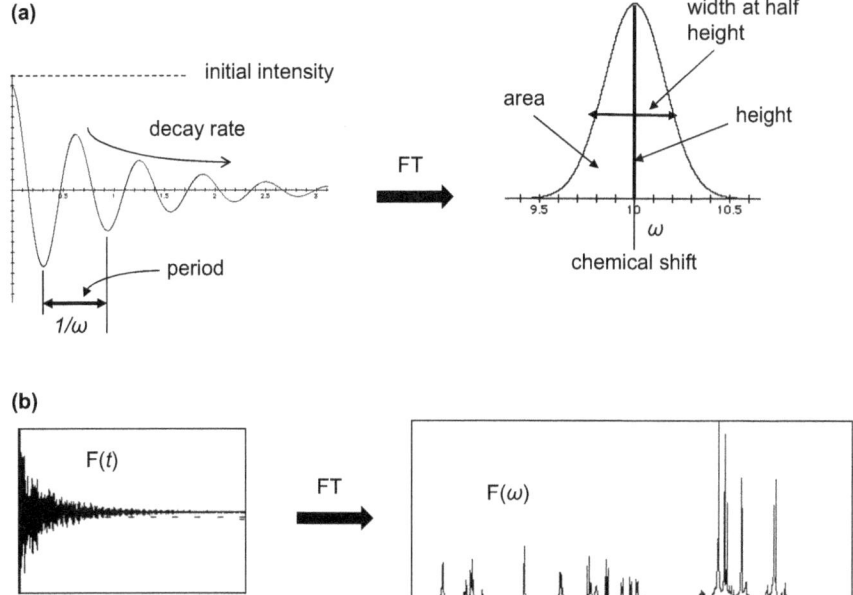

Figure 2.12 (a) Correspondence of signal characteristics in the time domain (FID) and frequency domain (spectrum). The amplitude of the cosine wave controls the signal intensity, the decay rate the line width, and the period the resonance frequency (chemical shift). (b) Fourier transform of a complex FID.

molecule, the NMR parameters mentioned previously can be easily extracted from the FID as shown in Figure 2.12(a), which in addition depicts the correspondence of the spectral parameters in the time and frequency domain.

For molecules, however, with a significant number of non-equivalent nuclei and signal multiplicities, the FID is much more complicated constituted by the superposition of the FIDs of the individual resonances including noise (Figure 2.12(b)). In this case, Fourier transform of the spectrum in the time domain cannot be avoided.

The unique advantages inherent in the pulsed NMR experiment with Fourier transform opened new possibilities of applying special pulse sequences that offer a large range of new NMR experiments. 2D and 3D NMR would not be possible without the introduction of pulses in NMR.

2.7 Pulses

Apart from the excitation of nuclei within a frequency range, pulses have been employed to obtain specific information from NMR spectra. Pulses of different strengths, widths, phases, frequencies and shapes, as well as composite and multiple pulses were invented and applied in 1D and 2D NMR experiments.

2.7.1 Non-Selective Pulses

Non-selective or hard pulses, which constitute an indispensable feature of all 1D and 2D multi-pulse experiments are usually 90° and 180° pulses of variable phases, and pulse lengths of the order of a few μs. The 90° pulse maximises the signal in the (xy) plane, whereas the 180° pulse inverts the spin-population of the energy levels. Figure 2.13 depicts the effect of the 90° and 180° pulse on M_0. The phase of the pulse depends on the direction of the applied RF field. For instance, a 90°_x pulse is applied along the x-direction of the RFR, and flips the magnetisation clockwise on the y-axis. This pulse is referred as an x-pulse or pulse about x.

2.7.2 Selective Pulses

Selective or soft pulses have bandwidths of 10 to 50 Hz, and durations of 10–100 ms. They are used to suppress unwanted signals or to selectively excite a narrow region in the spectrum, which can be so narrow involving just one

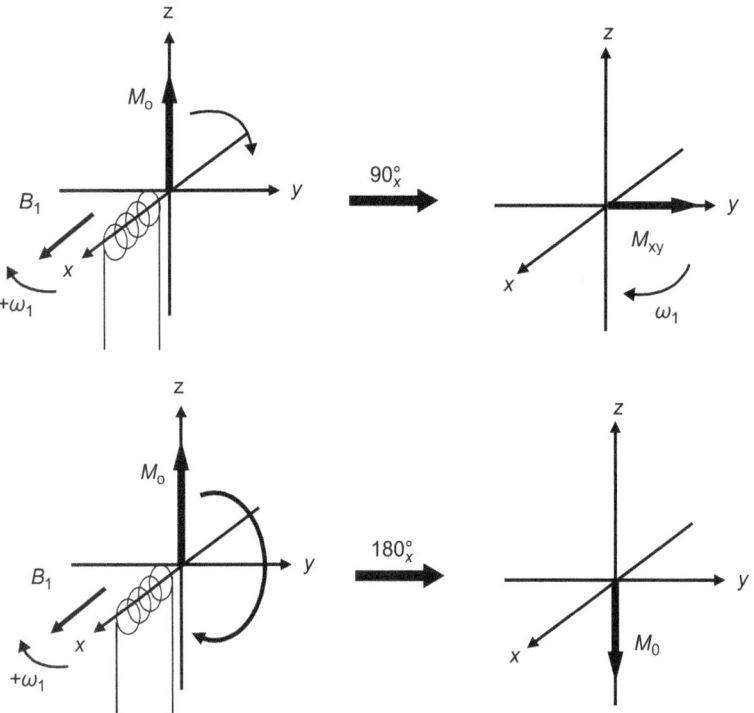

Figure 2.13   The 90°_x and 180°_x pulses. The 90°_x pulse deflects clockwise the magnetisation vector onto the (xy) plane along the y-axis, whereas the 180°_x pulse brings about inversion of the magnetisation, *i.e.* inversion of the spin population between the energy states.

component of a multiplet. Also, selective 90° pulses are used for proton decoupling experiments upon irradiating specific resonances in ¹H NMR spectra, removing the splitting pattern caused by homonuclear or heteronuclear spin–spin coupling. The pulse is applied selectively to the Larmor frequency of the proton to be decoupled during the acquisition of the free induction decay. The irradiated ¹H spins can be envisioned as rapidly flipping between α and β states, so the ¹³C nucleus (or a nearby coupled proton) sees only the average dipolar interaction (which is zero).

Another use of selective pulses is the measurements of the steady state NOE upon irradiation of specific resonance and observation of the results (intensity increase or decrease) on another signal in the spectrum connected with the irradiated nucleus *via* dipole–dipole interactions. Selective 90° and 180° pulses were used in polarisation transfer experiments (see below) and in 1D variants of the 2D pulse sequences, such as COSY, TOCSY, NOESY, and ROESY.

2.7.3   Composite Pulses

A composite pulse[4] is simply a sequence of single pulses with different amplitudes and phases. They were used initially to correct pulse imperfections and inefficient pulse strength to excite the whole spectral width in question, resulting in loss of S/N and creating artefacts. Many composite pulse schemes have been developed and proposed to obtain more accurate pulses. The composite pulse sequence is written in chronological order from left to right. For instance the composite pulse $90^\circ_x 180^\circ_y 90^\circ_x$ replaces the 180°_x pulse for inversion of the magnetisation vector. Figure 2.14 shows the effect of the composite pulse $90^\circ_x 180^\circ_y 90^\circ_x$ on the magnetisation vector that fully aligns this vector along the $-z$-axis.

The trajectory of the $90^\circ_x 180^\circ_y 90^\circ_x$ composite pulse, when the effective pulse was 80° is depicted in the same Figure. Even with this significant error, the net magnetisation still winds up very close to $-z$-axis.

Composite pulses found extensive application in broadband (BB) proton decoupling, which eliminates all heteronuclear, *e.g.* ¹³C–¹H splitting in ¹³C NMR spectra, or at least make them smaller than the line width. The earlier applications of BB decoupling were based on the white noise decoupling. White noise is a random signal containing a range (bandwidth) of frequencies of equal power that succeeds in the removal of all ¹³C–¹H couplings. However, this decoupling technique becomes ineffective when irradiation should cover a far greater spectral bandwidth especially at high-field NMR intensities for protons. In order to irradiate larger frequency range higher power is needed. Unfortunately, higher power decoupling is more likely to destroy the probe and sensitive sample (*e.g.* biological samples) by creating sample overheating.

The first full decoupling experiment using non-selective pulses consisted of a sequence of effective 180° pulses separated with delays of equal length that rapidly exchange α and β spin states and decouple ¹H from the heteronucleus, *e.g.* ¹³C. However, this pulsing technique was inefficient, since errors in accurately measuring a pulse length lead to cumulative errors in a series. Pulse

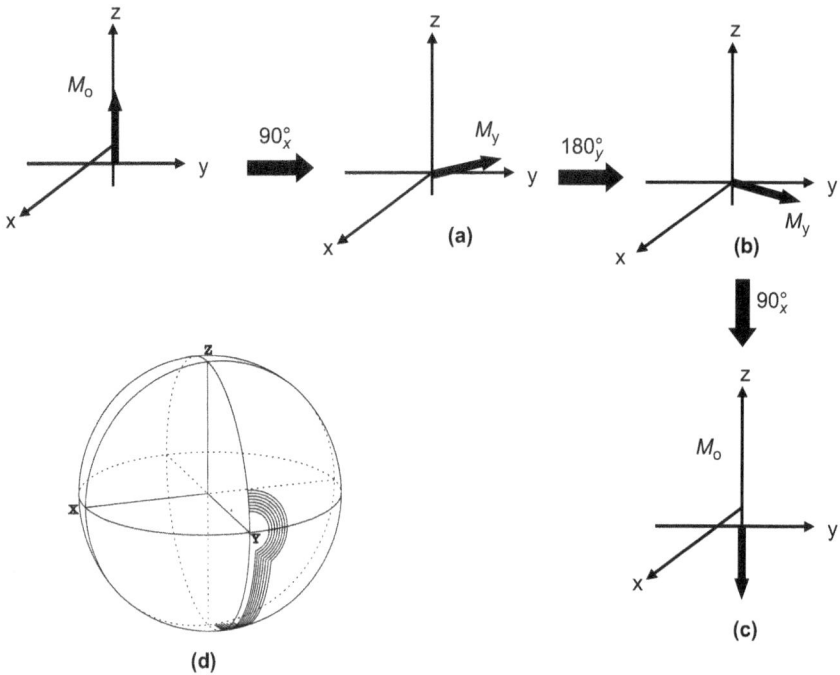

Figure 2.14 The self-compensating ability of the composite pulse $90^\circ_x 180^\circ_x 90^\circ_x$ for inverting the nuclear magnetisation vector. (a) Application of a pulse less 90° along the *x*-axis deflects the magnetisation vector above the (*xy*) plane. (b) The second 180° pulse along the *y*-axis flips the vector below the (*xy*) plane in a symmetrical position. (c) The final missed 90° pulse terminates the vector along the $-z$-axis. (d) A family of trajectories calculated for pulse lengths between 80 and 90% of the nominal values.

errors are minimised by a combination of different pulse lengths and phases. This led to a number of composite pulses with characteristic acronyms like MLEV and WALTZ. These schemes effectively allow decoupling to be efficient over a range of proton frequencies with minimum power. MLEV-16 makes use of $(90^\circ)_x (270^\circ)_y (90^\circ)_x$ composite pulse united into the 16 step cycle $R = 90^\circ_x 270^\circ_y 90^\circ_x$ and the reverse step (180° phase shift) $R' = 90^\circ_{-x} 270^\circ_{-y} 90^\circ_{-x}$. The pulse repetition in 4 cycles with phase permutation[4,5] is:

$$MLEV\text{-}16 = RRR'R' | R'RRR' | R'R'RR | RR'R'R$$

This composite pulse decouples efficiently a frequency bandwidth of ± 4.5 kHz. Better offset compensation is offered by **WALTZ** decoupling scheme.

WALTZ-16 can be obtained by permutation of the basic composite pulse $R = 90^{\circ}_{x}180^{\circ}_{-x}270^{\circ}_{x}$ and its reverse $R' = 90^{\circ}_{-x}180^{\circ}_{x}270^{\circ}_{-x}$; the pulse repetition[6,7] is:

$$WALTZ\text{-}16 = RRR'R'|R'RRR'|R'R'RR|RR'R'R$$

which decouples efficiently over $\pm 6\,kHz$, and corrects imperfections in MLEV. An alternative broadband decoupling scheme is GARP that is a computer-optimised decoupling technique using non-selective 90° flip angles; effective decoupling frequency range $\pm 15\,kHz$.[8]

2.7.4 Shaped Pulses

The shape of the pulse is tailored to give an excitation profile other than that for rectangular pulses shown in Figure 2.10. The latter profile follows a sinc function ($sinc\,x = \sin x / x$), which is not uniform across the whole frequency spectrum. Beyond the frequency range corresponding to $\pm 1/t_{p}$ of maximum effect, there are null points where nuclei remain unperturbed or frequency regions where inversion may occur. These features become a serious problem at high magnetic field strengths (800 and 900 MHz) for ^{13}C spectra that have a large chemical shift range ($> 200\,ppm$) or in cases where selective pulses are required. Also, complex experiments including multiple pulses depend on the accuracy and consistency of pulse widths. These problems were solved by the introduction of shaped pulses.[9] Several shaped pulses with different excitation profiles have been suggested, especially for selective excitation. The most common are the Gaussian, the half-Gaussian and the Gaussian 270° pulses shown in Figure 2.15.

The Fourier transform of a Gaussian is a Gaussian and therefore a narrow Gaussian shaped excitation bandwidth is obtained. Although the excitation is not flat, the phase is constant along the excitation profile, and in addition the total duration of the pulse is about ten times shorter than the rectangular pulse. A careful examination, however, of the profiles in Figure 2.15 advocates in favour of the half Gaussian pulse, since the simple Gaussian and the Gaussian 270° pulses retain some undesirable oscillatory behaviour.

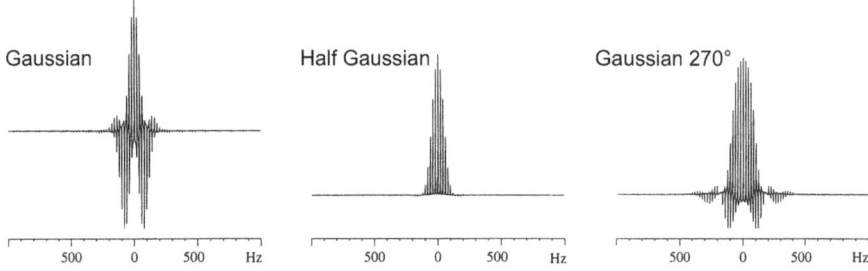

Figure 2.15 Simulated excitation profiles of Gaussian, half-Gaussian and Gaussian shaped pulses.
(Reprinted from ref. 4. Copyright (1999), with permission from Elsevier.)

2.7.5   Field Gradient Pulses (PFG)

Static field gradients have been used long ago in NMR imaging in order to spatially encode images, and only recently they have been applied to high-resolution 1D and 2D NMR experiments. PFG have been used to select the desired signals and suppress all others (*e.g.* suppression of solvent signals, purging of unwanted magnetisation, selection of coherence transfer pathways).[10] The benefits arising from the use of PFG in high-resolution NMR is to obtain NMR spectra of good quality in much shorter time than the traditional methodology of field cycling.

The application of a linear field gradient pulse, G_z along the *z*-direction for a discrete period of time τ_g in addition to the static field B_0 introduces an inhomogeneous magnetic field environment for all nuclei in the sample that were previously excited with an RF pulse. The result of the PFG application is that the nuclear spins will precess with different frequencies, $\omega_z = \gamma G_z$ because they experience different magnetic field strengths along the field gradient (Figure 2.16(a)).

The magnetisation vectors of the nuclei in the (*xy*) plane rotate through a spatially dependent phase angle, $\Phi_{(z)} = \gamma G_z \tau_g$. As a result, the net transverse magnetisation, which is the sum of all transverse components of the nuclear spins, is zero (Figure 2.16(b)). The dephasing effect of the PFG on the magnetisation is removed by applying a second PFG immediately after the first, of equal intensity and duration, but in the opposite sense, *i.e.* along the $-z$-axis. The individual vectors rotate in the opposite direction (Figure 2.16(c)) and after the end of the second PFG they refocus to produce a gradient echo and an observable net signal. This defocusing and refocusing effect on the

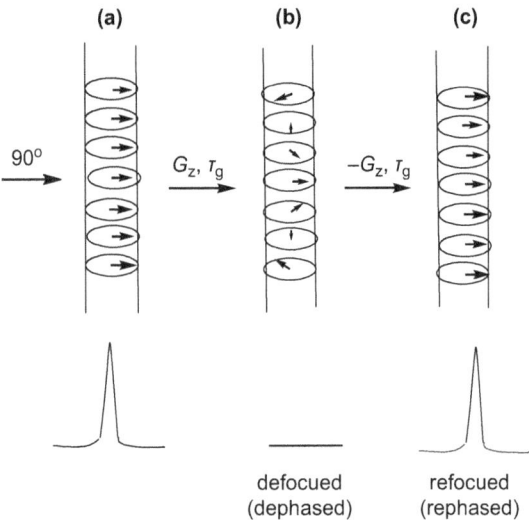

Figure 2.16   Graphic representation of the effect of a pulsed-field gradient on a collection of spins (see text).
(Reprinted from ref. 4. Copyright (1999), with permission from Elsevier.)

magnetisation vectors can be performed selectively, *i.e.* suppressing unwanted signals and refocusing only those that are of interest, making PFG the most important accessory for every gradient selected experiment.

2.8 Useful Pulse Sequences

The availability of the Fourier transform technique made possible the development of many experiments that use more than one pulse per acquired FID. These pulse sequences composed of a series of RF pulses, delays, gradient pulses and phases; give information that is difficult or impossible to obtain by the one pulse technique.

The multi-pulse sequences that will be described briefly in this paragraph belong to the family of 1D NMR techniques, inasmuch the various time delays are kept constant during the experiment (constant time sequences); We will start presenting pulse sequences that are used as building blocks in more complicated 1D and 2D NMR experiments.

2.8.1 Spin-echo

Figure 2.17 Panel A illustrates the spin-echo sequence and a diagrammatic summary of the spin echo formation described by the vector model. The initial 90_x° pulse (Figure 2.17a, b) is followed by a delay t_D during which the transverse magnetization M_y is resolved (loss of phase coherence) into its spin components as a result of differences in chemical shifts and therefore they precess at different rates ($\omega \neq \omega_0$) on the (xy) plane of the rotating frame of reference (Figure 2.17c). Placing a 180_x° pulse in the middle of the delay period (Figure 2.17d) will reverse the direction of the spins precession, bringing them all back to the origin after the second half of the pulse delay t_D (Figure 2.17e). The result is the refocusing of the chemical shifts, but obtaining a reverse signal after Fourier transformation (Figure 2.17f).

The spin-echo sequence has been used to measure the spin–spin relaxation time T_2 by applying a train of 180_x° pulses interrupted by time intervals τ and acquiring the successive spin-echoes (Carr–Purcell–Meiboom–Gill sequence). Spin-echo is used to compensate field inhomogeneity (see above) and as a building block for complicated sequences to refocus the distribution of chemical shifts and coupling constant effects. Also, it is an important method to obtain magnetic resonance images (see section 2.11).

2.8.2 Spin-Lock

This is a modified spin-echo sequence $(90_x^\circ - t - 180_y^\circ)n$ where the time interval t is very short and the 180° pulses are repeated n times (n is a very large number). Continuous application of the B_1 field along the *y*-axis results in locking the magnetisation along the *y*-direction. This means precession around B_1. Spin-lock pulse sequences have been used in major 2D NMR experiments, such as TOCSY and ROESY.

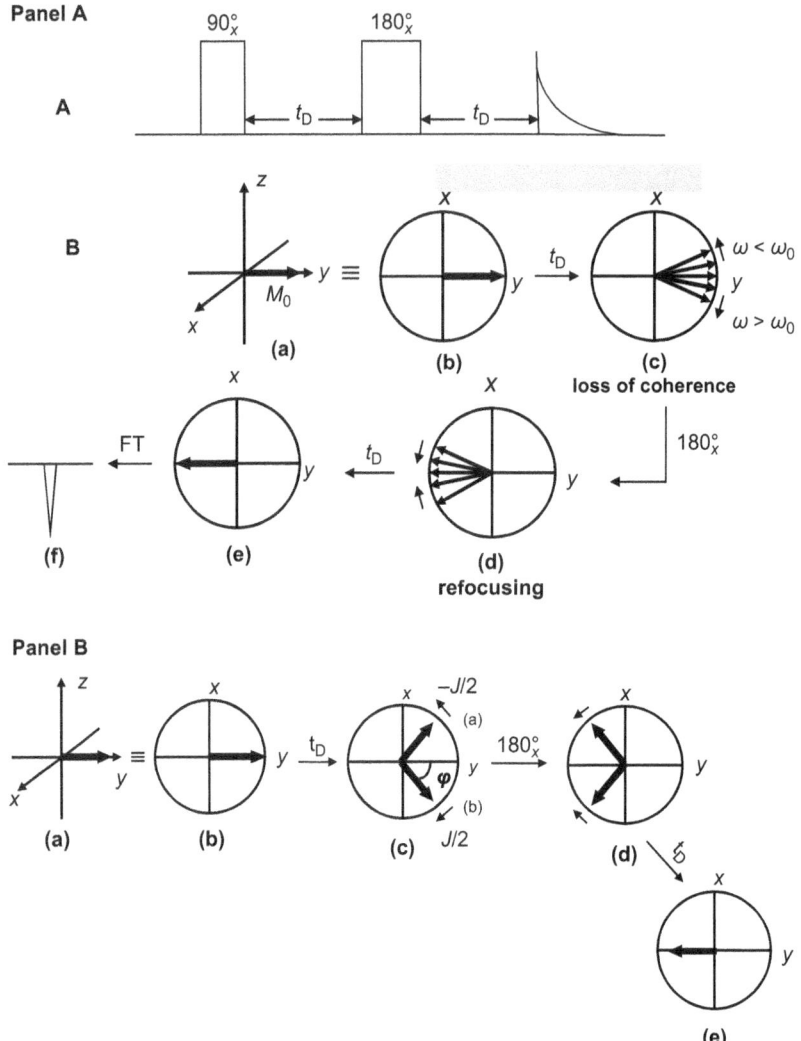

Figure 2.17   Panel A: (a) the diagram of the SE sequence. (b) Illustration of the effect of the SE sequence on the magnetisation vector M_0 in the rotating frame of reference. Panel B: refocusing effect of the SE on the heteronuclear spin-coupling (see text).

2.8.3   Polarisation Transfer

An important group of multiple-pulse experiments includes the polarisation transfer methods for sensitivity enhancement of nuclei with poor sensitivity because of low natural abundance and small magnetogyric ratio, such as ^{13}C, ^{15}N. The term polarisation denotes changes in the spin populations in the occupied energy levels of coupled nuclei induced by pulses. The polarisation transfer involves the transfer of the large excess population (polarisation) from a sensitive nucleus like proton to an insensitive nucleus like ^{13}C by using selective or non-selective pulses. One way to obtain polarisation transfer is to employ selective pulses, either to saturate (Selective Polarisation Transfer, SPT) or to invert (Selective Population Inversion, SPI) one ^1H components of a multiplet by a 90° or 180° pulse, respectively, and observe the results on the ^{13}C NMR spectrum. A two-fold and four-fold increase in the carbon lines' intensity is achieved by SPT and PSI methods, respectively. Nevertheless, the utility of these methods is limited by the obvious disadvantage that only one multiplet at a time can be polarised. Moreover, the ^{13}C spectrum is proton-coupled and has up and down peaks. No proton decoupling can be applied because the sensitivity enhancement is induced by population differences in the ^1H energy levels, which would be gone by decoupling.

The use of non-selective pulses to manipulate the polarisation from protons to carbons avoids the limitations involved in the SPT and PSI experiments. This is accomplished by the pulse sequence called INEPT (Insensitive Nuclei Enhanced by Polarisation Transfer). The INEPT pulse sequence and the vector description of the ^1H magnetisation in the rotating frame for an X–^1H system with coupling constant J are shown graphically in Figure 2.18.

The components of the ^1H doublet, rotating in the (xy) plane move through an angle $\varphi = \pi^* t_D {}^* J = \pi^*(1/4J)^* J = \pi/4$ from the y-axis (Figure 2.18). The end result of this sequence is a spectrum with an anti-phase doublet. The sensitivity for X-nuclei is enhanced by the ratio γ_H/γ_X. For instance, for the ^{13}C nucleus, INEPT offers intensity enhancement by a factor ~ 4, *i.e.* twice as much as that from NOE. Another important aspect of INEPT is the fact that nuclei with negative γ values (*e.g.* ^{15}N, ^{29}Si, ^{119}Sn) is not a disadvantage as in NOE, because the later is determined by $1 + \gamma_H/\gamma_X$, while polarisation transfer is governed by γ_H/γ_X. Sensible limitation of INEPT is the presence of anti-phase lines in the spectrum (Figure 2.18), precluding thus any proton decoupling. Any attempt to removing the J-splitting will cause a collapse of the anti-phase lines. The latest version of INEPT (Refocused INEPT) involves a refocusing period (delay–180°(^1H,^{13}C)–delay) in order to obtain in-phase ^{13}C magnetisation. Now, it is possible to apply broadband proton decoupling during acquisition in order to obtain a ^1H-decoupled ^{13}C spectrum.

2.8.4 Spectral Editing

Application of proton decoupling during the second t_1 period of the spin-echo sequence provides useful information about spin multiplicity, identifying carbon atom types in a molecule, such as CH, CH_2, CH_3 groups. This experiment is the so-called Attached Proton Test (APT). However, the poor editing accuracy of spin-echoes due to a wide range of the one-bond CH coupling constant ($^1J_{CH}$) values is the major limitation of this technique. Indeed, $^1J_{CH}$ varies dramatically (*e.g.* aliphatic CH_n *ca.* 125 Hz; HC= *ca.* 160 Hz, HC≡*ca.* 200 Hz), making the choice of a single value for the delay that matches $^1J_{CH}$ for

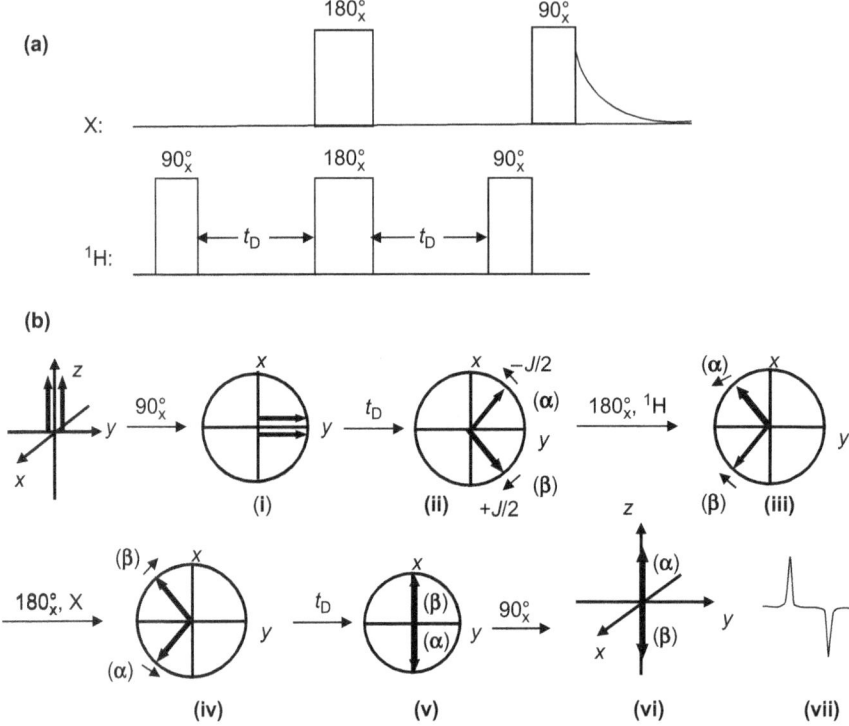

Figure 2.18 (a) Pulse sequence for the INEPT experiment. (b) The effect of the INEPT sequence on a X–^{1}H spin system. (i) The 90_x° pulse flips the ^{1}H magnetisation onto the (xy) plane. (ii) The delay t_D is set to $1/4J$, so the doublet components of ^{1}H spins labelled α and β, precessing in the rotating frame with frequencies $\pm J/2$, and each moving through and angle φ of $\pi/4$ from the y-axis up to the 180_x° pulses applied to both ^{1}H and X nuclei. (iii) The 180_x° pulse on ^{1}H spins rotates these components into the second half of the (xy) plane, while (iv) the 180_x° pulse on the X-nuclei ensures that reversing the sense of their precession does not refocus the coupling. During the second interval t_D, chemical shifts and field inhomogeneity are refocused. (v) By the end of the second interval the two components of ^{1}H spins doublets are aligned along the $\pm x$-axis. (vi) The final 90° pulse along the y-axis flips the components into the $\pm z$-direction (one up and one down). (vii) FT of the FID provides two anti-phase lines.

all carbons in a molecule problematic. Refocused INEPT can be used to produce edited spectra. Although INEPT is more efficient than APT, is unable to edit quaternary carbon nuclei. Nowadays, INEPT is used mostly as a building block in heteronuclear 2D sequences.

The most common spectral editing methods belong to the DEPT (Distortionless Enhancement by Polarisation Transfer) family. The relevant pulse sequence involves the generation and manipulation of multiple quantum coherences (MQCs) after the polarisation transfer. These coherences are associated with double-quantum and zero-quantum transitions met in section 2.3.2, which are forbidden and therefore cannot be seen or described with vectors. In the context of the vector model, it could be said that MQC of a simple spin system IS is composed of a group of evolving anti-phase vectors which have zero net magnetisation. For complete understanding of the full meaning and consequences of MQC, quantum mechanical analysis is necessary. In DEPT, the multiplicity editing is not done *via* a delay but through the pulse angle θ (Figure 2.19(a)).[4]

DEPT is much less dependent on the correct choice of the time intervals than INEPT. For the determination of the number of attached protons it is sufficient to take two DEPT spectra, one with $\theta = 90°$ (only CH, positive) and one with $\theta = 135°$ (CH and CH_3 positive, CH_2 negative). Full spectral editing, *i.e.* separate spectra with CH_n, ($n = 1$, 2, and 3) multiplicities, involves an additional spectrum with $\theta = 45°$; linear combination of the three spectra are then used to generate the edited versions. The CH, CH_2, and CH_3 edited spectra of the terpene andrographolide are shown in Figure 2.19(b) along with the normal ^{13}C spectrum.[4] A disadvantage of DEPT is its inability to detect quaternary carbons, and therefore always a regular ^{13}C spectrum is needed to be recorded in addition, increasing thus the time to acquire all the necessary chemical shifts information. Multiplicity information for all carbons including quaternary carbons has been accomplished by performing DEPTQ[11] (DEPT quaternary) and/or PENDANT[12] (Polarisation Enhancement During Attached Nucleus Testing) experiments. Both sequences use polarisation transfer and therefore have the sensitivity advantage of INEPT and DEPT. Comparing DEPTQ and PENDANT, DEPTQ gives better S/N than PENDANT.

2.9 Two-Dimensional (2D) NMR Spectroscopy

2.9.1 The Principle

The introduction of an additional time period, the evolution period, between the preparation period and detection (acquisition) period of the 1D experiment (Figure 2.11) is the basis for the 2D NMR spectroscopy. This additional time period is created, for instance, by application of a second 90° pulse (Figure 2.20).

The FID in such an experiment is now dependent on two time variables $F(t_1, t_2)$, the evolution time, t_1 and the detection time, t_2. Consequently, the 2D spectrum is a signal that depends on two frequencies $F(\omega_1, \omega_2)$ obtained after a double Fourier transformation of the signal in the time domain. The first 90° pulse creates a non-equilibrium condition for the spin system, which after the pulse evolves freely during the evolution period under certain physical condition (*e.g.* chemical shifts and/or spin–spin coupling, dipolar coupling). The

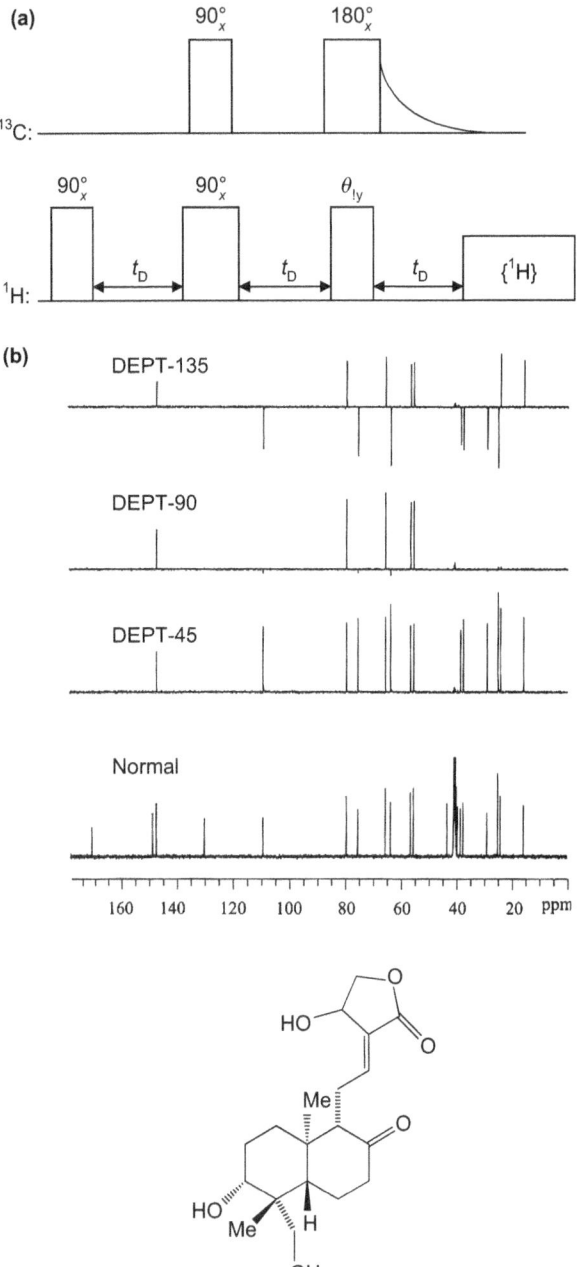

Figure 2.19   (a) The DEPT pulse sequence. (b) The conventional ^{13}C and DEPT-
edited spectra of the terpene andrographolide.
(Reprinted from ref. 4. Copyright (1999), with permission from Elsevier.)

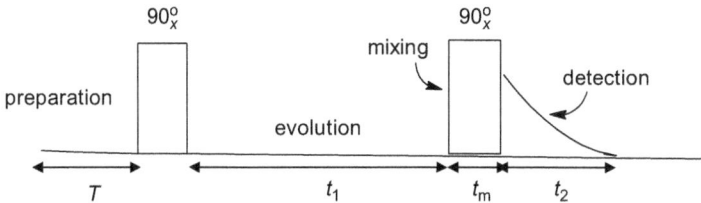

Figure 2.20 A simple 2D pulse sequence. This is also the basic COSY experiment.

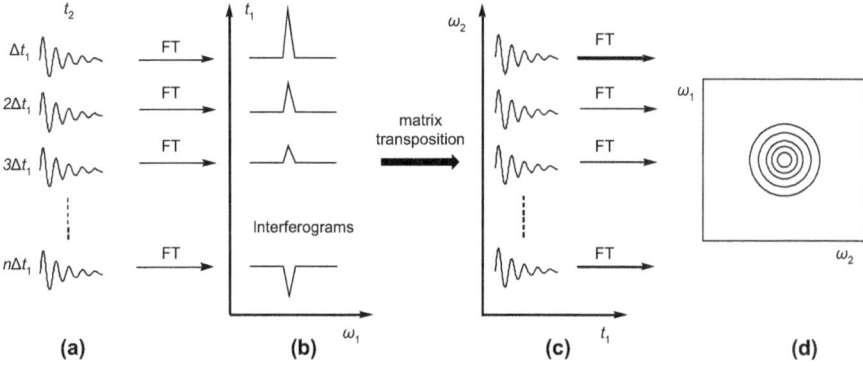

Figure 2.21 The general data flow for any 2D NMR experiment.

nature of the spin interactions during the evolution period characterises the type of the 2D experiment. The evolution period ends up with the second 90° pulse followed immediately by the sampling of the FID during the detection or acquisition period, t_2. What happens to the spin system during the evolution period cannot be detected directly, but the memory of the spin system during t_1 is transferred and recorded during time t_2. To map out the history of the spin system during the evolution period, and measure the dependence of the signal on two time variables, the experiment is repeated by making the evolution time t_1 systematically longer by adding consecutively time intervals (increments), Δt_1, as shown in Figure 2.21(a).

For each Δt_1 increment, a separate FID is detected during t_2. The complete data set can be regarded as a matrix with rows representing spectra obtained in individual experiments and columns representing the modulation of the intensity of each discrete signal. After the first Fourier transformation of the data points in t_2 domain, a series of ω_2 spectra are attained. The set of the signals obtained resemble a 2D spectrum, but it is not, since the second dimension is the time variable t_1. Each signal, $F(t_1, \omega_2)$ is called an interferogram to distinguish from the FID (Figure 2.21(b)). As mentioned previously, the signal's intensity is modulated in t_1 at a frequency corresponding to the correlated chemical shift and/or coupling constant. After the data matrix transposition (Figure 2.21(c)), so that each sequence of data points for transformation appears sequentially in the computer memory, a second Fourier transform of all interferograms produces the second dimension, ω_1, of the 2D NMR spectrum, $F(\omega_1, \omega_2)$ (Figure 2.21(d)). In most 2D NMR experiments there is an additional period, the mixing period (combination of pulses and/or time intervals), placed between the evolution and detection period. The mixing period ensures that information from the evolution period is transferred to the detection period. The mixing period of the pulse sequence of Figure 2.20 is the second 90° pulse. The 2D NMR spectrum is thus a representation of the transfer processes occurring during the mixing time amongst the different types of the spin systems.

2.9.2   The Appearance of 2D NMR Spectra

There are two general ways to present 2D spectra; the stacked-trace plots and contour plots with their pros and cons. A stacked-trace plot is obtained by drawing a series of spectra $F(\omega_2)$ one behind the other in order of increasing frequency ω_1, or *vice versa* by plotting $F(\omega_1)$ spectra in the order of frequency ω_2. The net result of both cases is a 3D arrangement (Figure 2.22(a)).

The 3D impression is enhanced further by leaving out those parts of the traces that have lower intensities than previous traces (whitewashed spectra). However, the spectrum in Figure 2.22(a) is actually a 3D spectrum, since the third dimension is simply the intensity of the signals. The risk of using this presentation is the loss of signals that are hidden behind signals in the preceding traces. Moreover, the exact location (chemical shifts) of the signals in the spectrum is difficult to find, as well as the distance between two signals. Plotting is also time

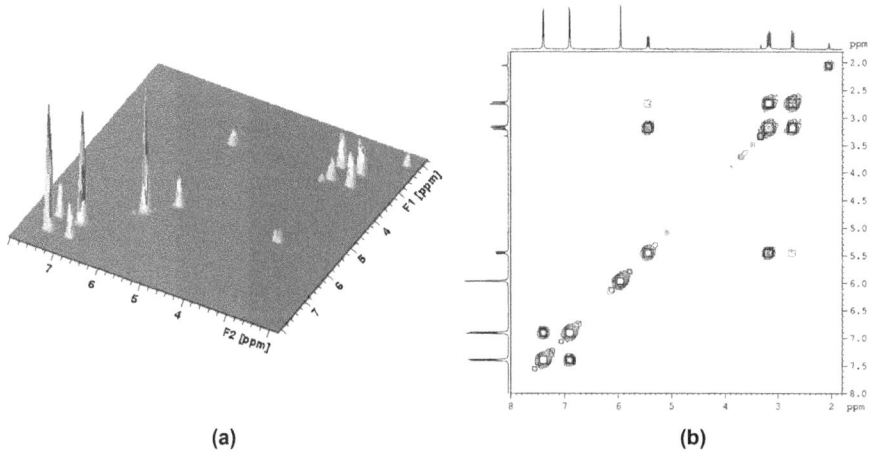

(a)   (b)

Figure 2.22   The 2D spectrum of naringenin (phenolic compound found in oregano) obtained by the pulse sequence of Figure 2.20. (a) Stacked plot, (b) contour plot.

consuming. Contour plot presentation avoids the aforementioned dangers and is fast to plot, although the precision of the chemical shift assignment depends on the diameter of the contour level. The contour plot depicted in Figure 2.22(b) shows the distribution of signal intensities, and resembles the geographical maps indicating the altitude of a region or a mountain. The problem with this presentation is when weak and strong signals are present in the spectrum. Increasing the level-threshold to observe weak signals, the high-intensity signals extend over a large area complicating the interpretation of the spectrum.

Other presentations of 2D NMR spectra involve cross sections and projections along certain directions usually through the two axes ω_1, and ω_2 of the spectrum, although cross sections can be obtained as well at an angle of 45° with respect to the ω_1 axis. Both presentations are suitable for simplifying congested spectral regions, detecting hidden peaks, and performing quantitative analysis. Also, this type of spectral presentation saves time, since it speeds up the plotting duration.

2.9.3 Description of a Simple 2D NMR Experiment

A rigorous analysis of a pulse sequence requires the use of either the density matrix formalism or the newly developed quantum mechanical theory of the product operators. These methodological approaches demand advanced knowledge of physics and mathematics in order to fully understand the specific mechanisms operating on the spin system during the application of a pulse sequence. This is not so important for the food scientist, who needs to apply 2D NMR for food analysis, and gather pertinent information from it. For the interested scientist, there are several wonderful books dedicated to the mathematical description of pulsed NMR experiments. On the other hand, a full understanding of the spin physics during the evolution and the mixing periods is not a prerequisite to perform successful 2D NMR experiments, provided that the experimenter is knowledgeable about setting up the experiments, as well as analysing and interpreting the results.

As an example, the experiment in Figure 2.20 will be described. This experiment is the well-known ^1H, ^1H-COSY-90 experiment that shows all existing connectivities amongst all coupled protons in the molecule. This experiment that takes no more than 10 min can replace successfully the laborious 1D decoupling experiments irradiating selectively each resonance at a time in the spectrum. During the preparation period the spin system described by the magnetisation vector M_0 (see section 2.7.2) is in its Boltzmann equilibrium along the positive z-axis of the rotating frame (Figure 2.23(a)).

The first 90° pulse deflects M_0 into the (xy) plane along the positive y-axis (Figure 2.23(b)). Then the spins evolve during the variable t_1 period; the magnetisation vector M_{xy} is analysed to its constituent nuclear magnetic moments that precess under the influences of both chemical shifts and spin–spin couplings (Figure 2.23(c)). It is said that the magnetisation for each type of proton nucleus is modulated (in a sense 'labelled') by the chemical shifts and spin–spin couplings. Following the evolution period, the second 90° pulse occurs that distributes the magnetisation among the various spin states of the

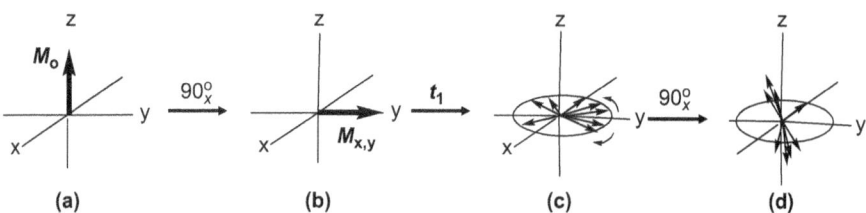

Figure 2.23 Approximate description of COSY experiment using the vector model. For a detailed analysis of this experiment the use of the product operators technique is required.

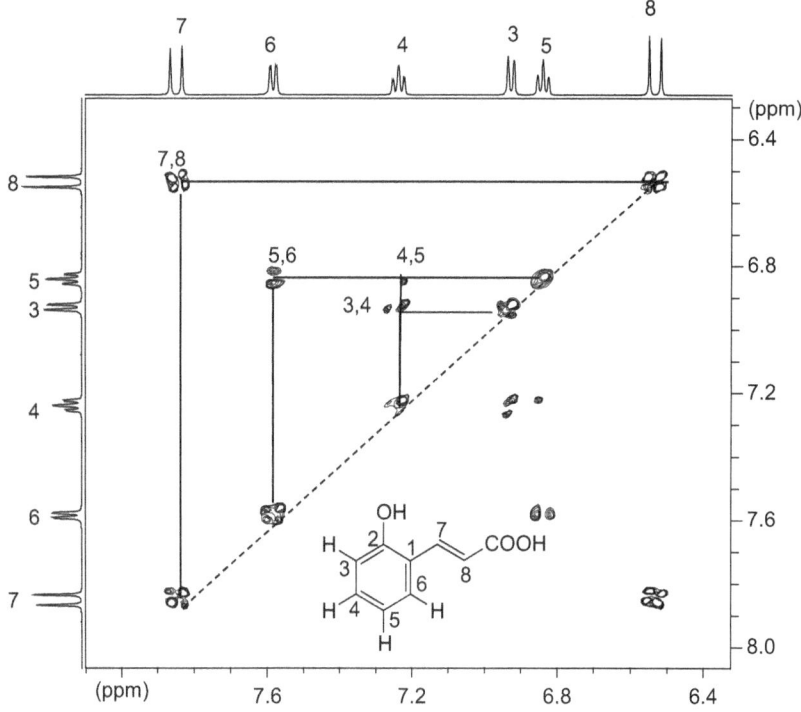

Figure 2.24 The 500 MHz COSY contour plot of o-coumaric acid in pyrine-d_5/chloroform-d solution shown with the 1D ^{1}H NMR spectrum. Connectivities between coupled nuclei are shown with solid traces above the diagonal (dotted line).

coupled nuclei (Figure 2.23(d)). The effect of the mixing pulse is detected in the period t_2, and after the double Fourier transform a contour plot is obtained with both axes representing the region of the chemical shifts (in ppm) of the proton nuclei within the molecule. This is shown clearly in the COSY-90 spectrum of o-coumaric acid (Figure 2.24).

This 2D plot is symmetrical about the diagonal and signals appear on the diagonal and off the diagonal. The diagonal signals result from the contribution of magnetisation that has not been changed by the mixing pulse, *i.e.* these signals reflect the distribution of magnetisation among nuclei of the same chemical shift. The coordinates of the diagonal signals are the same with respect to both axes and occur at the Larmor frequencies (chemical shifts) of the various protons in the sample. The projection on each axis is the corresponding 1D ^{1}H NMR spectrum. The off-diagonal signals or cross-peaks reflect the exchange of magnetisation during the mixing period indicating the interaction (correlation) among different types of nuclei through their spin – spin couplings, and show correlations between coupled nuclei. The cross-peaks are those that contain the information in 2D NMR spectra. The coordinates of the cross-peaks are the chemical shifts of the nuclei that are coupled to each other. There are several variants of the COSY experiment with several applications, *e.g.* COSY with delay for the detection of small couplings, the Double Quantum Filtered COSY (DQFCOSY) for the determination of *J* values, β-COSY for the determination of the relative sign of *J*, *etc.* The food scientist may find more effective the most powerful variant of COSY, the Total Correlation Spectroscopy (TOCSY).

2.9.4 Types of 2D NMR Experiments

The multitude of 2D NMR experiments has its origin on the properties of the spin system, the type of the mixing sequences and the nature of the various transfer processes. From the huge number of 2D pulse sequences that have been published, and mentioned in several books, practically, only 10–15 are helpful in the analysis of the NMR spectra of foodstuffs in the liquid state. Old and recent developments in multi-dimensional NMR pulse sequences with references to the original publications and comments are collected Berger and Braun.[13]

Through bond scalar coupling (*J* coupling) and through space dipolar coupling, mentioned in previous paragraphs, are two of the mechanisms that are responsible for two different classes of 2D experiments. A third mechanism for magnetisation transfer provided by exchanges processes, such as chemical reactions, may be useful for food scientists who are interested in kinetic studies of physicochemical changes, occurring in food materials, *e.g.* fruits ripening, freezing and melting processes, meat spoilage, *etc.* The through bond coupling underlies most of 2D NMR experiments. The pulse sequences of these experiments include a mixing period that transfers the coupling information from one spin system to another and the signals in the 2D spectra correlate the Larmor frequencies (chemical shifts) of the spin systems. These techniques are referred as correlation methods and are arranged as homonuclear and heteronuclear experiments, when correlations exist between homonuclear and heteronuclear nuclei, respectively. The most important homonuclear and/or heteronuclear pulse sequences for the food analyst are tabulated in Tables 2.2 and 2.3 along with their utility and relevant

Table 2.2 2D NMR experiments with mixing through bond scalar interactions.

Acronym/Name	Application	Comments
[H–(–)$_n$–H] homonuclear correlation		
H, H-COSY Correlation Spectroscopy Basic (90° pulses)	Chemical shift correlation of geminal and vicinal protons; identification of signals; spectral assignment.	An important technique, although it has been replaced by the robust TOCSY sequence; diagonal signals with mixed phases; variant with field gradients.
Long range COSY	Introduction of an additional delay time before and after the second 90° pulse in the basic COSY; detection of very small couplings.	The method succeeds even for couplings that are not observed in the spectrum; diagonal signals with mixed phases.
DQF-COSY Double Quantum Filtered COSY	Best standard technique; operates in the phase sensitive mode; determination of J values.	Diagonal peaks in the absorption mode, no long tailing that overlaps nearby off-diagonal signals; variant with field gradients.
E.COSY Exclusive Correlation Spectroscopy	Accurate measurements of J values.	Low intensity diagonal signals.
TOCSY Total Correlation Spectroscopy	Chemical shifts correlation of protons that are not directly coupled, but have common coupling partners; signals assignment within a scalar-coupled protons.	Generalised COSY experiment; variant with field gradients exists; Selective TOCSY is the 1D version; variant with field gradients.
(X–X) homonuclear correlation		
INADEQUATE Incredible Natural Double Quantum Transfer Experiment	It provides unequivocally X–X connectivities over one or two and rarely three bonds.	Poor sensitivity for the basic experiment; variant with field gradients; 1D variants are the basic INADE-QUATE with anti-phase signals and the refocused INADEQUATE.
INEPT-INADEQUATE	^{1}H detected INADEQUATE.	Slightly more sensitive than the basic INADE-QUATE, but it provides spectra (cross sections) with anti-phase signals; recent variants correct this drawback yielding in-phase doublets.

[H–(–)$_n$–X] heteronuclear Correlation		
HETCOR (H, C-COSY) Heteronuclear Correlation Spectroscopy	Heteronuclear shift correlation through one-bond coupling; spectral assignment.	No field gradients are required. Short set up. Good resolution at the carbon dimension, but low sensitivity.
(HMQC) Heteronuclear Multiple Quantum Coherence	Heteronuclear shift correlation through one-bond coupling; spectral assignment.	2–4 times more sensitive than HETCOR depending on the molecular size. Good resolution in proton dimension. Broader cross-peaks than HETCOR due to unsuppressed H–H couplings; variant with field gradients.
HSQC Heteronuclear Single Quantum Coherence	Heteronuclear shift correlation through one-bond coupling; spectral assignment.	Better resolution but lower sensitivity than HMQC. No signal broadening by H–H coupling.
X-edited HSQC	Structure elucidation; spectral assignment; simultaneous X chemical shift assignment and determination of XH_n multiplicity.	Combines the one X–H bond correlation (HSQC) together with carbon multiplicity selection similar to that obtained using the DEPT-135 experiment; variant with field gradients.
HSQC-TOCSY	Structure elucidation; spectral assignment; X-edited TOCSY spectrum; detects all ^{1}H signals within a coupled spin system.	This experiment is useful when overlap in the proton spectrum prevents analysis, while the corresponding X-nuclei are well resolved, or alternatively, the origin of proton signals is known and that of the X signals is sought. Cross-peaks are seen between the J-coupled protons in a spin network and each carbon involved in this network; variant with field gradients.
HMBC Heteronuclear Multiple Bond Coherence	Structure elucidation; spectral assignment; heteronuclear shift correlation through two- and three-bonds coupling.	Suppression of the unwanted one-bond coupling does not work equally well for all protons; variant with field gradients.

Table 2.3 2D NMR experiments with mixing through dipolar interactions and/or exchange.

Acronym/Name	Application	Comments
H–H NOESY Nuclear Overhauser Effect Spectroscopy	Distance information between protons, 3D structure for large molecules; molecular dynamics.	Besides COSY and TOCSY very important technique; major disadvantage is its dependence on the molar mass, viscosity and magnetic field strength, which can change its sign, and cause its disappearance under certain conditions; impossible to distinguish NOE from chemical exchange in large molecules; variant with field gradients.
zz-NOESY	Solvent suppression; separation of chemical exchange and NOE.	
H–H ROESY NOE in the rotating frame	Distance information between protons; 3D structure for small and medium sized molecules.	This pulse sequence is almost identical as the one used for TOCSY. To avoid TOCSY artifacts, the power used to achieve spin-lock is reduced; cross peaks are always positive.
H–X NOESY or HOESY Heteronuclear Overhauser Effect Spectroscopy	Provides information on the spatial relationship between heteronuclear spins (*e.g.* $^{13}C-^1H$ or $^{31}P-^1H$); provides distances between quaternary carbons and protons.	Very helpful when information from spin–spin coupling is not available; relatively poor sensitivity compared to NOESY and/or ROESY.
EXSY Exchange Spectroscopy	Study of chemical and conformational exchange processes; it is applicable equally well for two-sites and multi-sites exchange processes.	The NOESY experiment used for chemical and conformational exchange. EXCY and NOESY pulse sequences are identical. The cross-peaks due to exchange are often much more intense than those due to NOE, depending on the experimental conditions.

applications short description, and useful comments regarding their advantages and disadvantages. Application of these methods will be found in subsequent chapters. Table 2.3 contains the second family of 2D NMR experiments, where the dipolar coupling and/or chemical exchange are effective during the mixing period, while it need not be operative in the evolution or detection period.

2.10 Solid-State NMR

2.10.1 Line-Broadening Factors

Conventional liquid-state NMR cannot be used to study food samples in the solid state. As a matter of fact, the single pulse experiment mentioned previously results in a broad featureless spectrum. This is shown in Figure 2.25, which compares the liquid-state and solid-state ^{13}C NMR spectra of alanine obtained using a single pulse experiment.

There are two major mechanisms that contribute to such a broad signal; dipole–dipole (DD) interactions, and chemical shift anisotropy (CSA). Both mechanisms do not average by the molecular motion in the solid state. The first interaction is the direct coupling of the dipole moments of two homonuclear or heteronuclear spins through space. This interaction involves a local magnetic field generated by the magnetic moment of a particular nucleus at the location of a neighbouring nucleus. For two neighbouring nuclei, I and S, e.g. two protons or a ^{13}C nucleus with a proton, the local magnetic field B^{IS} depends on the internuclear distance r and the orientation of the angle θ formed by the internuclear vector with the external magnetic field B_0 (eqn (2.17) and (2.18)).

$$B^{IS} = d(1 - 3\cos^2\theta) \qquad (2.17)$$

$$d = \left(\frac{\mu_0}{4\pi}\right)\frac{\gamma_I\gamma_S}{r_{IS}^3} \qquad (2.18)$$

where γ_I and γ_S ($\gamma_I = \gamma_S$ for homonuclear spins) are the gyromagnetic ratios of the I and S nuclei, respectively, and d is the dipolar coupling, which provides

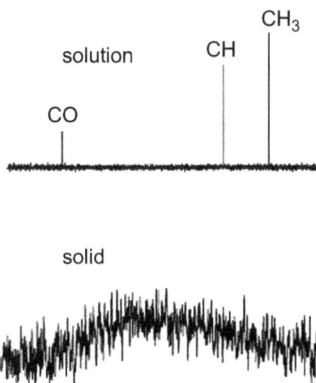

Figure 2.25 Comparison of the single pulse 1D liquid- and solid-state ^{13}C NMR spectra of the amino acid alanine. Strong dipolar ^{13}C–^{1}H coupling and CSA broaden considerably the line width in the solid. The rapid molecular motion in solution eliminates both mechanisms.

distance information. The total interaction is the summation of all possible pairwise interactions leading to broad featureless envelopes of unresolved splitting. Dipolar interactions typically range from 0 to 10^5 Hz in magnitude. In solution, however, where the molecules rapidly tumble, the angle θ assumes all values from 0 to 360°. Therefore, the angular term $(1 - 3\cos^2\theta)$ in eqn (2.17) is averaged out to zero, and the effect of dipolar interactions becomes minimal giving narrow lines. In solids, molecules are fixed with respect to B_0 and thus the angular term does not approach zero.

The second line-broadening factor is the CSA. As mentioned previously (see section 2.3.1), the chemical shifts are determined by the electron cloud shielding the nucleus from an applied magnetic field. Shielding will have a definite directional component because electron densities are not arranged spherically around nuclei. For molecules tumbling in solution the shielding constant is isotropic and determines the isotropic chemical shift (eqn (2.9)). This averaging mechanism is not available in solids. In the solid state, the shielding constant, σ, is anisotropic. Actually, σ is a tensor with nine directional components. This tensor can be diagonalised with respect to a molecular, principal axis frame into three principal components, namely σ_{xx}, σ_{yy}, σ_{zz}. The isotropic value of this tensor is given by $\sigma_{iso} = (1/3)(\sigma_{xx} + \sigma_{yy} + \sigma_{zz})$ or in the form of the chemical shift, δ: $\delta_{iso} = (1/3)(\delta_{xx} + \delta_{yy} + \delta_{zz})$ The CSA can be described by the so-called anisotropy parameter $\Delta\delta$ (eqn (2.19)):[14]

$$\Delta\delta = \tfrac{1}{2}(\delta_{xx} + \delta_{yy}) - \delta_{zz} \qquad (2.19)$$

and the asymmetry parameter η (eqn (2.20)):

$$\eta = \frac{3}{2}\frac{\delta_{xx} - \delta_{yy}}{\Delta\delta} \qquad (2.20)$$

The resulting line shapes are shown in Figure 2.26.

Figure 2.26(a) corresponds to an axially symmetric shielding tensor: $\delta_{xx} = \delta_{yy} \neq \delta_{zz}$ (*i.e.* η is zero). If no axial symmetry is present in the electronic environment, *i.e.* for $\delta_{xx} \neq \delta_{yy} \neq \delta_{zz}$, the line shape changes characteristically as shown in Figure 2.26(b) and (c). For axially symmetric molecules (with two equivalent axes), the anisotropic part of the shielding constant (eqn (2.21)) can also be written as:[15]

$$\delta = (1/3)(\delta// - \delta\bot)(3\cos^2\theta - 1) \qquad (2.21)$$

where $\delta//$ and $\delta\bot$ are the chemical shift tensor at the nucleus parallel and perpendicular to the symmetry axis of the molecule, and θ is the angle B_0 makes with the symmetry axis. The angular dependence of CSA (eqn (2.21)) is exploited to suppress this line-broadening factor, as we shall see below.

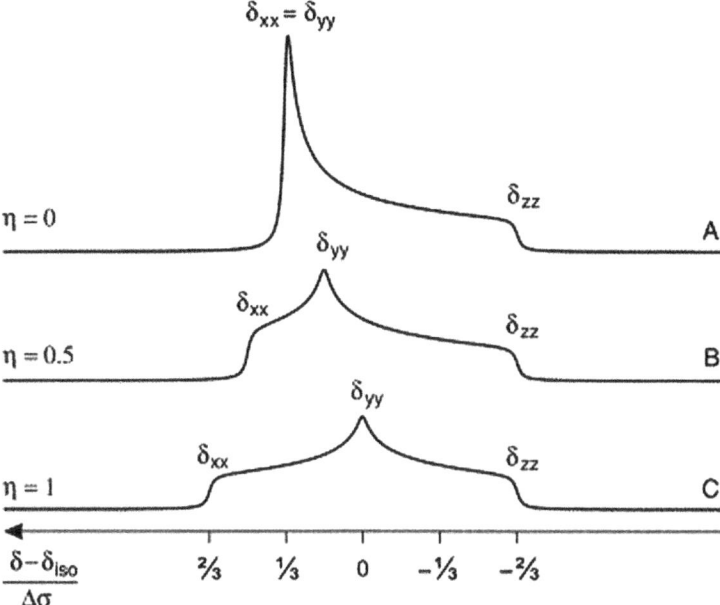

Figure 2.26 Calculated powder line shape of an isolated ^{31}P nucleus for various values of the asymmetry parameter, η.
(Reprinted from ref. 14. Copyright (2007) with permission from Elsevier.)

2.10.2 Removal of Dipolar Coupling and CSA

Both line-broadening factors are common for $\frac{1}{2}$ nuclei. However, the full suppression of these two factors requires different techniques for homonuclear decoupling (^{1}H–^{1}H, ^{19}F–^{19}F) as compared to heteronuclear decoupling with protons (^{13}C–^{1}H, ^{31}P–^{1}H).

2.10.2.1 Heteronuclear Spins

High-power proton decoupling and magic-angle spinning (MAS) can be used to eliminate the effects of dipolar coupling and CSA. Heteronuclear dipolar interactions, such as ^{1}H–^{13}C (or ^{1}H–^{31}P) can be reduced significantly with MAS and the residual dipolar coupling can be further eliminated using high power decoupling. MAS is used to average the chemical shift anisotropy. When a sample is spun rapidly at 54°74′ (the magic angle) relative to the static magnetic field, the angular part in eqn (2.21) and thereby the anisotropic component of the shielding constant is either partially or completely averaged to the isotropic value, yielding high-resolution solid-state NMR spectra. This angle dependence also reduces the residual dipolar interaction (see eqn (2.17)). Figure 2.27 shows the effect of MAS and dipolar decoupling on the ^{13}C spectrum of solid alanine.

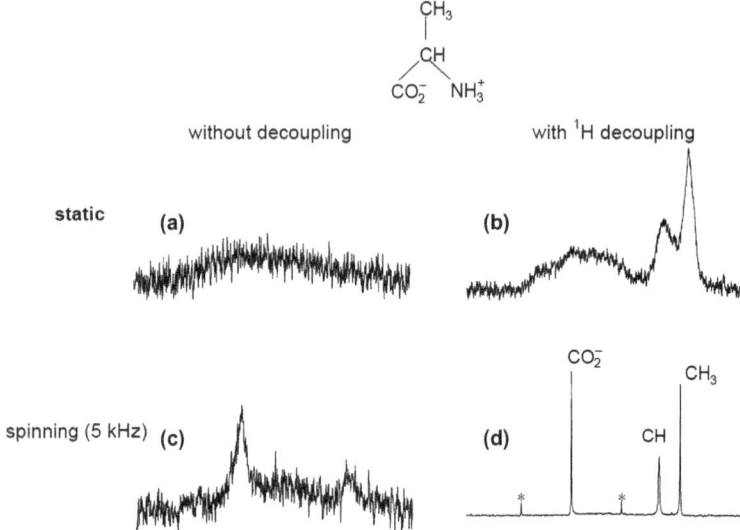

Figure 2.27 ^{13}C NMR spectra of solid alanine recorded with (a) the single pulse experiment as in solution, (b) proton decoupling, (c) MAS technique, and (d) MAS and proton decoupling. The asterisks denote spinning sidebands.

The combined effect of both techniques appears to be much more effective. The spinning speed must be higher than the static line width of the powder pattern otherwise spinning sidebands are observed at multiples of the spinning speed making thus difficult the interpretation of the spectra. At any rate, total suppression of spinning sidebands (TOSS) present in MAS has been accomplished by performing the homonymous experiment.

A further development is the cross polarisation (CP), which provides a four-fold signal enhancement for ^{13}C experiments. It also leads to shorter acquisition delay times, because the recycle delay is determined by the ^{1}H relaxation rate, which is much shorter than that of ^{13}C. The basic CP experiment can be rapidly repeated to obtain a satisfactory S/N. The concept is similar to INEPT, although CP uses a different pulse sequence. The basic CP experiment is shown in Figure 2.28.[15]

A 90° pulse provided by the magnetic field $B(^{1}$H$)$ generates ^{1}H transverse magnetisation $M(^{1}$H$)$ onto the y-axis. A subsequent 90° phase shift of $B(^{1}$H$)$ along the y-axis keeps (spin-locks) the ^{1}H magnetisation in the (xy) plane, precessing around $B(^{1}$H$)$ at a frequency $(\gamma_H/2\pi)B(^{1}$H$)$ Hz for a period of time, t_1, during which the proton polarisation decays with a specific time $T_{1\rho H}$ referred as relaxation time in the rotating frame for protons.

The spin locking process is better visualised in Figure 2.29. As the ^{1}H spins are spin-locked along the y-axis, application of an on-resonance ^{13}C pulse afforded by the magnetic field $B_1(^{13}$C$)$ generates transverse magnetisation $M(^{13}$C$)$ that precesses at a frequency of $(\gamma_C/2\pi)B_1(^{13}$C$)$ Hz.

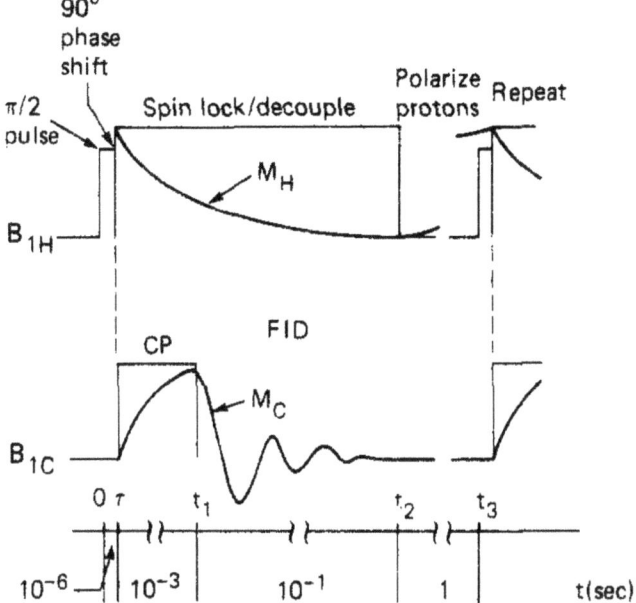

Figure 2.28 Single-contact spin-lock cross-polarisation pulse sequence, showing the time dependence of the carbon and proton magnetisation. Time scales (in s) for various parts of the sequence are shown. For CPMAS, the sample is continuously rotated about the magic axis.
(Reprinted from ref. 15. Copyright (1982) with permission from the American Chemical Society.)

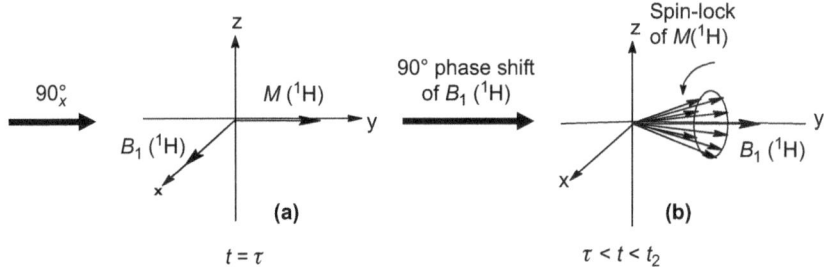

Figure 2.29 Proton spin-locking. (a) After the 90° pulse and (b) after the 90° phase shift of the field B_1 (^1H). The individual proton magnetic moments precess around the field. Spin-locking decouples the protons from carbons through the time during which the carbon FID is acquired $(t_1 < t < t_2)$. The time periods refer to the pulse sequence of Figure 2.28.

The spins ^{13}C are also locked along the y-axis and decay with the time $T_{1\rho C}$ known as the relaxation time in the rotating frame for carbons. During the spin-lock period or contact period, t_1, polarisation transfer occurs from the

^{1}H to the ^{13}C spins, if the Hartmann-Hahn condition (eqn (2.22)) is satisfied, *i.e.*

$$\gamma_H B_1\left(^1H\right) = \gamma_C B_1\left(^{13}C\right). \tag{2.22}$$

After the build-up of ^{13}C magnetisation at the end of the contact period t_1, $B_1(^{13}C)$ is switched off and the carbon FID is sampled at time t_2 under the dipolar decoupling by the field $B_1(^1H)$ (Figure 2.28). Finally, both spins are allowed to relax back to their equilibrium states during the time period t_3. A theoretical maximum magnetisation enhancement of carbon equal to the ratio γ_H/γ_C can be achieved by the CP sequence, although the increase in the ^{13}C signal depends on the strength of the dipolar interaction and the duration of the contact time.

The combination of CP and MAS (CP-MAS) with high-power decoupling affords high-resolution solid-state NMR spectra for the less sensitive nuclei. Figure 2.30 shows the effect of each technique to the ^{13}C spectrum of the of 4,4′-bis[(2,3-dihydroxypropyl)oxy]benzoyl diacetonide, separately and in combination.

The best result is obtained by combining these three techniques, although each of them is much more effective than the conventional one-pulse experiment.

2.10.2.2 Homonuclear Spins

Application of MAS without dipolar decoupling can be used in cases where the dipolar-coupling network is relatively weak and/or if using a fast spinning rate or an ultrafast spinning rate. Commercial MAS probes are now available with rotors of an external diameter of 1.3 mm that have MAS spinning rates of up to 67 kHz. In most cases, homonuclear dipolar interactions are very large (> 100 kHz) and appear very difficult to remove it by MAS. The highest spinning rates can reduce a 10 kHz line width to 1.5 kHz. Therefore, dipolar decoupling is attempted by pulsed techniques that simulate the MAS spinning, *i.e.* average to zero the Hamiltonian describing the dipolar coupling. Simulation can be performed by a series of RF pulses with different acronyms, such as the older MREV-8, WAHUHA and CRAMPS,[16] and the most recent pulse sequences FSLG, PMLG, DUMBO and variants.[17] The latter sequences are more effective if combined with MAS to suppress incompletely averaged dipolar couplings (high-resolution MAS).[18] Finally, most of 2D experiments used in solution NMR can be applied to solid samples with the appropriate experimental modifications.[19]

2.11 MRI

2.11.1 Spatial Encoding and Decoding

MRI is essentially the technique of using a homogeneous magnetic field, B_0, in combination with magnetic field gradients, G to provide a well-defined spatial

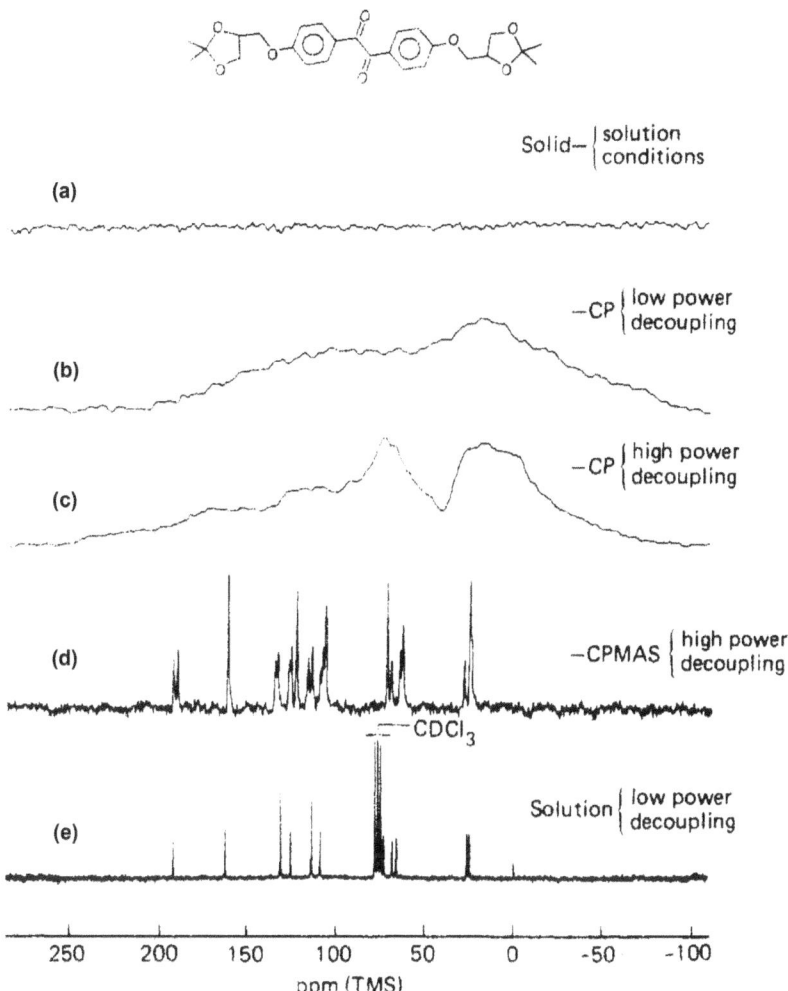

Figure 2.30 ^{13}C NMR spectrum of solid 4,4'-bis[(2,3-dihydroxypropyl)oxy]benzoyl diacetonide. (a) Conditions as in solution, (b) CP + high power decoupling, (c) CPMAS + high power decoupling, (d) sample dissolved in CDCl$_3$ with the same NMR parameters as in (a).
(Reprinted from ref. 15. Copyright (1982) with permission from the American Chemical Society.)

dependence to nuclear spin dynamics. A spatial image of the sample can then be reconstructed from the NMR signals acquired. At the most elementary level, the image can be thought of as a map of the spatial nuclear spin density, $\rho(r)$. Imaging is the technique by which the behaviour of the spins can be modulated by using cogent combinations of delays, RF pulses and gradients to obtain information on $\rho(r)$. There are three common ways (see below) by which a gradient can be used to impart a spatial dependence into the acquired NMR signal.

2.11.1.1 Frequency Encoding

The basic task of MRI is to give information on the spatial distribution of the nuclear magnetisation within a sample. This is achieved by recording the magnetic resonance signal in the presence of linear magnetic field gradients G. The influence of the field gradient on the image generation is depicted in Figure 2.31.[20]

Upon application of the 1D field gradient, say along the z-axis, G_z, in a homogeneous magnetic field B_0, the nuclei in say two different regions of the sample experience different magnetic fields. The result is an NMR spectrum with two separate signals instead of one. The signal intensity will depend on the spin population or density $\rho(r)$ in each region of the sample. This procedure,

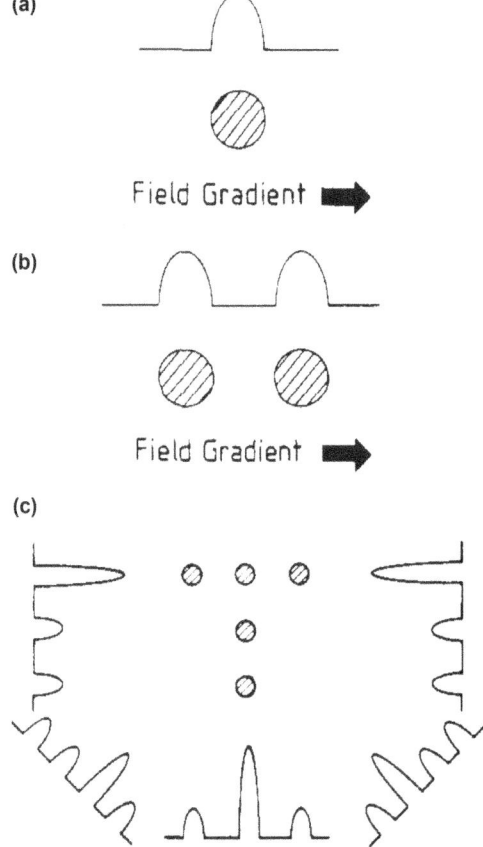

Figure 2.31   NMR spectrum of a structured specimen in a linear field gradient gives an 1D projection of proton density along the gradient direction: (a) a single tube of water, (b) two tubes, and (c) five tubes arranged as a T. (Reprinted from ref. 21. Copyright (1998) with permission from Elsevier.)

which is the basic principle of MRI, is called *frequency encoding* and is described by eqn (2.23):

$$\omega(z) = \gamma B_0 + \gamma G_z z = \omega_0 + \gamma G_z z \quad \text{and} \quad z = \frac{(\omega(z) - \omega_0)}{\gamma G_z} \qquad (2.23)$$

where ω_0 is the Larmor frequency, γ is the gyromagnetic ratio and z is a space-vector. γB_0 and γG_z are the homogeneous and inhomogeneous frequency gradients. Gradient G_z can be applied at several angles (from 0 to 359°) to the z-axis, and at each angle a spectrum will be recorded following a 90° pulse. Once the data has been acquired in the computer memory, they can undergo back-projection through space to obtain the one-dimensional image.

In conventional imaging sequences, instead of varying the angle of a single gradient, *e.g.* G_z, the required frequency encoding can be accomplished by linear combinations of two gradients G_x and G_y. Apart from frequency encoding, there are two more common ways in which a gradient can be used to impart a spatial dependence into the acquired NMR signal; these are the slice selection and phase encoding.

2.11.1.2 Slice Selection

Slice selection means the excitation of selected spins in a plane through the object. The slice selection is achieved by applying a 1D linear field gradient across the sample while applying a 180 or 90° selective or semi-selective pulse. Selection is obtained either by modifying the strength of the field gradient across the sample, while using a pulse of a constant excitation frequency, or by adjusting the offset frequency of the pulse for a constant gradient strength. Both gradient strength and excitation bandwidth $\Delta\omega_{B_1}$ of the pulse determine the thickness Δz of the slice selected (eqn (2.24)).[21]

$$\Delta z = \frac{\Delta\omega_{B_1}}{\gamma G_z} \qquad (2.24)$$

Regardless of the method used to select the slice, the spins within the selected slice are excited by the 90°_x pulse and rotate onto the (xy) plane. When the selection pulse is switched off, a frequency encoding gradient G_z (or linear combinations of G_x and G_y) is turned on and the FID representing the spectrum of the slice is collected and Fourier transformed to produce the frequency domain spectrum. The concept of slice selection with the application of selective 90 or 180° pulses is shown schematically in Figure 2.32.[21]

2.11.1.3 Phase Encoding

The third method of obtaining spatial information is to consider a spatially dependent phase change of the transverse magnetisation vectors of spins located at different regions of the sample. Assuming that the spins in the different regions have the same chemical shift, they will rotate with the same

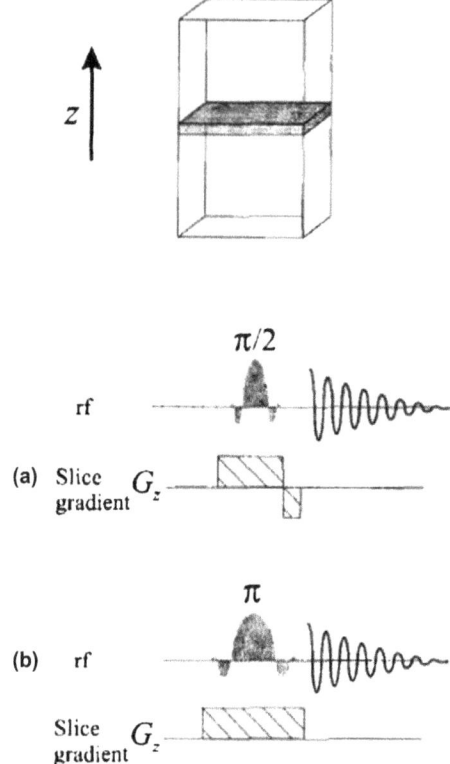

Figure 2.32 Two possible slice selection schemes using selective RF pulses together
with a gradient applied perpendicular to the slice of interest. (a) The
magnetisation in the slice is rotated into the (xy) plane by a 90° pulse.
Since the transverse magnetisation is de-phased during the slice selection,
a short, negative gradient is applied immediately after the RF pulse to
refocus the magnetisation. (b) Slice selection using a 180° pulse. The
magnetisation is automatically refocused.
(Reprinted from ref. 21. Copyright (1998) with permission from
Elsevier.)

frequency under the influence of the homogeneous field B_0. If a gradient is
applied within B_0 along the z-axis, the vectors in the various regions will precess
with frequency given by eqn (2.23), *i.e.* each vector will have its own unique
Larmor frequency. This situation is not very different from that described for
the frequency encoding. However, instead of collecting the FID, the gradient is
turned off, and the vectors now experience only the external field B_0, and
precess with the same frequency, but with different phase angles with respect to
the y-axis. In this case the gradient is referred to as a phase-encoding gradient
(G_φ). In contradistinction to frequency encoding, in phase encoding it is
necessary to observe an appropriately phase-encoded signal multiple times
(usually, 128 or 256 steps).

2.11.2 Basic Imaging Techniques

Using the different spatial encoding techniques presented in previous paragraph, it is possible to construct many imaging schemes for obtaining 2D (slices) and/or 3D images. Due to content restrictions imposed by this book, a few of these techniques will be discussed briefly. For further information on MRI, the reader should consult the pertinent sources at the end of this chapter.

2.11.2.1 Gradient-Echo (GE) Imaging

This MRI sequence is used for slice selection and corrects artefacts imparted by the rapid switching of the gradient, which distorts the FID and consequently the resulting image. The gradient-echo sequence is illustrated in Figure 2.33.

A slice selective 90° RF pulse is applied simultaneously with a slice selection gradient G_S. Next, a phase encoding gradient is applied, whose phase is varied between, say G_φ and $-B_{g(\varphi)}$ in 128 or 256 equal steps. A dephasing frequency encoding gradient G_F is applied at the same time as the phase encoding gradient. This gradient is negative in sign from the frequency-encoding gradient turned on during the acquisition of the signal. The last gradient refocuses the dephasing of the spins occurred during the dephasing gradient. As a result of this gradient reversal, an echo is formed. The signal from a gradient-echo sequence depends on the spin-echo time (T_E), *i.e.* the time between the start of the RF pulse and the maximum in the signal, and the repetition time of the sequence T_R (eqn (2.25)):[22]

$$S = k\rho(1 - \exp(-T_R/T_1)) \exp(-T_E/T_2^*) \tag{2.25}$$

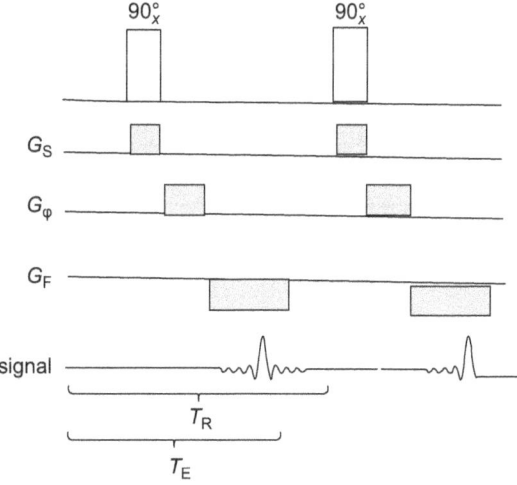

Figure 2.33 The MRI GE sequence.
(Reprinted from ref. 22. Copyright (1996–2011) with permission from J.P. Hornak.)

where k is a proportionality constant, which depends on the sensitivity of the detector of the MRI instrument, and ρ is the spin density.

2.11.2.2 Spin-Echo (SE) Imaging

This sequence is superior to the gradient-echo, because not only the gradient-induced dispersion is refocused, but also the chemical shifts and frequency dispersion due to residual *Bo* inhomogeneity effects in heterogeneous samples are refocused. A slice selective 90° RF pulse is applied simultaneously with a slice selection gradient (G_S) (Figure 2.34).[22]

The phase encoding gradient is applied between the 90° and 180° pulses, whose phase is varied between G_φ and $-G_\varphi$. Finally, a frequency encoding gradient G_F is applied after the 180° pulse during the time that the echo is collected. The recorded signal is the echo. The signal produced by the spin-echo sequence is given by eqn (2.25) upon replacing T_2^* with T_2, since field inhomogeneities are refocused.

2.11.2.3 Inversion Recovery (IR) Imaging

This pulse sequence starts with a slice selective 180° pulse in conjunction with a slice selection gradient G_S. The remainder of the sequence is equivalent to a spin-echo sequence, which is applied after a period of time T_I.

The phase encoding gradient G_φ could not be applied after the first 180° pulse because there is no transverse magnetisation to phase encode at this

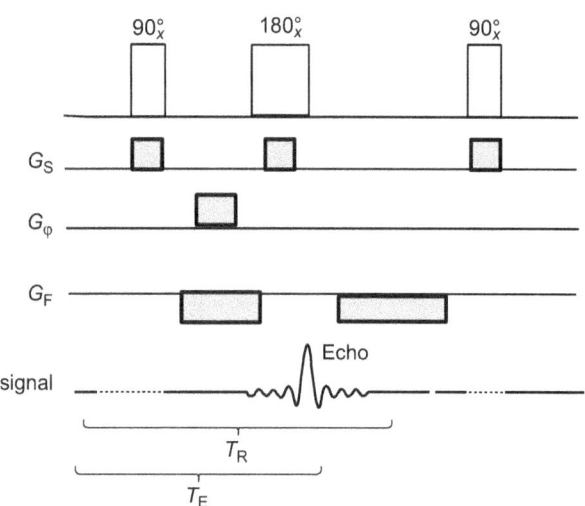

Figure 2.34   The MRI SE sequence.
(Reprinted from ref. 22. Copyright (1996–2011) with permission from J.P. Hornak.)

point. The frequency encoding gradient G_F is applied after the second $180°$ pulse during the time that the echo is collected. The entire sequence is repeated every T_R s and the recorded signal is shown in eqn (2.26).[22]

$$S = k\rho(1 - 2\exp(-T_I/T_1) + \exp(-T_R/T_1)) + \exp(-T_E/T_2) \qquad (2.26)$$

The inversion recovery sequence works equally well with a gradient-echo signal detection substituting the spin-echo sequence.

2.11.3 Basic MRI Scans

The aforementioned techniques use preferably the most sensitive proton to generate images. They can be optimised to obtain image contrast distinguishing differences between similar but not identical objects (*e.g.* tissues) using the resonance of the protons to generate images. For example, with selected values of the echo time T_E and the repetition time T_R, fluid-containing tissues are bright and fat-containing tissues are dark. Additional information about the MRI scans are given in the reference cited.[8–10,21,22]

2.11.3.1 *T_1-Weighted MRI*

T_1-weighted MRI is one of the basic types of MR contrast and is a commonly clinical scan. It uses a gradient-echo sequence, with short T_E and short T_R, differentiating fat from water with water being darker and fat lighter. The T_1 weighting can be increased (improving contrast) with the use of an inversion pulse. Due to the short repetition time (T_R) this scan can be run very fast allowing the collection of high-resolution images.

2.11.3.2 *T_2-Weighted MRI*

T_2-weighted scans are another basic type. As with T_1-weighted scan, fat is differentiated from water, but in this case fat shows darker, and water lighter. These scans are therefore particularly well suited to imaging oedema, with long T_E and long T_R. This MRI method is preferred from T_1-weighted, because the spin-echo sequence is less susceptible to inhomogeneities in the magnetic field.

2.11.3.3 *Spin Density Weighted MRI*

Spin density, also called proton density, weighted scans try to have no contrast from either T_2 or T_1 decay, the only signal change coming from differences in the population of available spins (hydrogen nuclei in water). It uses a spin echo or sometimes a gradient echo sequence, with short T_E and long T_R.

2.12 Time Domain (TD) NMR

Food materials show time-dependent phenomena; phase transitions, matrix and water distribution, compositional changes during processing and storage are a few representative examples. The study of these time-dependent evolutions require TD NMR experiments as opposed to frequency domain measurements that are suitable for assessing compositional and structural features of foods. Suitable parameters for monitoring time-dependent phenomena are the relaxation times (T_1 and T_2) and the self-diffusion coefficients, D, of the various food components.[23] TD NMR instruments are usually small in size, bench-top, low resolution and low field, and designed to detect hydrogen or fluorine nuclei. In most cases, TD NMR analysis is quantitative and rapid (normally within seconds or minutes). It is a low cost, easy to use, non-destructive and non-invasive method of analysis. Because of all of these benefits, TD NMR is widely used for routine analysis in agriculture, food, chemical, pharmaceutical, petroleum, and medical industries.[23–25]

2.12.1 TD NMR Methods

2.12.1.1 *Relaxometry*

The most basic experiment in TD NMR is the single pulse experiment and the use of the FID for a rapid assessment of the relaxation behaviour of materials with low molecular mobility (*e.g.* crystalline and glassy phases). The operation principle of the FID method[24] for the protons of a two-component mixture is illustrated in Figure 2.35.

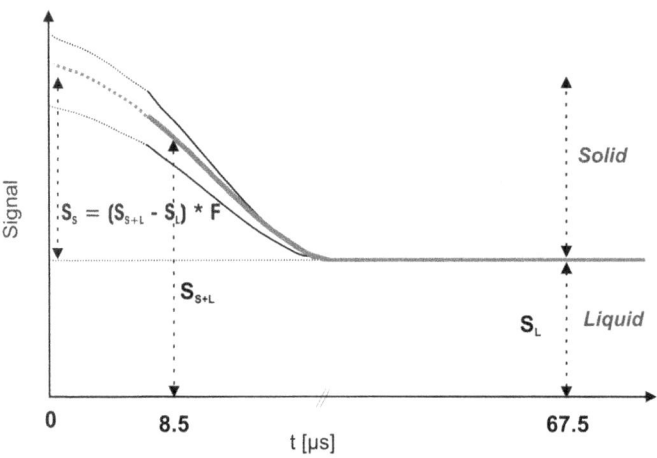

Figure 2.35   Time scheme of the FID and its analysis in case of the solid fat content (SFC) determination. S_S and S_L are samples on the fast and slow decaying parts of the FID, representing the solid and liquid component of fat, respectively; F is a correction factor determined from the calibration standard samples.
(Reprinted from ref. 24. Copyright (2008), with kind permission from Springer Science and Business Media.)

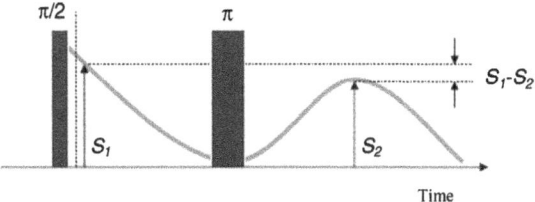

Figure 2.36   Hahn-echo sequence and data processing for simultaneous determination of moisture and oil content in seeds.
(Reprinted from ref. 24. Copyright (2008), with kind permission from Springer Science and Business Media.)

The initial rapid decay of the FID corresponds to the fast relaxing nuclei of the solid fraction, and the last slow decaying part of the time domain signal belongs to the slow relaxing nuclei of the liquid fraction. Sampling the signal just after the pulse and after the full decay of the signal of the solid fraction provides a measure of the solid content. This method has been used to determine the oil content in oilseeds and the fat content in animal fats (see chapter 7). However, this method is unable to simultaneously determine the content of both the slow and fast relaxing component (*e.g.* oil and water). This was achieved by the introduction of the Hahn spin echo sequence in food analysis.[24] This method, which is depicted schematically in Figure 2.36 allows the measurement of both the FID signal amplitude, S_1, and the echo amplitude S_2.

The first amplitude S_1 represents the sum of the fast and slow relaxing component, *e.g.* oil and water, respectively, whereas the spin echo sequence can be tailored in such a way that the second amplitude S_2 is only due to the slow relaxing component, the oil. Therefore the difference S_1-S_2 gives the content of the slow relaxing component, the moisture. Further advancement in the spin-echo technique was the Carr–Purcell–Meiboom–Gill (CPMG) pulse sequence, which, in addition to refocusing the magnetic field inhomogeneity, minimises the diffusion and exchange processes.[25]

Besides the TD NMR experiments based on transverse relaxation, experiments that examine the longitudinal relaxation (T_1) behaviour of foodstuffs have been used. Specific pulse sequences for T_1 measurements, such as the Inversion Recovery (IR) and the less time consuming Saturation Recovery (SR) have been proposed for time dependent measurements of foodstuffs.[1] These methodologies have been replaced by the Fast field Cycling (FFC) relaxometry,[26,27] which offers the possibility of rapid assessment of the T_1 dependence on static magnetic field, from very low field strengths (0.01 MHz proton Larmor frequency) up to about 45 MHz proton Larmor frequency; measurements can be extended to include magnetic fields of conventional NMR spectrometers. In order to cover a wide range of magnetic field values without changing the spectrometer, the excitation/detection system works at some static field, and after the excitation of the nuclei by a pulse sequence for relaxation measurements, and during the relaxation period, the sample is subject to another, easily variable magnetic field. The principle of FFC sequence for measuring T_1 is depicted in Figure 2.37.[27]

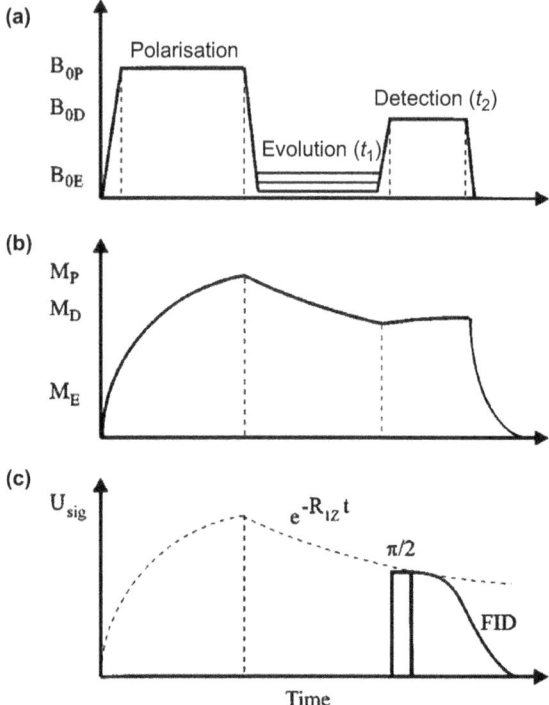

Figure 2.37   Summary of principles of magnetic field-cycling sequence for measurement of longitudinal (T_1) relaxation time indicating the polarisation (P),
evolution (E), and detection (D) periods. (a) Magnetic field (B_{0X}), where
$X = P$, D, or E; (b) longitudinal magnetisation M_X, and (c) nuclear
induction signal U_{sig} as a function of time. During the *polarisation period*
the magnetisation is increased to a relatively high value given a sufficiently
long polarisation interval. Following the downward switch to the selected
magnetic field level B_{0E}, the magnetisation relaxes during the *evolution
period* towards the equilibrium value within the B_{0E} field, with a time
constant T_1 (B_{0E}). Immediately after the upward switch to the field B_{0D},
the free induction decay (FID) is monitored during the *detection period* by
means of a switchable radiofrequency pulse scheme; here $B_1(t)$ corresponds to the Larmor frequency at the larger detection field strength B_{0D}.
Measurements of the magnetic field dependence of the longitudinal
relaxation time T_{1Z} are among the typical forms of field-cycling studies.
(Reprinted with permission from ref. 27. Copyright (1996) from the
American Institute of Physics.)

Relaxation measurements at low frequencies are useful, since they probe
long-range motions associated with rheological properties, including the elas-
ticity, viscosity, and other characteristic macroscopic properties of food
materials. The field cycling experiment generates a dispersion curve of relaxation time *versus* frequency, which yields information about molecular dynamics
and identifies the occurrence of different contributions that can be assigned to
various components of the foodstuff. By comparing the characteristic

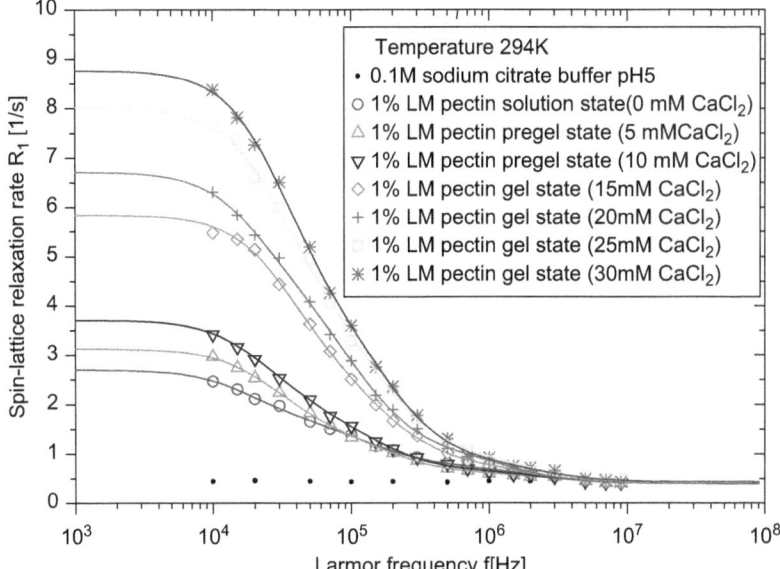

Figure 2.38 Dispersion of the water ^{1}H spin–lattice relaxation rate in solution of 1% w/w pectin at pH 5 and 21 °C without added calcium chloride and after addition of 5, 10, 15, 20, 25 and 30 mM CaCl$_2$.
(Reprinted from ref. 30. Copyright (2004) with permission from Elsevier.)

relaxation dispersions of various substances, one can gain a better understanding of their microscopic behaviour in relation to the observed macroscopic food properties. Nevertheless, FFC has not yet been applied in food analysis as it would be expected.[28,29] Fast Field Cycling Relaxometry was applied to study the gelation process in aqueous low methoxyl pectin solutions in the presence of divalent cations from calcium chloride.[30] Low methoxyl pectins are complex polysaccharides used in food industry as textural ingredients because of their thickening and gelling properties. Figure 2.38 shows the ^{1}H NMR dispersion profile of the water ^{1}H spin-lattice relaxation rates, R_1, at different concentrations of the added calcium chloride.[30]

The calculated mean correlation time through a dynamic model indicated an increase in water mobility with increasing salt concentration, but remained unchanged for concentrations above 15 mM. The abrupt increase of the mean correlation time in the range from 10 to 15 mM corresponds to the sol–gel transition.

2.12.1.2 Diffusometry

The introduction of PFG combined with spin-echo[31] or stimulated echo[32] pulse sequences allowed the separation of the various components of mixtures on the basis of differences in their self-diffusion coefficients. Figure 2.39 illustrates the basic (PFG + spin-echo) PFGSE and (PFG + stimulated echo) PFGSTE experiments.[33]

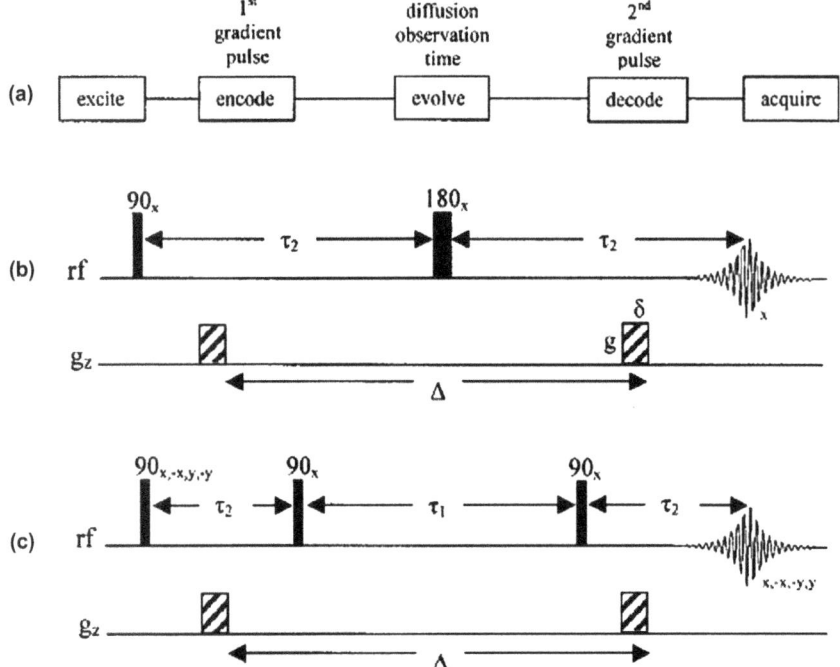

Figure 2.39   (a) The essential components of a PGSE NMR diffusion experiment
include, besides an excitation and acquisition process, three key elements:
an encoding element performed by a gradient pulse that labels the
position of the spins with a position-dependent phase angle, an evolution
period long enough to allow for sufficient translational displacement of
the spins to occur, and a decoding element performed by a gradient pulse
identical to the first that refocuses those spins that have not changed
position during evolution. (b) The PGSE pulse sequence. (c) The pulsed
gradient stimulated echo pulse sequence.
(Reprinted from ref. 33. Copyright (2002) with permission from Wiley
Interscience.)

Apart from the excitation by a 90° pulse and the detection of echo, both
sequences are composed by three elements: (a) an encoding element performed
by a gradient pulse G_z applied along the z-axis of duration δ (in ms) that labels
the position of the nuclei with a position-dependent phase angle $\varphi = \delta \gamma G_z z$ (γ
and z are the gyromagnetic ratio and the position of the nucleus, respectively);
(b) an evolution period long enough to allow sufficient translational displacement for the nuclei; and (c) a decoding element performed by a gradient pulse
identical to the first that refocuses those nuclei that have not changed position
during evolution. The signal intensity generated by the PFGSE pulse sequence
is shown in eqn (2.27).[33]

$$E_{(t-2\tau_g)} = E_{(t=0)} \exp[-D\gamma^2 G_z^2 \delta^2 (\Delta - \delta/3) - R] \qquad (2.27)$$

The parameter $R = 2\tau_g/T_2$ accounts for the relaxation rate of the nuclei during the pulse sequence. In the PFGSTE pulse sequence, two 90° pulses essentially replace the 180° pulse. This minimises the time the magnetisation spends in the (xy) plane. After the second 90° pulse the magnetisation is placed back into the z-direction where the relaxation is governed by T_1. This is generally more favourable because the ratio T_1/T_2 is commonly greater than unity for ^1H nuclei and the J-modulation effects are also minimised. The signal intensity generated by the PFGSTE pulse sequence (Figure 2.39) is given in eqn (2.28).[33]

$$E_{(t-2\tau_g)} = \frac{1}{2} E_{(t=0)} \exp[-D\gamma^2 G_z^2 \delta^2 (\Delta - \delta/3) - R] \qquad (2.28)$$

The factor $\frac{1}{2}$ designates the fact that the second 90° pulse returns only one of the y- or the x-components of the transverse magnetisation back to the z-axis, not both. This means that half of the intensity is lost and this constitutes the main drawback of the PFGSTE pulse sequence. In restrictive media, such as seeds, the self-diffusion coefficient, D, can be derived from the attenuation of the signal in a field gradient as a function of the parameters of the product $G_z\delta$ and the delay time Δ. The order of the magnitude of the measurable self-diffusion coefficient ranges from 10^{-5} to 10^{-10} m^2 s^{-1}.

2.12.1.3 2D TD NMR

The application of 2D TD NMR experiments was necessitated by the fact that the corresponding 1D TD NMR experiments failed to resolve the components of heterogeneous materials with similar relaxation times. 2D NMR experiments in the time domain correlating; either relaxation times (*e.g.* T_1–T_2), or relaxation times with self-diffusion coefficients (*e.g.* D–T_2) were generated by combining the appropriate pulse sequences. For instance, the combination of saturation or inversion recovery (SR or IR) and CPMG sequences is able to measure 2D T_1–T_2 correlations, whereas the combination of PFGSE and CPMG sequences is used to correlate the self-diffusion coefficient and the spin–spin relaxation time, $(D$–$T_2)$. Recent review articles[23,34] report overviews of 2D time domain sequences built up from 1D relaxometric and diffusometric building blocks and their applications in food science. The potential of combining two dimensions for resolving different components in foods was facilitated by the application of a fast 2D Laplace inverse algorithm.[35] The potential of the 2D methodology in TD NMR is depicted in Figure 2.40.

This Figure shows the Laplace transformed ^1H T_1–T_2 spectra of the fresh and the D_2O exchanged xylem tissues of carrots obtained by the combination of the IR method (T_1 measurements) and the CPMG pulse sequence (T_2 measurements).[36] In Figure 2.40(a), the signal designated as (w) is due to water in xylem vessels, whereas signal 7 corresponds to water inside the carrot starch granules; signal 5 is assigned to the non-exchangeable protons of proteins, whereas signals 2, 3, 4, and 6 are attributed to the cell wall biopolymers, pectin,

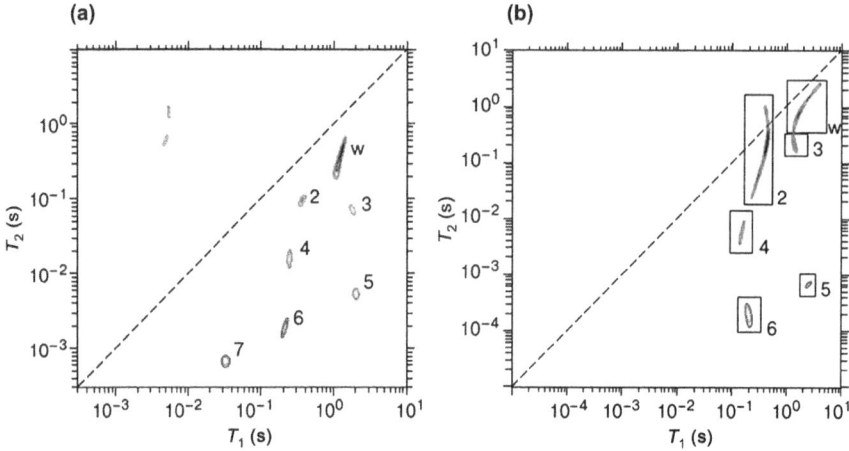

Figure 2.40   (a) ^{1}H T_1–T_2 spectra of (a) fresh xylem tissue and (b) D_2O-exchanged
xylem tissue of carrots acquired at 100 MHz.
(Reprinted from ref. 36. Copyright (2009) with kind permission from
Springer Science and Business Media.)

hemicelluloses and cellulose. These assignments are tested in Figure 2.40(b),
which shows the T_1–T_2 spectrum of carrot tissue after replacement of H_2O
with D_2O. Signals that are due to exchangeable protons either vanish or
decrease in intensity, whereas signals of non-exchangeable protons increase in
intensity.

Further Reading

Liquid-State NMR

The following books are recommended for beginners in NMR spectroscopy.

1. R. S. Macomber, *A Complete Introduction to Modern NMR Spectroscopy*,
   John Wiley & Sons, Inc., 1998.
2. H. Günther, *NMR Spectroscopy. Basic Principles, Concepts, and Applications in Chemistry*, John Wiley & Sons, New York, 1998.

For advanced descriptions of NMR theory and practical considerations for
doing NMR experiments with relevant examples the following books are
recommended.

3. A. D. Derome, *Modern NMR Techniques for Chemistry Research*,
   Pergamon Press, Oxford, 1997.
4. T. D. W. Claridge, *High-Resolution NMR Techniques in Organic
   Chemistry*, Pergamon Press, Oxford, 1999.

Theoretical and experimental aspects of NMR spectroscopy, including the quantum mechanical treatment of the experiments and the product operator formalism, are included in the following books.

5. J. Keeler, *Understanding NMR Spectroscopy*, John Wiley & Sons Ltd., Chichester, 2005.
6. R. R. Ernst, G. Bodenhausen and A. Wokaum, *Principles of Nuclear Magnetic Resonance in One and Two Dimensions*, Oxford Science Publications, Oxford, 1987.
7. M. H. Levitt, Spin Dynamics: Basics of Nuclear Magnetic Resonance, 2nd edn, John Wiley & Sons Ltd, 2008.

Solid-State NMR

8. M. J. Duer, *Introduction to Solid State NMR Spectroscopy*, 2nd edn, Blackwell Publications, Oxford, 2004.
9. E. O. Stejkal and J. D. Memory, *High resolution NMR in the solid State*, Oxford University Press, Oxford, 1994.
10. R. W. Vaughan, *Ann. Rev. Phys. Chem.*, 1978, **29**, 397–419.

Magnetic Resonance Imaging

11. M. A. Brown and R. C. Semelka, *MRI: Basic Principles and Applications*, 4th edn, Wiley–Blackwell, 2010.
12. B. Hills, *Magnetic Resonance Imaging for Food Science*, John Wiley & Sons Inc., New York, 1998.
13. J. P. Hornak, *The basics of MRI*, www.cis.rif.edu/htbooks/mri, 1996–2011.

References

1. J. R. Lyerla Jr. and G. C. Levy, in *Topics in Carbon-13 NMR Spectroscopy*, ed. G. C. Levy, John Wiley & Sons, New York, vol. 1, ch. 3, p. 100.
2. D. J. Craik and G. C. Levy, in *Topics in Carbon-13 NMR spectroscopy*, ed. G. C. Levy, John Wiley & Sons, New York, vol. 4, ch. 9, p. 239.
3. T. C. Farrar and E. D. Becker, *Pulsed and Fourier Transform*, Academic Press, New York, 1971, ch. 2, p. 22.
4. M. H. Levitt, *Prog. NMR Spectrosc.*, 1986, **18**, 61.
5. M. H. Levitt, R. Freeman and T. Frenkiel, *J. Magn. Reson.*, 1982, **47**, 328.
6. A. J. Shaka, J. Keeler and R. Freeman, *J. Magn. Reson.*, 1983, 53.
7. A. J. Shaka, J. Keeler, T. Frenkiel and R. Freeman, *J. Magn. Reson.*, 1983, **52**, 335.
8. A. J. Shaka, P. B. Keeler and R. Freeman, *J. Magn. Reson.*, 1985, **64**, 547.
9. R. Freeman, *Prog. NMR Spectrosc.*, 1998, **32**, 59.
10. W. S. Price, *Concepts Magn. Reson.*, 1997, **9**, 299; W. S. Price, *Concepts Magn. Reson.*, 1998, **10**, 197.

11. R. Burger and P. Bigler, *J. Magn. Reson.*, 1998, **135**, 529.
12. J. Homer and M. C. Perry, *J. Chem. Soc., Perkin Trans.*, 1995, **2**, 533.
13. S. Berger and S. Braun, *200 and More NMR Experiments. A Practical Course*, Wiley-VCH, Weinheim, 2004.
14. A. Iuga, C. Ader, C. Gröger and E. Brunner, *Ann. Rep. NMR Spectrosc.*, 2007, **60**, 145.
15. C. S. Yannoni, *Acc. Chem. Res.* 1982, **15**, 201.
16. R. W. Vaughan, *Ann. Rev. Phys. Chem.*, 1978, **29**, 397.
17. P. Hogdkinson, *Ann. Rep. NMR Spectrosc.*, 2011, **72**, 185.
18. P. K. Madhu, *Solid State NMR*, 2009, **35**, 2.
19. S. P. Brown, *Solid State NMR*, 2012, **41**, 1.
20. E. R. Andrew, *Acc. Chem. Res.*, 1983, 16, 114.
21. W. S. Price, *Ann. Rep. NMR Spectrosc.*, 1998, **35**, 139.
22. J. P. Hornak, *Basic Imaging Techniques*, The basics of MRI, www.cis. rif.edu/htbooks/mri, 1996–2011.
23. J. van Duynhoven, A. Voda, M. Witek and H. Van As, *Ann. Rep. NMR Spectrosc.*, 2010, **69**, 145.
24. H. Todt, G. Guthausen, W. Burk, D. Schmalbein and A. Kamlowski, in *Modern NMR Spectroscopy*, ed. G. A. Webb, Springer, Dordrect, vol. 3, 2008, pp. 1739–1743.
25. P. N. Gambhir, *Trends Food Sci. Technol.*, 1992, **3**, 191.
26. E. Anoardo, G. Galli and G. Ferrante, *Appl. Magn. Reson.*, 2001, **20**, 365.
27. C. Job, J. Zajiceka and M. F. Brown, *Rev. Sci. Instrum.*, 1996, **67**, 2113.
28. P. S. Belton and Y. Wang, in *Magnetic Resonance in Food Science. A View to the Future*, ed. G. A. Webb, P. S. Belton, A. M. Gil and I. Delgadillo, Royal Society of Chemistry, Cambridge, UK, 2001, p. 145.
29. S. Baroni, *Magnetic Resonance in Food Science: Challenges in a Changing World*, Royal Society of Chemistry, Cambridge, UK, 2009, p. 65.
30. M. Dobies, M. Kozak and S. Jurga, *Solid State NMR*, 2004, **25**, 188.
31. E. O. Stejskal and J. E. Tanner, *J. Chem. Phys.*, 1965, **42**, 288.
32. J. E. Tanner, *J. Chem. Phys.*, 1970, **52**, 2523.
33. B. Antalek, *Concepts Magn. Reson.*, 2002, **14**, 225.
34. B. P. Hills, *Ann. Rep. NMR Spectrosc.*, 2006, **58**, 178.
35. Y.-Q. Song, L. Venkataramanan, M. D. Hürlimann, M. Flaum, P. Frulla and C. Straley, *J. Magn. Reson.*, 2002, **154**, 261.
36. M. E. Furfaro, N. Marigheto, G. K. Moates, K. Cross, M. L. Parker, K. W. Waldron and B. P. Hills, *Appl. Magn. Reson.*, 2009, **35**, 521.

CHAPTER 3
Instrumentation

3.1 The NMR Spectrometer

Early NMR experiments were conducted by the so-called CW technique (section 2.6.2). In the early 1970s the first FT instruments became available. As noted in section 2.6.2, the pulsed NMR experiment consists of a very short high power pulse, of the order of μs applied to the sample to excite a band of frequencies of nuclei about the Larmor frequency. Pulsed NMR spectrometers are more sophisticated than CW instruments, and require more complex circuitry for power transmission and signal detection. The block diagram in Figure 3.1 shows[1] the main components of a modern NMR spectrometer for liquid samples that will be described in the following paragraphs.[2,3]

3.1.1 The Magnet

The NMR magnet[1,4] is considered the most important (the heart) and most expensive part of the NMR spectrometer. Early NMR magnets were iron core permanent or electromagnets producing magnetic fields no more than 2.3 T. Today, NMR magnets for high-resolution NMR experiments are of the superconducting type, ranging in field strength from approximately 6 to 23.5 T, or in frequency units 250 to 1000 MHz for the proton nucleus, respectively. Figure 3.2 shows the basic design of the superconducting magnet for high-resolution liquid NMR.[5]

The superconducting wire of the magnet is wound into a multi-turn solenoid or coil. The coil has a resistance equal to zero, when it is cooled to a temperature close to absolute zero. This is achieved by immersing the wire in liquid helium. Once current flows in the coil it will continue to flow for as long as the coil is kept at liquid helium temperatures. The coil and liquid helium are kept in a large Dewar surrounded by a second Dewar that contains liquid nitrogen (77.4 K), which acts as a thermal buffer between the room temperature

RSC Food Analysis Monographs No. 10
NMR Spectroscopy in Food Analysis
By Apostolos Spyros and Photis Dais
© Apostolos Spyros and Photis Dais 2013
Published by the Royal Society of Chemistry, www.rsc.org

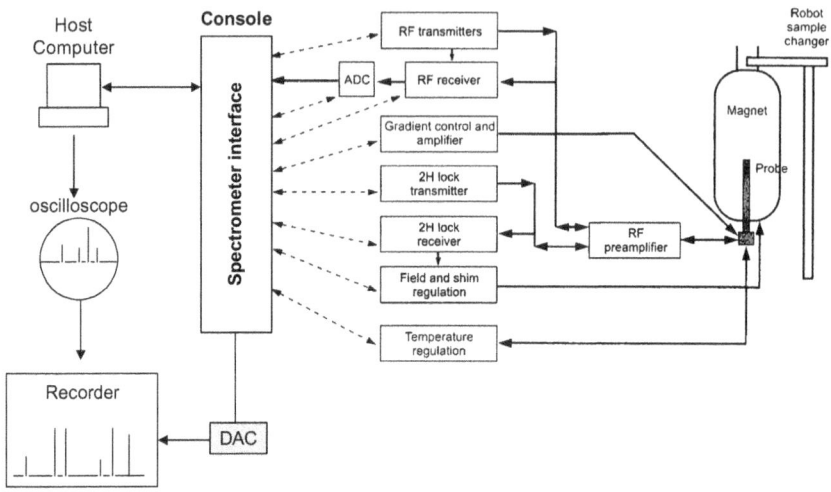

Figure 3.1   A schematic illustration of the Fourier transform NMR spectrometer.
(Reprinted from ref 1. Copyright (1999), with permission from Elsevier.)

air and the liquid helium. A longer liquid helium hold time (about a year) can
be achieved by inserting a thermal reservoir between the liquid helium and
liquid nitrogen that has a temperature less than the liquid nitrogen. Today,
NMR spectrometers provide active magnet shielding from the stray magnetic
field surrounding the magnet, avoiding thus field shifts inducing loss of the
magnet homogeneity, and eventually causing anxiety to people experimenting
with the NMR spectrometer. Finally, a room-temperature bore is inserted with
the shim coils, and a spinner assembly, which contains the turbine system for
spinning the NMR sample tube.

The magnet for solids is not different than that used for liquids, except
perhaps the extra circuitry for cross-polarisation, dipolar decoupling, and the
MAS rotor that gives bulkiness to the bore. In this respect, special wide-bore
(89 mm) magnets were required for solid-state NMR experiments. However,
this is no longer the case and most standard-bore spectrometers can be easily
adapted by the addition of a probe, a MAS controller, and possibly some
strong amplifiers to do solid-state NMR. Using a standard-bore machine is
advantageous for food analysis, since these machines are more common and
less costly.

The MRI apparatus used for food research is a special design for small
objects. It is equipped with permanent magnets of 1.0–4.7 T field strength. The
temperature around the magnet is regulated with a controller to prevent any
shift of resonating radio frequency and to stabilise the signal intensity against
influences of the ambient temperature. Shim coils are used to reduce inhomogeneity within the static magnetic field and to narrow the spectral width, which
is necessary for obtaining good-quality images for food materials. The direction
of the magnetic field is horizontal denoted as the z-axis.

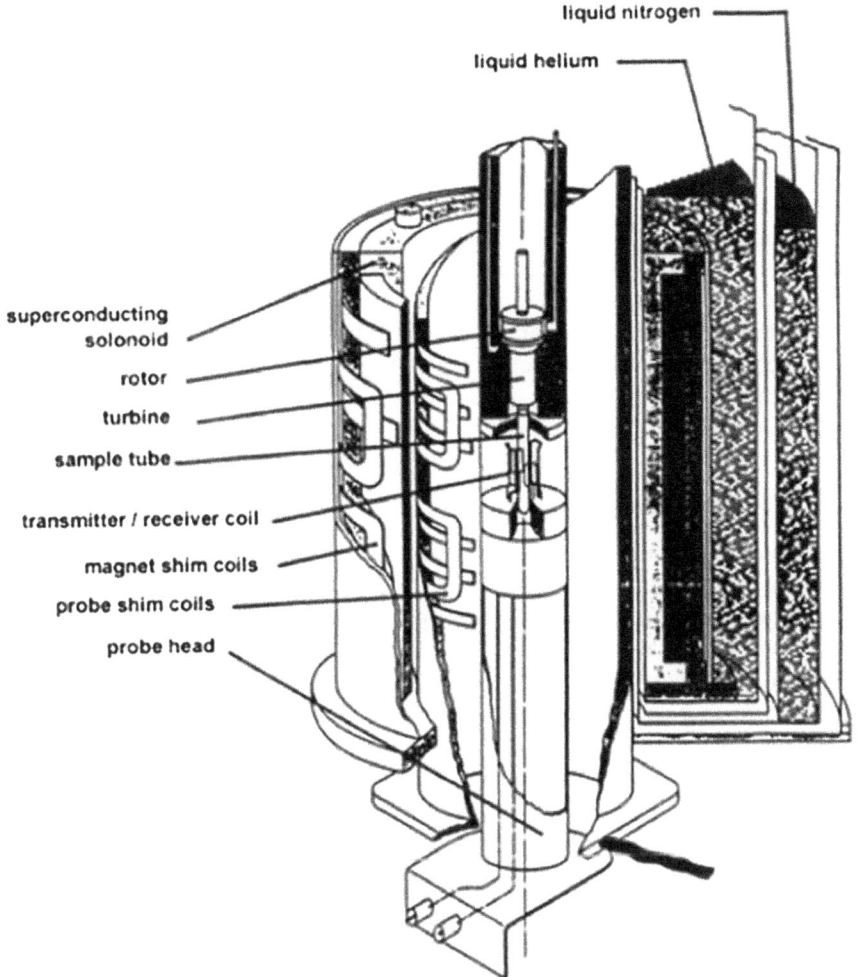

liquid nitrogen ——

liquid helium ——

superconducting
solonoid

rotor

turbine

sample tube

transmitter / receiver coil

magnet shim coils

probe shim coils

probe head

Figure 3.2 Illustration of the interior of a cryostat, showing the essential parts of the
superconducting magnet and probe.
(Reprinted from ref 5. Copyright (1998) with permission from John Wiley
& Sons.)

3.1.1.1 Shim Coils

For high-resolution NMR spectroscopy, the magnetic field homogeneity
should be adjusted to better than 1 ppb (part per billion) over the sample
volume. Minor spatial inhomogeneities may be created by the magnet design,
variations in the thickness of the sample tube, sample permeability, and
ferromagnetic materials around the magnet.

The homogeneity requirements are usually lower for MRI. The homogeneity of the magnetic field created by the magnet is adjusted by using a set
of shim coils. Superconducting magnets usually have at least eight

superconducting shim coils placed symmetrically to the centre of the main coil, and can be mounted either on the outside of the coil, between two sections of the main coil, or inside the innermost section of the coil. Simplest is a placement on the outside, although this placement requires shim coils with a larger number of windings. The Cartesian coordinate description of the gradients is normally used to name the shim coils. They are usually referred to as the z shim, the y shim, the zz shim, $2xy$ shim, and so forth. Shimming is obtained by passing the appropriate amount of current through each shim coil until a homogeneous magnetic field is achieved. The optimum shim current settings are found either by minimising the line width, maximising the size of the FID, or maximising the signal from the field lock. On most spectrometers, the shim coils are manageable by the computer by means of an appropriate algorithm, whose task is to find the best shim value by maximising the lock signal.

3.1.1.2   Field-Lock

After obtaining a satisfactory homogeneity using the shim coils, the next step is to secure the stability of the homogeneous field. The field strength might vary over time due to the magnet aging, and for a new magnet owing to the movement of metal objects near the magnet, or temperature fluctuations. Stabilisation is achieved by the so-called field-lock. It could be said that the field lock system is a self-contained mini-NMR spectrometer within the spectrometer, which measures most often the resonance position of deuterium. The deuterium signal comes from the deuterium solvent used to prepare the sample. The frequency of the lock transmitter is swept over a range of about 1 MHz as it searches for the resonance frequency of deuterium. When it first detects a resonant response, it stores this value and then jumps to a higher frequency and approaches resonance from the other direction. When it detects a resonance response as it approaches from the other direction, it stores this frequency and then jumps to a frequency at the midpoint between the two stored values and then 'locks' on this frequency as the resonance frequency of deuterium. Any drift from this resonance frequency, caused by field inhomogeneities, is corrected by minor changes in the B_0 magnetic field to keep the resonance frequency constant.

3.1.2   The Probe

The probe where the sample sits is introduced into the magnet from the bottom and contains the RF coils, which creates the B_1 field for the ^{1}H- and X-nucleus, and the ^{2}H lock. Also, the probe detects the transverse magnetisation as it precesses in the (xy) plane. RF coils can be divided into three general categories; transmit and receive coils, receive only coils, and transmit only coils. Modern NMR spectrometers are equipped with transmit and receive coils that serve as the transmitter of the B_1 fields and receiver of RF energy from the relaxed nuclei.

Probe selection depends primarily on the nature of the experiment(s) that will be conducted. The present state of the art in probe design offers a wide range of probes for almost any application; from liquid and solid NMR to imaging and diffusion experiments.

3.1.2.1 Dual Probe

A dual probe is one that is optimised for maximum sensitivity of a single nucleus, *e. g.* ^{13}C. There is always some loss of sensitivity when a coil is tunable to more than one nucleus. One of the coils will be tuned to an X-nucleus and the other coil will be tuned to ^1H. Similarly, a quadruple nucleus probe (QNP) has a coil that can be tuned to three different nuclei and a coil that is tuned to ^1H.

3.1.2.2 Broadband Probe

A broadband probe is one that has an RF coil that can be tuned to many different nuclei. These will always be nuclei other than ^1H or ^{19}F nuclei. This is because the frequencies of ^1H or ^{19}F are much higher than all the others and require separate hardware from the lower frequency nuclei. These lower frequency nuclei are referred to as X-nuclei. The broadband coil is the inner coil. Broadband probes also have an outer coil that is tuned to the ^1H frequency. Since the sample is closest to the inner coil, broadband probes offer high sensitivity to the X-nuclei.

3.1.2.3 Triple Resonance Probe

In this probe, the inner coil may be double tuned for two nuclei X and Y, *e.g.* ^{13}C and ^{31}P, or ^{13}C and ^{15}N, so that three channels (nuclei) ^1H, X, Y experiment are feasible. The primary nucleus used for detection (usually proton) should be assigned to the inner coil.

3.1.2.4 Inverse Probe

In this broadband probe, the inner coil is optimised for ^1H observation while the outer coil is tunable over a frequency range that permits decoupling of X-nuclei, usually between ^{31}P and ^{15}N depending on probe configuration. Since the ^1H coil is closest to the sample, the ^1H sensitivity is maximised. The X nucleus in an inverse broadband probe has much less sensitivity than in a broadband probe. Nevertheless, inverse probes offer high sensitivity for the less sensitive X-nuclei in 2D heteronuclear experiments, *e.g.* ^1H–^{13}C or ^1H–^{15}N HSQC, HMQC and HMBC, which allow ^1H observation.

3.1.2.5 Cryogenic Probe

Cryogenic probes[6] or cryoprobes have the RF coils and preamplifiers cooled to $\sim 10\,K$, whereas the sample area will still be at room temperature. As will be discussed below, the advantage of a cryoprobe over a conventional probe is the substantial increase in sensitivity because the S/N is increased by reducing the thermal noise at very low temperature. Cryoprobes can be made in any of the above configurations. However, a cryoprobe requires a large support system and is a very expensive probe to operate and complex to service. Currently, efforts are being made to manufacture cryoprobes for solid-state NMR.

3.1.2.6 Microcoil Probe

These probes[7-9] are manufactured with small RF coils, which have active volumes in the microliter to nanolitre range. They have several advantages and interesting properties relative to ordinary probes. Small volume probes need smaller amount of deuterated solvents, increasing thus the concentration of the solute, which may be difficult to obtain. Also, the small amount of deuterated solvents makes signals from impurities less prominent. In addition, small coils reduce dramatically the need of deuterated solvents for microseparations, decreasing thus the cost of the analysis, since most deuterated solvents are expensive. Small coils are simpler to shim and show significant sensitivity gain relative to conventional probes. The sensitivity improvement offered by microcoil probes, as well as they contribution to the design of multiple-coil probes will be discussed below.

3.1.2.7 Flow Probes

A flow probe[10,11] is used to inject liquid samples into NMR probes with an integrated sample chamber called flow cells, which is centred within the core of the RF coil. Both the bottom and the top ends of the flow cell are connected to PTFE tubes. The liquid sample is transferred from the reservoir to the bottom of the cell, and after analysis, it is retrieved in a serial fashion. The flow probes have sample volumes ranging from 100 to 480 μL and active volumes ranging from 40 to 250 μL. The overall recycle time is approximately 2–5 min per sample, which for 1D NMR analysis is the limiting time factor. The concentration sensitivity of the flow probes is good and can be improved using a cryogenic flow probe. Moreover, they provide high reproducibility of magnetic-field homogeneity from sample to sample, thus facilitating applications where clean difference spectra are desired. They increase the throughput of the NMR methodology and constitute an indispensable accessory of the hyphenated liquid chromatography-NMR (LC-NMR) technique.

3.1.2.8 Solids Probes

Solids probes are in many aspects different compared to probes for liquids, because they are designed to perform under CPMAS conditions. Often a solids probe is double-tuned (DT) to permit CP and high-power decoupling for increased sensitivity and resolution. Multiple-pulse line-narrowing techniques may be used to remove dipolar broadening for ^1H nuclei (section 2.10.2.2). The CPMAS probe adds high-speed sample spinning at the magic angle to the other capabilities of the solids probe for effective averaging of susceptibility broadening, chemical shift anisotropy, and often dipolar broadening, thereby greatly improving resolution for most spin 1/2 nuclei. Apart from probes dedicated to a single X-nucleus, there exist triple resonance solids probes, especially $^1H/^{13}C/^{15}N$ that are extremely useful in solving structures of larger molecules (*e.g.* polymers, proteins). High-resolution MAS (HR-MAS) probes for solid ^1H NMR experiments can be carried out on a conventional commercially available liquid-state NMR spectrometer. The only extra equipment required is a probe capable of MAS and a pneumatic unit for controlling the sample spinning.

3.1.2.9 MRI Probe

The main constituents of the MRI probe are the field gradients and amplifiers. Since the size of the food samples is relatively small, so must be the size of the probe head with the gradient coils inside. The working diameter is usually less than 25 mm for a nominal 89 mm wide bore magnet. A typical arrangement for performing imaging is to use a single turn coil to generate a read gradient in *x*-direction, and a detection gradient in the *z*-direction. The gradient probe with $G_x = 40\,mT/m$, $G_y = 35\,mT/m$ and $G_z = 52\,mT/m$ could image materials of 3.0 cm in diameter at most (the maximum field of view) with a spatial resolution (the smallest distance between two points in the object that can be distinguished as separate details in the image) of 10 μm.

3.1.3 The Console

Next to the magnet is what is called the console, which provides all the necessary parts that are needed to record an NMR spectrum. It supplies at least three radiofrequency channels, the observe channel, the lock channel, and another channel for decoupling. Usually these frequencies are derived from digital frequency synthesisers, which are phase-locked to a central quartz oscillator. These frequencies are controlled, amplified, pulsed, and transmitted to the probe-head. Depending on the spectrometer design, the various NMR signals are pre-amplified then mixed with the local oscillator frequency to yield the intermediate frequency. In other spectrometers no mixing stage is used. The signal is further amplified, and then in a second mixing stage the NMR audio signal is obtained after quadrature phase detection (detection of M_x and M_y components on the *xy* plane). The two signal components are digitised in the analog-to-digital converter (ADC) and fed into the computer memory or, in the case of the lock signal, used for field/frequency regulation.

3.1.3.1 ADC

ADC is an important part of the detection system of the spectrometer. This electronic device samples the free induction decay at regular time intervals (sampling rate) and converts the analog signal (voltage or current) to digital numbers proportional to the magnitude of the analog signal, which is stored in the computer. An important property of the ADC is its resolution, which indicates the number of discrete values it can produce over the range of the analog values. These values are usually stored electronically in binary form, so the ADC resolution is usually expressed in bits (binary digits). Consequently, the number of discrete values available is a power of two. For instance, an ADC with a resolution of 12-bits can encode $2^{11}-1$ analog signals; one bit being reserved for the sign. We will return to ADC and its function in section 5.1.1.

3.1.3.2   The Computer

All modern spectrometers are driven by the computer. Most of the activities formally carried out by analogue electronics such as filtering are now carried out digitally giving much greater precision than analogue devices. Some activities may be delegated to slave processors for specific functions such as the timing and generation of pulses.

The NMR computer is characterized by two important parameters: the number of memory locations and the word length. Memory locations are counted in multiples of K (kilobytes); each K is consisted of $2^{10} = 1024$ bytes. Older NMR instruments were equipped with computers of 64K or 128K, with storage capacity of 65536 and 131072 data points, respectively. New computers, however, have vast RAM capacity and can use solid state storage with very rapid read/write times so memory limitations is not a problem anymore. However, not all memory locations are used for data storage. The computer memory is occupied by the so called acquisition system, which controls the overall execution of the NMR experiment. The acquisition system communicates with the computer via Ethernet, receiving instructions and uploading data. Timing for the delays and other events within the pulse sequence making up an NMR experiment is either handled by the acquisition or by a separate timing control microprocessor. Part of the computer memory is reserved for the software including the FID transformation and processing routines.

3.1.3.3   Variable Temperature (VT) Unit

It is an important accessory for any fully equipped NMR spectrometer for conducting experiments at low or high temperatures. The VT unit affords the temperature selection device and is connected with the probe. Temperature increase is achieved through a heated resistance located beneath the probe, whereas sample cooling is obtained by immersing the whole probe in a Dewar containing liquid nitrogen. An indicator shows the current temperature of the experiment. However, for an accurate determination of the sample temperature, the chemical shift difference between the methyl and hydroxyl protons of methanol is measured for low temperature experiments. For high temperature measurements, the chemical shift difference between the methylene and hydroxyl protons of ethylene glycol is used.

3.2 Sensitivity of the NMR Experiment

One of the major drawbacks of the NMR spectroscopy is its low sensitivity compared with other spectroscopic techniques. For instance, the limit of detection (LOD) of Fourier transform infrared (FTIR) and Raman spectroscopy is as low as 10^{-12} to 10^{-15} mol, and that of mass spectrometry 10^{-19} mol, whereas the LOD for NMR is several orders of magnitude lower. This is particularly annoying when limited amount of a food sample is available or minor constituents in food sample need to be detected and eventually quantified.

Sensitivity in spectroscopy is defined as the ratio of signal to noise *i.e.* S/N, where S is the signal height and N is the average thickness or height (the maximum peak-to-peak noise amplitude) of the base line measured at a distance from the signal as shown in Figure 3.3.

The noise is determined by the relation $N =$ noise height/2.5. The computer of the NMR spectrometer gives the S/N ratio automatically. The usual methodology to increase the S/N of an NMR experiment is the accumulation of several transients or scans averaged in the computer of the NMR spectrometer. It is known that the signal intensity (S) increases with the number of scans, say n scans, whereas thermal noise (N) increases as $\sqrt{n}$; therefore, S/N increases as $\sqrt{n}$. For instance, in a single pulse NMR experiment, doubling the S/N of one scan, from its initial value of 2, requires the accumulation of 4 FIDs, (eqn (3.1)) *i.e.*

$$(S/N)_f = \sqrt{n} \times (S/N)_i \rightarrow n = \left(\frac{(S/N)_f}{(S/N)_i}\right)^2 = \left(\frac{4}{2}\right)^2 = 4 \qquad (3.1)$$

To improve the sensitivity by 4 or 8, 16 and 64 scans would be required, and so forth. Apart from signal averaging, which may be time-consuming, especially for the less sensitive nuclei, or samples of low concentration, efforts have been made in recent years to improve the NMR sensitivity by other means. These attempts to increase the sensitivity of the less sensitive nuclei (*e.g.* ^{13}C, ^{15}N) include recent technological advances in the strength of static magnetic fields, B_0, the development of cryogenically cooled probes, the miniaturisation

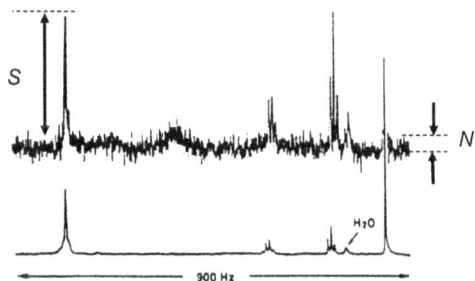

Figure 3.3   Definition of the S/N ratio as the quotient of average signal height, S, and the average noise level N, the latter value being divided by 2.5.

of the receiver coils, the application of pulsed field gradients, and the exploitation of polarisation experiments based on INEPT and DEPT pulse sequences mentioned in section 2.8.3.

3.2.1 Technological Advances to Improve NMR Sensitivity

According to theoretical considerations, the signal, *i.e.* the voltage, U_i, induced in the receiver coil of the probe, over the noise current, U_N, (thermal noise from the receiver, the preamplifier and the sample itself) is given by the following equation.[6,12]

$$\frac{S}{N} \propto \frac{U_i}{U_N} = \frac{V_s\, N\, \gamma_e \sqrt{\gamma_d^3\, B_0^3}\, (B_1/i_{coil})}{\sqrt{4 k_B T \Delta\omega [R_c(T_c + T_a) - R_s(T_c + T_a)]}} \tag{3.2}$$

Eqn (3.2) includes factors related to sample, such as the sample volume V_s, the number of nuclei N, the gyromagnetic ratios of the excited, γ_e, and detected nucleus, γ_d. Significant for sensitivity is the strength of the static magnetic field, B_0, and the coil design expressed by the ratio B_1/i_{coil}, *i.e.* the magnetic field B_1 providing the RF pulse over the intensity of current generated by the coil. Additional factors contributing to sensitivity is the resistance of the receiver coil, R_c, as well as the temperatures of the coil, T_c, and that of preamplifier, T_a. The sample resistance, R_s, induced in the coil and temperature, T_s, become important for solutions of high conductivity. $\Delta\omega$ is the receiver bandwidth, and k_B is the Boltzmann constant. From eqn (3.2), it appears that there are two ways to increase the S/N ratio, either upon increasing the factors affecting the signal voltage, U_i, and/or by decreasing the factors influencing the noise voltage, U_N. U_i increases by increasing B_0, or by conducting experiments with nuclei having high γ values. Optimising the coil design as expressed by the ratio B_1/i_{coil} can also augment the signal voltage.

Increasing the static magnetic field strength, the difference in the population between two energy levels is enhanced leading to an increase of the NMR signal (section 2.1). Commercial NMR spectrometers with superconducting magnets are quite expensive, and therefore one should be very cautious and purchase the instrument that meets the needs of his research. For the food scientist a fully equipped NMR spectrometer with a magnet of 14.1 T, operating at 600 MHz for the proton nucleus is proper choice for liquid samples. For solid samples, a 400 MHz wide-bore magnet (89 mm bore diameter) is sufficient for most NMR experiments.

Regarding the factor B_1/i_{coil} in eqn (3.2), it has been shown[12] that its value is inversely proportional to the diameter of the receiver coil. Therefore, the design of small coils increase this ratio and offer a higher sensitivity with little amount of sample.[13] This is particularly important for the quantification of minor compounds in foodstuffs, *e.g.* polyphenols, sterols, mono- and diacylglycerols, terpenic compounds, *etc.* Three basic geometries of small coils have been used, namely saddle, solenoid and planar, each with pros and cons. In general, solenoid coils offer a higher sensitivity compared to saddle coils, although the sample loading is quite involved. At any rate, the gain in sensitivity using small coils can only be estimated by comparison with similar experiments conducted with standard larger coils. For a sucrose sample in aqueous solution at 600 MHz, the sensitivity using 1 mm diameter capillary tube was found to be five times greater than a 5 mm conventional probe.

Further improvement of the S/N ratio can be accomplished by reducing the noise voltage (eqn (3.2)). The term $R_s(T_s + T_a)$ is immaterial for non-conducting solvents and/or samples, but the term $R_c(T_c + T_a)$, which is a measure of the thermal noise produced by the receiver coil and preamplifier, can be diminished by cooling the probe, preamplifier and other probe electronics. In cryogenically cooled probes, R_c, and T_c, are low and thermal noise is reduced.[6] The coil assembly and preamplifier are cooled using cold helium gas in a closed-loop cooling system. Thermal isolation in the probe allows samples to be measured at room temperature only a few millimetres from the cold coil assembly. The overall reduction of thermal noise in a cryogenic probe is typically about a factor of four.

3.3 High-Throughput NMR

Another aspect of NMR spectroscopy that may interest the food scientist, who intends to use NMR as an analytical tool, is the possibility of analysing a large number of samples within a reasonable analytical time, *i.e.* the ability of NMR spectroscopy for high and fast throughput screening. The food scientist expects through this process to rapidly perform multiple compositional analysis to identify changes in chemical and physical properties of foods occurring under certain conditions, such as storage, packaging, or farming; to detect abuse of overuse of food additives or other illegal compounds in food products; to detect and characterise veterinary drug residues in animal food products, and many other applications.

3.3.1 Technological Advances

Current approaches to high-throughput NMR use automatic sample changers with robotic liquid handlers. Automatic sample changers are limited by a relatively high failure rate mainly due to the use of glass NMR tubes, which can break, and unattended variations in automatic routines such as spinning the sample and failure of finding the ^2H lock. However, current NMR spectrometers have overcome the last two obstacles for automation. In addition, robotic systems do not allow any increase in speed and efficiency, limiting further the sample throughput. To avoid filling glass tubes, an alternative approach is to use a flow probe[10] in which direct transfer of a sample is possible into the NMR detector cell itself. This injection NMR detection system is simple and could be linked to an automatic sample-handling device in which samples are transferred to the probe from a rack of sample vials or well plates. Notwithstanding this methodology has the disadvantage of a thorough sample preparation before injection and it has been replaced by the closely related technology of LC-NMR, *i.e.* the directly coupled HPLC-NMR spectroscopy in which the output from a chromatographic column is fed to a flow probe (see below).

Another approach to NMR flow probe design is the development of microcoil NMR probes.[11] The microcoil flow probe has been used with several hyphenated techniques, such as microbore LC-NMR and capillary electrophoresis (CE)-NMR. A recent advance in hardware technology to increase the NMR throughput screening is the design of conventional or flow probes containing multiple RF receiver microcoils.[7–9] Such probes are capable of performing a number of 2D NMR experiments for a given sample, reducing significantly the experimental time. The basis of the proposed method is the splitting of a sample into R separate fractions placed into each of the R coils which, with the appropriate pulse sequence, allow each fraction to experience a different t_1 increment, thus reducing the data acquisition time by a factor of R. Each receiver channel has an independent preamplifier, transmit/receive switch and ADC, whereas pulses are conveyed to the sample fractions from a common transmitter that is connected to the receiver coils *via* switches. The position of the switch is controlled within the pulse program. Figure 3.4 shows diagram for an eight-coil probe used to run COSY experiments for eight different samples.[8]

The 2D spectra are obtained by reconstruction of the data set from each coil. This methodology is now in its infancy and has not been exploited commercially, although it requires minimal modifications to the spectrometer hardware.

3.3.2   Hyphenated NMR (LC-NMR)

LC-NMR was proposed over 30 years ago, but only recently has this versatile technique found extensive application in the field of food science. It had first to overcome several obstacles related to the interfacing of LC to NMR, the flow through probe design and issues, such as NMR sensitivity, solvent suppression, NMR and LC compatible solvents, and large differences in the volume of the chromatographic peak *versus* the volume of the NMR flow cell.[14–16] Figure 3.5 shows schematically the LC-NMR assembly and the various modes of operation.[16]

The sample is pumped to pass through the LC-column, and the eluent passes through a flow cell for UV (or DAD) detection. Finally, the fractions from the LC column pass to the probe of the NMR magnet. The status of the sample during the measurement characterises the mode of the LC-NMR operation. There are three general modes of operation: the on-flow mode (continuous flow), the stop-flow mode, and loop collection. In the on-flow mode, all the LC-column eluent flows without stopping to the NMR flow-cell during the LC run, and NMR spectra are recorded over the whole time. The result is a two-dimensional time-frequency plot consisting of a set of one-dimensional spectra (frequency axis, ppm) *versus* retention time (RT axis, min). Slices along the time axis provide the one-dimensional spectra of the various fractions eluted from HPLC. The on-flow mode is a rapid screening with ^{1}H NMR of all components in a mixture, including those that do not give UV signals, but suffers from low

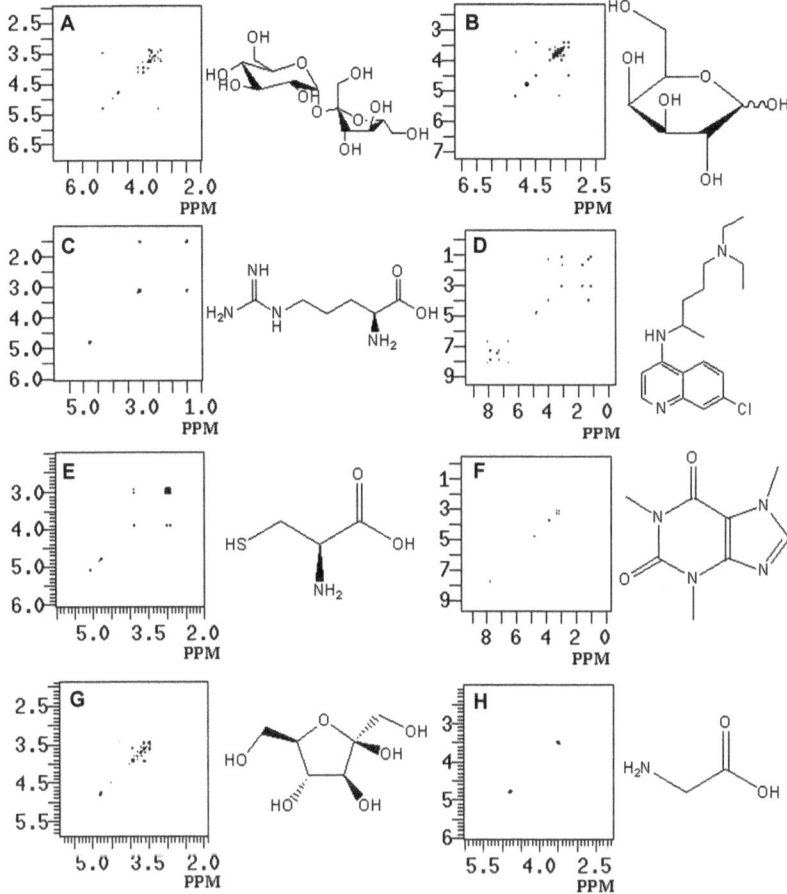

Figure 3.4 COSY spectra acquired with the 8-coil probe and the chemical struc-
tures of the compounds used. (A) Sucrose, (B) galactose, (C) arginine,
(D) chloroquine, (E) cysteine, (F) caffeine, (G) fructose, and (H) glycine.
(Reprinted from ref 8. Copyright (2005), with permission from Elsevier.)

sensitivity. In stop-flow mode, the LC flow is stopped while NMR spectra are
recorded for a particular LC peak, and then is turned on again. A series of
peaks may be measured this way during the run. The UV detector or prior
knowledge of LC retention times is generally used to stop the flow. The stop-
flow mode provides longer NMR acquisition times, advanced NMR techniques
(such as 2D NMR), detection of less sensitive nuclei (*e.g.* ^{13}C), while smaller
component concentrations can be accommodated. Stopping the flow does not
usually affect chromatographic resolution. Nevertheless, frequent stops may
disturb the quality of separation. In this respect, this mode could be better used
with mixtures containing a relatively small number of constituents. In the third
mode, the loop collection, the LC fractions are collected in their own capillary
loops during the LC run. After the run is completed, the individual loops are

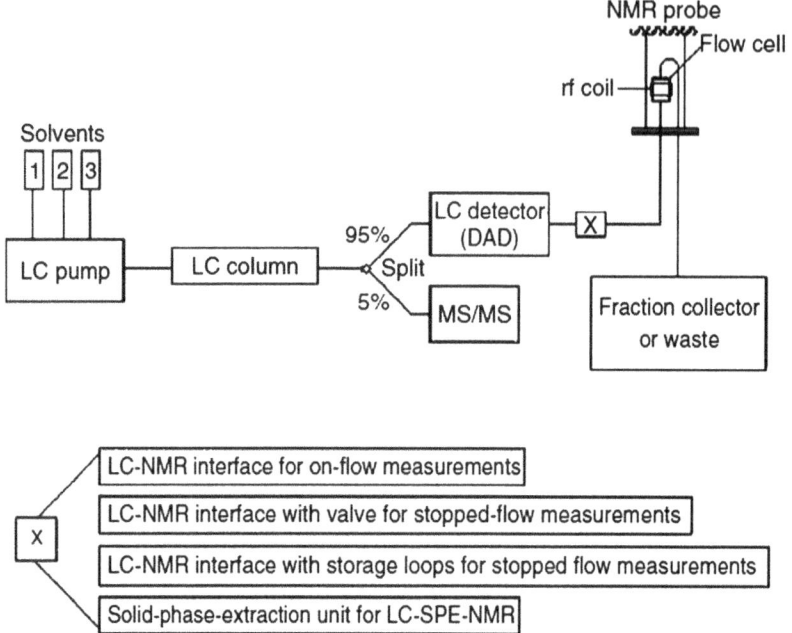

Figure 3.5   Illustration of the LC NMR assembly giving emphasis to the various
                 modes of operation. The splitter directing the HPLC fractions to the
                 NMR probe and/or to mass spectrometer can be placed after the LC
                 detector.
                 (Reprinted from ref. 16. Copyright (2005), with permission from Wiley
                 Interscience.)

eluted for NMR analysis at a later stage. The loops can also be removed and
stored for later or repeated analysis. This mode is useful when long analysis
times are required, for example with small amounts of sample (micrograms
scale), and it is preferred for 2D homonuclear and heteronuclear experiments,
although it cannot be used with unstable samples that may decompose during
the extended period of the analysis.

The eluent from the HPLC column can be split to give two parallel flows.
One goes to the NMR probe and the other to the mass spectrometer.
The splitter is adjusted in such a way that the split ratio is 95:5 for NMR *versus*
MS. This is because NMR is much less sensitive than MS. The LC-NMR/MS
technique can be used in both on flow and static conditions.

Further improvement in the LC-NMR performance was attained by integration of a solid-phase extraction (SPE) trap between the LC unit and the
NMR flow probe (Figure 3.5).[17] The use of the SPE technique is to trap the
eluted fractions from LC in a similar way to loop collection. In this respect,
repetitive LC runs can increase the concentration of the eluted fractions, and
thereby the sensitivity of the LC-NMR experiments. The SPE cartridges can be
flushed with D_2O, dried by nitrogen gas, and eluted with pure deuterated

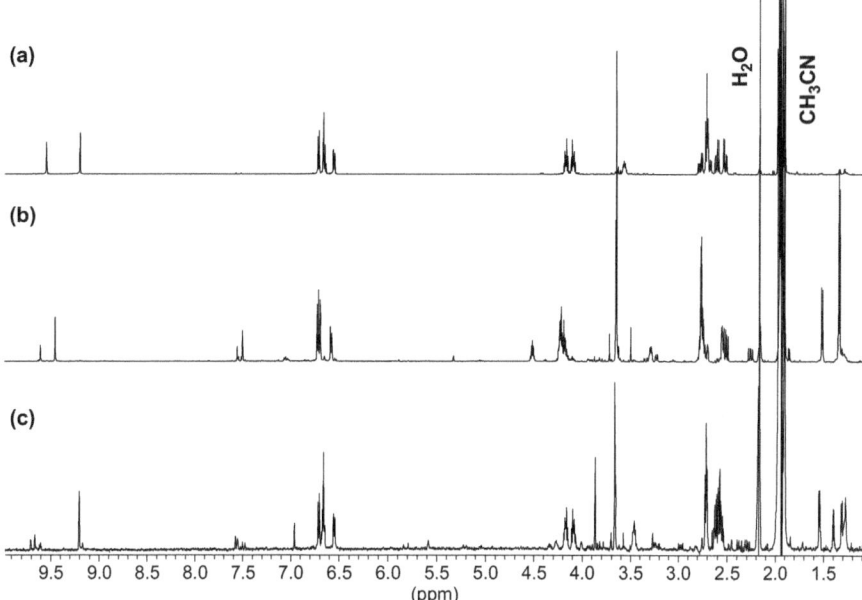

Figure 3.6 600 MHz LC-SPE ^{1}H NMR spectra of oleuropein derivatives: (a) dialdehydic form of oleuropein lacking a carboxymethyl group; (b) two
co-eluted isomers of the aldehyde form of oleuropein; (c) hemiacetal of
the dialdehyde form of oleuropein. The suppressed signals of H_2O and
CH_3CN solvents give spikes at δ 1.95 and δ 2.18 ppm.
(Reprinted from ref. 18. Copyright (2005), with permission from the
American Chemical Society.)

solvents, to minimise interfering solvent signals. The use of deuterated solvents
only during the SPE washing, allows the LC elution with protonated solvents,
which are much cheaper that the deuterated counterparts. Figure 3.6 shows the
first LC-SPE-NMR experiment on virgin olive oil for the detection of phenolic
compounds.[18]

Technical advances in magnets and probes design mentioned previously have
been incorporated in the manufacture of the NMR system coupled to LC.
Cryogenic technology in combination with the LC-SPE-NMR instrumentation
allowed the direct observation of ^{13}C spectra of natural products. Also, the
small detection volumes of microcoils are well suited to eluted volumes of
capillary separation techniques, *i.e.* capillary high-performance liquid chromatography NMR (capLC-NMR), capillary electrophoresis NMR (CE-
NMR), and capillary electrochromatography NMR (CEC-NMR).[15]

3.4 TD NMR

In most common TD NMR instrumentation, static magnetic fields are generated by electromagnets or permanent magnets with strengths B_0 ranging

between 0.12 to 1.4 T corresponding to Larmor frequencies 5 to 60 MHz, although higher magnetic field strengths (100–300 MHz) have been used. Such magnets are relatively cheap and can be installed on laboratory benches. These bench-top apparatuses are equipped with exchangeable RF parts for allowing fast and easy change of the measurement conditions and are most suitable for on-line measurements in industrial environments.

Regarding the FFC relaxometer, its heart is the magnet, which has to supply homogeneous magnetic fields of variable strength from a maximum of 0.5 T (20 MHz) to a minimum of 100 µT (10 KHz). This is obtained by using electronic modulation of the current flowing through the coil of an electromagnet. This technique allows fast variations of the field strength and makes the FFC approach capable of measuring very short relaxation times T_1 down to fractions of milliseconds. The instrumental aspects of FFC relaxometer concerning primarily the fundamental characteristics and the functional behaviour of those parts (magnet, power supply, data acquisition system, *etc.*), which are necessary for the FFC technique, are thoroughly discussed in reference cited.[19]

References

1. T. D. W. Claridge, *High-Resolution NMR Techniques in Organic Chemistry*, Pergamon Press, Oxford, 1999.
2. J. C. Lindon, *Encyclopedia of Spectroscopy and Spectrometry*, ed. J. Lindon, G. Tranter and D. Koppenaal, 2nd edn, Academic Press, Oxford, 2010, p. 1872.
3. J. C. Lindon and J. K. Nicholson, *Trends Anal. Chem.*, 1997, **16**, 190.
4. H. Nagai, A. Sato, T. Kiyoshi, F. Matsumoto, H. Wada, S. Ito, T. Miki, M. Yoshikawa, Y. Kawate and S. Fukui, *Cryogenics*, 2001, **41**, 623.
5. R. S. Macomber, A Complete Introduction to Modern NMR Spectroscopy, John Wiley & Sons, Inc., New York, 1998.
6. H. Kovacs, D. Moskau and M. Spraul, *Prog. NMR Spectrosc.*, 2005, **45**, 131.
7. M. A. Macnaughtan, T. Hou, J. Xu and D. Raftery, *Anal. Chem.*, 2003, **75**, 5116.
8. H. Wang, L. Ciobanub, A. S. Edisonc and A. G. Webb, *J. Magn. Reson.*, 2004, **170**, 206.
9. H. Wang, L. Ciobanub and A. Webb, *J. Magn. Reson.*, 2005, **173**, 134.
10. P. A. Keifer, *Ann. Rep. NMR Spectrosc.*, 2007, **62**, 1.
11. R. L. Haner, W. Llanos and L. Mueller, *J. Magn. Reson.*, 2000, **143**, 69.
12. A. G Webb, *J. Pharm. Biomed. Anal.*, 2005, **38**, 892.
13. G. Martin, *Ann. Rep. NMR Spectrosc.*, 2005, **56**, 1.
14. U. Braumann and M. Spraul, in *On-line LC–NMR and Related Techniques*, ed. K. Albert, Wiley, Chichester, 2002, p. 23.
15. M. Sandvoss, in *On-line LC–NMR and Related Techniques*, ed. K. Albert, Wiley, Chichester, 2002, p. 111.
16. V. Exarchou, M. Krucker, T. A. van Beek, J. Vervoort, I. P. Gerothanassis and K. Albert, *Magn. Reson. Chem.*, 2005, **43**, 681.

17. O. Corcoran, P. S. Wilkinson, M. Godejohann, U. Braumann, M. Hofmann and M. Spraul, *Am. Lab.*, 2002, **5**, 18.
18. S. Christophoridou, P. Dais, L.-H. Tseng and M. Spraul, *J. Agric. Food Chem.*, 2005, **53**, 4667.
19. G. Ferrante and S. Sykora, *Adv. Inorg. Chem.*, 2005, **57**, 405.

CHAPTER 4
Sample Preparation

4.1 Sampling

In general sampling is the process of selecting a small number of samples from a larger defined target group of samples such that the information gathered from the small group will allow judgments to be made about the larger group. There are two main categories of sampling (probability sampling and non-probability sampling), each including several types of sampling methods. These methods are intended to minimise any type of bias that is attributable to errors in either drawing a sample or determining the proper sample size. Although a full description of sampling methods is outside the scope of this book, a few remarks about sampling of foodstuffs will be given here. The interested reader is encouraged to get more information on sampling from the references cited,[1,2] and in particular from a collection of articles published in 2004 by a special issue of the journal *Chemometrics and Intelligent Laboratory Systems.*[2]

The food sample used for the analysis is a complex analytical system, and as such it must represent all sources of variability or at least the most important known variability factors that have an impact in food composition. For instance, the geographical and botanical origin of foods, such as honey, wine, olive oil, cheese must be known accurately. Samples purchased from the nearby supermarket should be avoided, except perhaps for a specific analysis of commercial benefit. Samples from supermarkets are inhomogeneous and most of the time they are simply repetitions. The next factor is the sample distribution, which is very important for both classification and multivariate calibration studies. In particular, calibration methods require nearly uniform distribution of samples amongst the location of harvesting, variety and harvest period. Otherwise, the calibration model will be highly influenced by too many similar samples, lowering significantly its prediction ability. Another factor that requires special attention is harvesting. Food composition of plant origin is affected by ripening processes reflected on rapid metabolite changes due to

RSC Food Analysis Monographs No. 10
NMR Spectroscopy in Food Analysis
By Apostolos Spyros and Photis Dais
© Apostolos Spyros and Photis Dais 2013
Published by the Royal Society of Chemistry, www.rsc.org

enzymatic degradation or oxidation. For example, the ripening process occurring in olive fruits induces a progressive deterioration of the sensory quality of the extracted olive oil. In this respect, samples of plant origin should be harvested at approximately the same time period of each year to guarantee similar ripening stage.

Sampling for animal tissues (meat and poultry) appears to be a bit more difficult mainly because of the limited access of the food scientist to the slaughterhouse. Nevertheless, meat samples can be obtained in collaboration with local supermarkets. The farm and its location where the animals were bred should be known accurately, as well as the animal type, gender, age and weight. The knowledge of breeding specifications would be useful information. For each type of animal, a number of cuts (usually six to eight) from different animal muscles can be abscised for further analysis.

4.2 Foodstuff Pre-treatment

In general, the preparation of food samples for the analysis requires the following preliminary steps:[3,4]

4.2.1 Conservation

A successful analysis requires the rapid sample pre-treatment followed by an immediate analysis. However, this is not really feasible in most cases, and therefore sample conservation for a short time period before the analysis is an important first step to avoid potential modifications in food composition. For instance, conversion of muscles to meat may induce post mortem changes, which alter its biochemical and physical properties. The metabolic changes that occur in muscles after death play an important role in meat quality. Metabolic and enzymatic changes occurring in food samples of plant and animal origin can be quenched by rapid freezing in liquid nitrogen. Specific conservation conditions must be used for certain food of plant origin that is sensitive to the extremely low temperature of liquid nitrogen. These include olive oil that is stored at –20 °C and honey kept at room temperature before the analysis. Also, specific conservation methods are used for milk and dairy products to preserve the quality and nutrient composition.

4.2.2 Removal of Extraneous Matter

The removal of irrelevant extraneous mater is a necessary step in food pre-treatment. Soil and sand that adheres to fresh fruits and vegetables can be removed by washing. However, care should be taken to avoid excessive washing that may induce leaching of useful soluble material. Flesh is usually separated from fruit kernels, whereas suspended materials or sediments from liquid samples (*e.g.* beer, wine, juice, edible oil) is removed by filtration or separated by centrifugation. The food analyst should be aware of possible changes in the composition of crushed plant and animal tissues. Rapidly enzymatic changes may result in appreciable changes of certain food constituents. To prohibit such compositional changes, inactivation food enzymes may be necessary, for example by denaturation using a boiling mixture of methanol–water or ethanol–water.

4.2.3   Reduction

The reduction of the sample size is more important for solid or semi-solid foodstuffs. Finely divided solid material facilitates the analysis, since samples of smaller size dissolve faster and are easier to extract because of their greater surface area. There are several approaches for reducing the size of solid and semi-solid foodstuffs such as mechanical grinding, crushing, mixing, rolling, agitating, stirring, mincing, pulverising, macerating or other reasonable means. Frequently, sample reduction in particular for solid foodstuffs offers a means for sample homogenisation. This process is necessary to reduce variations in composition throughout the food sample.

4.2.4   Drying

Moisture in food samples is an important factor and it should be recognised before the analysis. Lack of moisture determination through sample drying or by other methods will end up with erroneous analytical results. Moreover, the presence of moisture may cause a multitude of problems in subsequent analysis including the formation of emulsions, sample turbidity, and more importantly the masking of the sample signals in the NMR spectra by the huge water signal. Vacuum drying is preferred, since this technique reduces to a great extent the deterioration of sample during heating. Sample drying by lyophilisation is particularly useful for the analysis on non-volatile compounds.

4.2.5   Sample Extraction

Extraction is the process that is used to isolate the target components in food matrices or to remove co-extractive material (clean up). Since this procedure is of great interest for the food scientists, it will be presented next in more detail.

4.3   Extraction

Extraction of food material has three main objectives: (a) removal of unwanted matrix components that can interfere with subsequent analysis, increasing its specificity; (b) concentration of a particular analyte to increase the limits of detection and quantification; and (c) partition of the analyte from a polar to a non-polar solvent for further analysis. Sample preparation involves several techniques that can be divided into three general categories: sample digestion techniques, solvent extraction (liquid–liquid extraction, LLE) and sorbent extraction (solid-phase extraction, SPE). Sample digestion techniques are not appropriate for NMR analysis, since the sample is destroyed upon digestion with mineral acids (nitric acid is a common acid for food digestion). Apart from chemical modifications occurred in food samples, the acquired NMR spectra show reduced sensitivity and large variations in chemical shifts from sample to sample despite attempts to stabilise the pH of the solution.

4.3.1 LLE

LLE is used to separate substance(s) from food matrices by partitioning the compounds(s) between two immiscible liquids.[5] Solvents can be selected on the basis of their physicochemical properties such as polarity and selectivity, as well as of their toxicity and inertness. Since food matrices contain a wide variety of compounds with different polarities and solubilities, no single extraction is capable of extracting all metabolites in a single extraction. The LLE procedure of extracting phenolic compounds from edible oils is a good example of this separation method. The oil is dissolved in chloroform and the resulting solution is then vigorously mixed with several portions of an immiscible mixture of methanol: water (80 : 20 v/v). The polar phenolic compounds contained in the oil will prefer the polar medium and migrate to the mixture of solvents, whereas the hydrophobic lipids prefer to remain in the organic phase. The layer of mixture of solvents containing the largest part of polyphenols is separated, vigorously agitated with hexane to remove traces of lipids, and its content is acquired by solvent evaporation or better by lyophilisation.

Presentation and comparison of various LLE extraction methods for metabolites analysis of plant and animal tissues using various deuterated and non-deuterated solvents and/or mixtures of solvents has been reported in several articles.[6–9] An important stage while preparing foodstuffs for metabonomic analysis is the pH regulations, since changes in the pH values of the extracts are accompanied by changes in the chemical shifts. In this respect, a buffer solution has to be used instead of pure solvent. Phosphate (pH range 5.0–7.4) or acetate (pH range 3.7–5.6) buffers are possible choices, but phosphate buffer (pH 6.0) is preferable since it shows better NMR resolution. Another problem often encountered in metabonomic studies of plant extracts is the presence of the primary metabolites (*e.g.* sugars and organic acids), which are usually more abundant and their signals usually mask those of the secondary metabolites, especially when the latter are present in low quantities. It can thus be advantageous to remove interfering primary metabolites before the analysis. This can be accomplished either by using liquid–liquid partitioning or better SPE.[7] Most steps of sample preparation for metabonomic studies are similar, regardless of the analytical tools to be used. Eventually, what is important is that all steps of the metabonomic study are standardised in order to allow direct comparison

Another way of solvent extraction is the repetitive contact of food sample with the fresh solvent, as in the well-known Soxhlet method. However, this technique where both polar and non-polar solvents can be used suffers from the same disadvantages as the LLE method; it uses a large amount of solvent and takes hours to run although it has the benefit of automation. Recently, pressurised liquid extraction[3] (PLE) and supercritical liquid extraction[3,10–12]

(SLE) that use specific solvents at higher temperature (80–200 °C), and pressure (10–20 MPa) have emerged as viable improvements of the traditional Soxhlet method. Selected applications of PLE and SLE to food samples pre-treatment are tabulated in references cited.[3,12]

4.3.1.1 *Pressurized Liquid Extraction (PLE)*

This extraction method is similar to Soxhlet extraction, except that the solvents are used near their supercritical region where they have high extraction properties. In that physical region the high temperature enables high solubility and high diffusion rate of analytes in the solvent, while the high pressure, in keeping the solvent below its boiling point, enables a high penetration of the solvent into the sample. Thus, PLE allows high extraction efficiency with a low solvent volume (15–40 mL) and a short extraction time (15–20 min). The efficiency of extractions with pressurised solvents (hexane, methylene chloride, isopropanol, ethanol, or aqueous mixtures) of polar and non-polar lipids was examined in corn and oats kernels. The effects of solvent polarity and temperature were tested on the recovery of total lipids, triglycerides, glycolipids, and phytosterols. Also, PLE was used to isolate several food constituents, such as tocopherols from seeds and nuts, steroids, vitamins, flavours and fragrances, aroma compounds in fruits, vegetables and cereals, and it was shown to yield very clean extracts and recoveries similar to conventional techniques.

4.3.1.2 *Supercritical Liquid Extraction (SLE)*

The SLE has been applied only recently to sample preparation on an analytical scale. It has the advantage of substituting the variety of conventional solvents with a single supercritical fluid (above its critical temperature and pressure), thus reducing the problems of their storage and disposal in the laboratory. Carbon dioxide, which is the most adopted supercritical fluid has an equivalent solvating strength to several liquid solvents, depending on the pressure and temperature conditions, For instance, the supercritical CO_2 at 38 265 MPa and 40 °C is characterised by solvating strength similar to liquids such as methylene chloride, carbon tetrachloride, toluene or benzene. The SLE technique has been applied for the determination of fats and nutrients in several raw food material and food products.

Details and apparatuses for PLE and SLE can be found in several review articles and standards texts.[11,12]

4.3.2 Microwave Assisted Extraction (MAE)

The basis of this extraction technique is the use of the microwave energy to heat the traces of moisture inside the food sample. This generates tremendous pressure on the food tissues and cell walls causing their rapture and consequent exudation of the active constituents into the solvent. For instance, using this technique the ether linkage of cellulose, which is insoluble in all solvents, are hydrolysed and are converted into soluble material within a few minutes. Higher yields can be obtained by increasing the temperature, which facilitates faster penetration of solvent into the food interior. The dielectric constant of the solvent and food matrix components is the factor that dictates the heating process. MAE can be performed in a closed-vessel system when the solvents used have high dielectric constants, whereas solvents with low dielectric constants are used in open-vessel systems. In the closed-vessel system, the solvent absorbs almost the whole microwave energy and exudates constituents from the food matrix. With this extraction system higher temperatures can be reached due to the increased pressure inside the vessel that raises the boiling point of the solvents used; there is no loss of volatiles, and much less solvent is required. In the open-vessel method operating at atmospheric pressure is the sample, which collects the microwave energy. Components within food with high dielectric constants (water, *etc.*) are heated locally exciting the food analytes that are trapped by the surrounding cold solvent. The low pressure under which this method operates prevents sample degradation. However, the main advantages of the open-vessel technique are the efficiency to process large samples without previous cooling, and the capacity for full automation. Several MAE applications deal with the determination of minor food constituents, *e.g.* volatiles, flavour compounds.[3,13] Table 8 of ref. 3 lists a variety of foodstuffs extracted by the MAE technique.

Comparative studies[14,15] among PLE, SLE, MAE techniques, and the conventional Soxhlet extraction procedure have proved that the three extraction methods are superior than the classical method in decreasing the total time of the analysis and the consumption of much less organic solvents. Based on the same criteria, *i.e.* duration of extraction and solvent quantity, PLE was found to be more effective than MAE. However, one should be cautious in comparing extraction techniques, because comparative studies have been conducted on particular food matrices and the results obtained might not be applicable, when extraction is carried out with different foodstuffs.

4.3.3 SPE

SPE is a very powerful method for sample preparation, in which the analyte contained in a liquid phase comes into contact with a solid-phase extraction material that is loaded in a separate cartridge or disk.[3] The liquid is selectively absorbed onto the surface of the solid phase by chemical attraction. The other material not absorbed remains in the liquid phase and passes intact through the solid phase. The next step is to pass through the sorbent bed a wash solution that does not disturb the analyte, but removes adsorbed contaminants from the sample matrix. The final step is the selective elution, in which the analyte splits off the solid support into the solvent, for which the eluted substance shows a greater affinity than for the solid material. The decisive points for a successful use of SPE for food sample preparation involves:[16,17] (a) proper choice of the sorbent material, suitable for the extraction of the analyte; (b) sample pre-treatment to optimise sample conditions for the analyte retention; and

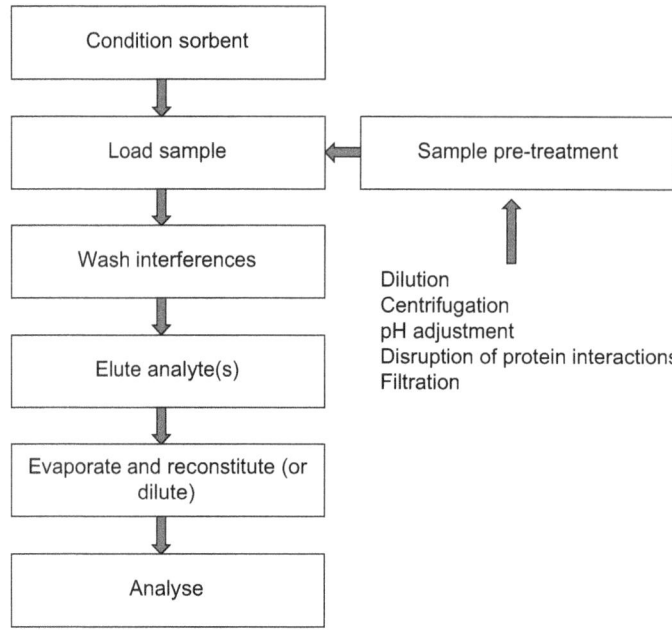

Figure 4.1 The general scheme for performing solid-phase extraction consists of several sequential steps. Sample pre-treatment is important to ensure analyte retention.
(Reprinted from ref. 4. Copyright (2003), with permission from Elsevier.)

(c) selection of the eluting solvent that should offer very good solubility to the analyte. There are three general classes of sorbents bonded to the analyte by dipole–dipole interactions, and hydrogen bonding (polar sorbents), van der Waals interactions (non-polar sorbents), and electrostatic interactions (ion-exchange sorbents). The choice of sorbents depends on the food matrix, the constituents of interests, and possible interferences. Sample pre-treatment and solvent selection is accomplished by investigating proper dilution of the analyte to the chosen solvent, pH correction, ionic strength, and polarity and flow rate of the elution solvent. In general, 4 to 6 steps are performed using SPE extraction, including conditioning of the sorbent bed, retention of the analyte, selective washing and elution.[16] Figure 4.1 depicts the general scheme for performing SPE.

For the extraction of tiny amounts of analytes that are especially useful for subsequent analysis with GC or HPLC, the solid-phase micro-extraction (SPME) procedure is recommended. The SPME process is simple and can be performed even by non-scientists without the need to have GC-MS or HPLC equipment. A sampling device, *i.e.* a fused silica fibre coated with an extracting solid phase material, concentrates different kind of analytes (both volatiles and non-volatiles) by adsorption and/or absorption and the extracted material may be released directly to the flow probe of the NMR spectrometer. The preference of using SPME lies on the facts that extraction is fast and simple and can be done usually without solvents, and detection limits can reach parts per trillion (ppt) levels for certain compounds. When properly stored, samples can be analysed days later in the laboratory without significant loss of volatiles.

Several standard texts and review articles provide in-depth coverage of SPE and/or MSPE,[3,16–19] and discuss strategies for automation and high-through-put on-line sample preparation.

4.3.4 LLE *versus* SPE

The food analyst is often faced with the dilemma of what extraction method should be used for a particular sample. Conventional LLE techniques are time-consuming, labour intensive procedure and multi-stage operations. In addition, each step of the analysis is prone to errors due to solvent or analyte (volatiles) evaporation. It requires large volumes of organic solvents, which can be expensive and requires disposal of hazardous waste increasing thus the cost of the analysis. Moreover, it may create health hazards to the personnel, and it is difficult to automate. Nevertheless, LLE is a straightforward technique, and well documented in standard texts and references. It is an easy and accurate method for sample preparation and could be used whenever large quantities of analytes are required for multiple experimentations. The other major liquid extraction methods PLE and SLE have been limited so far to toxicological and environmental analyses. The requirement for high pressures increases the cost compared to conventional liquid extraction, so these modern extraction will only be used where there are significant advantages, *e.g.* protection of food samples against any oxidative degradation. Carbon dioxide itself is non-polar, and has somewhat limited dissolving power, so cannot always be used as a solvent on its own, particularly for polar solutes. The use of modifiers increases the range of materials, which can be extracted. Food grade modifiers such as ethanol can often be used, and can also help in the collection of the extracted material, but reduces some of the benefits of using a solvent, which is gaseous at room temperature.

Most of the conventional LLE limitations are removed by using SPE cartridges or discs. SPE needs much less solvent and is the preparation method that should be preferred for small quantities of analyte, but it is still a time-consuming multi-step process, and it should be used with caution for the extraction of volatile components. In this case, the SPME technique should be applied.

4.4 NMR Samples Preparation

Sample preparation for NMR analysis comprises samples that do not need extra care or samples that have already undergone a suitable pre-treatment to isolate a specific substance or a particular family of substances.

4.4.1 Liquid-State NMR

To obtain high-resolution NMR spectra, it is important to use samples free of suspended material (precipitates, dust, *etc.*). Suspended material present in solution will produce local field inhomogeneity, increasing the line width of signals in the NMR spectrum. Broad spectral lines will reduce spectral resolution and shimming cannot correct this problem. Suspended particles can easily be removed from the NMR sample by constructing a filter using preferably cotton wool (the alternative glass wool does not filter out small particles) as a filtering medium. An effective filter can be made by placing a small amount of cotton wool pushed down inside a Pasteur pipette with the help of a second long pipette. The solute of interest can then be dissolved in a separate glass vial using less than the final volume of deuterated solvent required to make the NMR sample. After the solute has been dissolved it can be transferred directly to a NMR tube by passing the solution through the cotton filter. Finally, the sample volume can be adjusted by adding the remaining solvent to the NMR tube so that a final sample volume is reached followed by vigorous shaking of the sample to effectively mix its content. The amount of solute and the final sample volume depends on its availability and solubility to a particular solvent. The first factor determines, also, the size of the probe. Microprobes are suitable for scarce solutes, such as proteins or biofluids, whereas sample volume for microprobes goes down following the miniaturisation of the RF coils. Commercial microprobes as small as 1 mm are now available in the market with an active volume of 2–5 µL. However, for conventional 5 mm probes, the required sample volume for a good shimming should not exceed the coil dimensions. A short sample (*i.e.* volume less than that suggested by the manufacturer) can only be shimmed through tremendous effort. The volume and sample height for a 5 mm tubes is dictated by the probe specifications provided by the manufacturer.

In old times, one drop of the internal standard TMS was added in the NMR sample just before the insertion of the NMR tube to the probe. Nowadays, the presence of TMS is not needed to reference chemical shifts. Modern NMR spectrometers calculate directly the chemical shifts of signals in the spectrum on the basis of the NMR solvents chemical shifts relative to TMS; these values are known and are stored in the computer memory. This is feasible provided that the NMR user identifies the solvent in the acquisition parameters before the start of the experiment.

Some experiments, such as relaxation times and NOE measurements for molecular dynamic studies or for quantitative analysis, require removing the dissolved oxygen. Interaction of the excited nuclei with the lone electron pairs of the paramagnetic oxygen molecules accelerate their relaxation resulted in erroneously low experimental values. There are two ways for removing oxygen, either by bubbling an inert gas (nitrogen or argon) through the solvent in the NMR tube for a few minutes, or more effectively by using the so-called freeze–pump–thaw method. Some solvent may be lost through evaporation using the first method. The freeze–pump–thaw method includes the following steps: (1) freeze sample in liquid nitrogen or CO_2/acetone, (2) turn on the pump for evacuation of space above the frozen sample, (3) turn off the pump and leave sample to thaw, (4) repeat steps 1–3 at least three times, (5) seal the NMR tube.

4.4.2 NMR Solvent Selection

Solvents used for the NMR sample preparation must be deuterated. Protonated solvents should be avoided because their strong signals may overlap those of sample. Moreover, strong solvent signals receive extensive digitisation from ADC at the detriment of the sample signals, which may buried under the noise (section 5.1.1) Another benefit of using deuterated solvents is the deuterium nucleus, which provides the field-lock signal. The degree of deuteration for efficient experimentation should not be lower than 98.5%; usually varies from 98.5% to 100%. Fully deuterated solvents are more expensive, as well as solvents with several non-equivalent protons to be replaced by deuterium. For instance, deuterated chloroform ($CDCl_3$) is less expensive than deuterated methanol (CD_3OD). Non-polar samples or extracts, *e.g.* lipids, are commonly dissolved in non-polar solvents, such as chloroform-*d*, whereas polar solvents, such as D_2O, methanol-d_4, dimethylsulfoxide-d_6, or acetone-d_6 are used for polar samples (*e.g.* the polar extract of olive oil). Finally, it should be noted that chemical shifts of the same compound could be different in different solvents. Therefore, any chemical shifts assignment based on databases or literature values is valid only when the sample is dissolved in the same solvent reported in the literature.

There are experimenters that consider the water resonance as an ideal reference signal to quote chemical shifts. Because of the large variations of the chemical shift of the water signal with temperature and pH, it should never be used as an internal reference for chemical shifts. In aqueous solutions the use of trimethylsilyl propanoic acid (TSP) or 4,4-dimethyl-4-silapentane-1-sulfonic acid (DSS) (both at δ 0.00) or dioxane (at δ 3.75) or 1, 3, 5-trioxane (at δ 5.00) as internal references are recommended.

4.4.3 Solid-State NMR

Powdered and/or sliced solid samples are packed into cylindrical rotors made by Teflon (white) or Kel-F (transparent). The experimenter should use the material that best matches the susceptibility of the material under investigation. If samples are highly diamagnetic then Teflon must be used. For less diamagnetic samples Kel-F is the appropriate material. There are two types of caps clear (Kel-F) for temperatures not far from room temperature ($-20\,°C$ to $50\,°C$), and white (boron nitride or zircon) for more extreme temperatures ($-50\,°C$ to $120\,°C$). The boron nitride cap is very fragile and virtually unusable. The zircon cap is recommended for extreme temperatures because it is more durable that boron nitride. If there is not enough material, it is better to dilute the sample with something inorganic such as silica, alumina or calcium sulphate. Hard solid samples are ground up with a mortar and pestle, and the resulting powder is placed in the rotor with the aid a small spatula. If necessary, sample is compressed using a special tool. The rotor is filled up to the cap's place. For liquids and gels, formation of bubbles that deteriorates resolution should be avoided. If the density of the liquid is known then the rotor, insert and screw can be weighed before and after filling to confirm that there are no bubbles. For soft samples such as animal tissues, the inserts must hold the sample exactly in place. Solids that can be easily melted should be inserted with an automatic pipette or with a syringe. For resins and other swelling samples, the swelling ratio for the solvent (typically 5) is checked, and the amount of solid sample required filling up the rotor (usually 12 or 50 µL) after swelling, should be calculated. After filling up the rotor, the sample must be homogenised with a pin. Soft food and tissue samples are washed with D_2O and cut to the appropriate size with a scalpel. Flat samples such as leaves should be cut, rolled up, and inserted in the rotor.

4.4.4 MRI

Sample preparation for MRI experiments depends heavily on the food studied and the capacity of the MR probe. In this respect, the MRI user is advised to follow tested techniques for sample preparation of a particular foodstuff. For instance, meat samples obtained after slaughtering are excised and cut in parallelepipeds of appropriate dimensions to conform to those of the MR imager. Bread dough should be thawed for an overnight period at 4 °C and proofed at room temperature for 1 h prior to their placement to the MRI scanner. Another example is MRI measurements to monitor the ripening and decay of tangerine and orange fruits. Intact fruits are used, and put vertically in a 20 cm RF coil at the centre of the MRI magnet.

References

1. S. S. M. Curren and J. W. King, in *Sampling and Sample Preparation for Food Analysis, In Sampling and Sample Preparation for Field and Laboratory*, ed. J. Pawliszyn, Elsevier, Amsterdam, 2002, ch. 25.
2. Special Issue: 50 years of Pierre Gy's Theory of Sampling Proceedings: First World Conference on Sampling and Blending (WCSB1) Tutorials on sampling: Theory and Practice, *Chemometrics and Intelligent Laboratory Systems*, 2004.
3. P. L. Buldini, L. Ricci and J. L. Sharma, *J. Chromatogr. A*, 2002, **975**, 47.
4. D. A. Wells, *High-throughput Bioanalytical Sample Preparation. Methods and Automation Strategies*, Elsevier, Amsterdam, 2003, ch. 2, p. 41.
5. F. F. Cantwell and M. Losier, *Compr. Anal. Chem.*, 2002, **37**, 297.
6. C. Y. Lin, H. Wu, R. S. Tjeerdema and M. R. Viant, *Metabolomics*, 2007, **3**, 55.

7. K. A. Kaiser, G. A. Barding, Jr and C. K. Larive, *Magn. Reson. Chem.*, 2009, **47**, S147.
8. H. Wu, A. D. Southam, A. Hines and M. R. Viant, *Anal. Biochem.*, 2008, **372**, 204.
9. H. K. Kim and R. Verpoorte, *Phytochem. Anal.*, 2010, **21**, 4.
10. E. Anklam, H. Berg, L. Mathiasson, M. Sharman and F. Ulberth, *Food Additives and Contaminants*, 1998, **15**, 729.
11. M. A. McHugh and V. J. Krukonis, *Supercritical Fluid Extraction: Principles and Practice*, 2nd edn, Butterworth-Heinemann Series in Chemical Engineering, London, 1994.
12. M. Valcárcel and M. T. Tena, *Fresenius J. Anal. Chem.*, 1997, **358**, 561.
13. P. Tatke and Y. Jaiswal, *Res. J. Med. Plant*, 2011, **5**, 21.
14. V. Pères, J. Saffi, M. I. S. Melecchi, F. C. Abadc, R. de Assis Jacques, M. M. Martinez, E. C. Oliveira and E. B. Caramão, *J. Chromatogr. A*, 2006, **1105**, 115.
15. M. P. Lompart, R. A. Lorenzo and R. Cela, *J. Chromatogr. A*, 1997, **774**, 243.
16. D. A. Wells, *High Throughput Bioanlytical Sample Preparation Methods and Automation Strategies*, Elsevier, Amsterdam, 2003, ch. 11, p. 361.
17. C. F. Poole, *Compr. Anal. Chem.*, 2002, **37**, 341.
18. J. Pawliszyn, *Compr. Anal. Chem.*, 2002, **37**, 389.
19. G. Vas and K. Vékey, *J. Mass Spectrosc.*, 2004, **39**, 233.

CHAPTER 5

Experimental Conditions and Processing

5.1 Qualitative Liquid NMR

After sample preparation, one has to design the NMR experiment. Choosing the probe and the pulse program are critical factors for a successful NMR experiment. As described in section 3.2.1, the probe size and its operation are dependent of the quantity of the target substance(s) in the food matrix, the sensitivity of the experiment, the type of the experiment and the nature of the nucleus studied. Pulse sequences are computer programs that control all hardware aspects of the NMR measurements. Each pulse sequence is specific and the choice of the particular pulse program depends on the required information. Most common 1D and 2D pulse sequences are stored in the computer memory of the NMR spectrometer, whereas others can be found free of charge in pulse sequence libraries of the NMR manufacturers or in private collections. The operators are allowed to write and use their own pulse sequences to accomplish whatever NMR task is required, provided that they are familiar with the computer language of the spectrometer. Following the choice of the appropriate probe and pulse sequence, insertion of the NMR tube into the bore of the probe, tuning and shimming of the magnet is performed, either manually or automatically as in current NMR spectrometers. Pulse calibration and the entry of the proper parameters in the pulse sequence acquisition menu precede the start of the experiment. Calibration is not so important for the simple single pulse experiments, and an approximate 90° value is satisfactory. Pulse calibration appears to be crucial for multi-pulse experiments, which use more than one pulses separated by various time intervals, such as those used in 2D NMR experiments. In addition, the pulse width and power levels depend on the NMR probe being optimally tuned and

RSC Food Analysis Monographs No. 10
NMR Spectroscopy in Food Analysis
By Apostolos Spyros and Photis Dais
© Apostolos Spyros and Photis Dais 2013
Published by the Royal Society of Chemistry, www.rsc.org

matched. When the probe is not optimally tuned and matched, the pulse width and power level stored in the parameter files of the spectrometer are no longer calibrated correctly. Another important factor of the experimental design is the temperature range at which the experiment will be carried out. High temperatures may be required for viscous solutions or amorphous material to avoid signal broadening due to slow molecular motions (section 2.10.1). Rapid rotational motions at high temperature average to zero the dipolar interactions, which constitutes a significant factor for line broadening. Also, VT measurements are mandatory in order to perform kinetic studies. The following sections will be devoted to describe the effect of the acquisition parameters on the quality (sensitivity and resolution) of high-resolution 1D and 2D NMR experiments in the liquid and solid state. Additional sources of information about experimental conditions and effective NMR data manipulations to obtain good quality 1D and 2D NMR spectra are given in the list of references.[1–5]

5.1.1 Setting Up the High-Resolution 1D NMR Experiment

5.1.1.1 Pulse Width (PW)

As mentioned in section 2.6.2, a pulse is simply a high-power RF signal transmitted for a short period of time in the range of μs for non-selective excitation. The pulse duration, or *PW* should not be very long to protect the amplifiers from burning. In modern spectrometers operating with quadrature detection (see below), the frequency of the pulse constituting the reference frequency of the RFR is set at the centre of the spectral width (*SW*), *i.e.* within the range of resonant frequencies expected. This ensures that the pulse is homogeneous and capable of exciting the sample nuclei within the selected *SW*, which is now divided into two halves (*SW*/2).

5.1.1.2 Spectral Width (SW)

The *SW* depends primarily on the type of nuclei studied. ^{1}H nuclei resonate in a narrower frequency range ($\sim$15 ppm) than the heavier ones that show large paramagnetic shielding, *e.g.* ^{13}C ($\sim$250 ppm), ^{15}N ($\sim$900 ppm), ^{19}F ($\sim$400 ppm), ^{31}P ($\sim$1000 ppm). The second factor that governs the choice of *SW* is the decay of the FID. Too short a *SW* does not allow the full decay of FID. Truncation of the FID results in distorted baselines. On the other hand, too long *SW* makes the digitiser to sample noise after the complete decay of FID; which may have a defective impact on the sensitivity of the experiment. A third factor is the sampling rate of FID. Sampling rate is determined by the highest frequency signal, *i.e.* the signal farthest from the centre of *SW* where the pulse is applied. According to the fundamental result in the field of information theory as described by the Nyquist sampling theorem (eqn (5.1)), the highest frequency needs to be sampled at least twice during each cycle of its sine wave. This means that the number of samples or data points (DP) has to be twice the number of cycles of the highest frequency within the FID.

$$\text{No. of DP} = 1/DW = 2 \times \text{No. of cycles} = 2 \times \text{highest frequency} \qquad (5.1)$$

DW is the time at which each sample value is obtained, or the time between two samples and is called dwell time. The reciprocal of dwell time represents the sampling rate, implying that the maximum frequency that can be accurately measured is $1/(2 \times DW)$. DW is expressed in µs and the frequency in Hz. If the frequency of a signal exceeds the sampling rate, *i.e.* it is outside the spectral width, then it is misinterpreted as a signal of lower frequency, and it appears in wrong position in the NMR spectrum. It is usually recognised by its lower intensity and/or the wrong phase that cannot be corrected using the phasing routine of the spectrometer. This process is called folding or aliasing. Figure 5.1 shows the phenomenon of folding or aliasing.

Using a low-pass filter to remove frequencies above the sampling rate or increasing the spectral width to include all resonances may suppress aliasing. In conclusion, the spectral width of the experiment is determined by the sampling rate, and should contain the highest frequency signal. Any region of spectral width can be expanded for a better scrutiny in the screen, but is unchanged as a whole once the experiment has started.

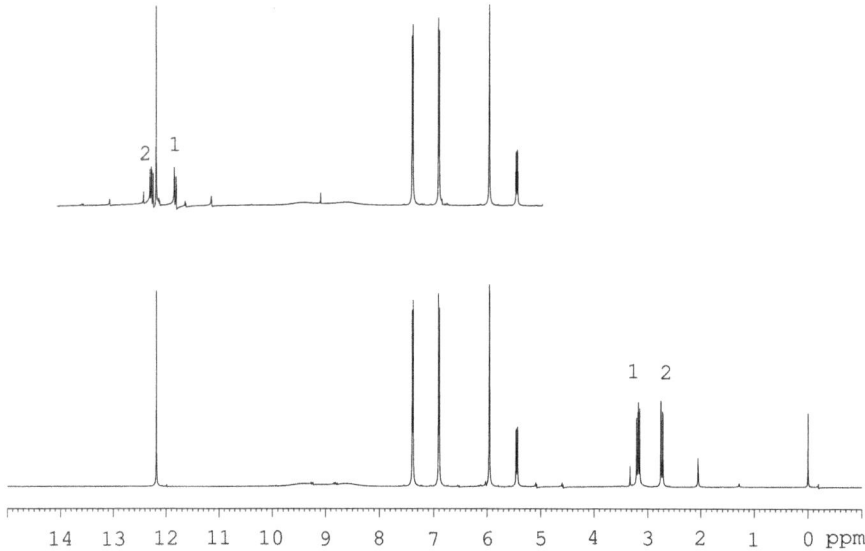

Figure 5.1   Aliasing with the 500 MHz ^{1}H NMR spectrum of naringenin in DMSO-d_6 solution observed with quadrature detection. The normal spectrum (bottom) recorded with an offset of 6 ppm and $SW = 18$ ppm (aromatic region). The spectrum (top) was recorded with an offset of 9.5 ppm and $SW = 9$ ppm. The spectral window of the top spectrum is too small for sampling the FID according to the Nyquist theorem, and thus aliasing of signals 1 and 2 occurs. These signals appear in the wrong chemical shifts and phase.

5.1.1.3 Number of Data Points (DP)

The decay of FID induces an analogue signal (voltage) in the receiver coil of the spectrometer. Sampling this electrical signal means recording the voltage at regular intervals of time. Since the computer cannot understand anything but numbers, this analog signal is converted to a digital number proportional to the magnitude of the voltage and is stored in the computer memory locations. Each such digital number (bit) is called a data point (*DP*), and the device that does the FID sampling and the analogue to digital conversion is the Analog-to-Digital Converter (ADC) or digitizer. The number of data points is inserted in the acquisition parameters as multiples of K, which stands for $2^{10} = 1024$ data points. For each time sampled two measurements need to be made a real part and an imaginary part, because of the quadrature detection. Quadrature detection is a clever way to tell the difference between positive frequencies and negative frequencies corresponding to clockwise and counter-clockwise rotations, respectively, relative to the rotating frame of reference as shown in Figure 2.9a.[6] The direction of rotation is monitored by two channels of the detection system of the receiver differing in phase by 90°. One channel detects signals along the *x*-axis, and at the same time the second channel distinguishes signals along the *y*-axis. Combination of the Fourier transforms of these signals gives a true representation of the actual positions of the resonance frequencies Figure 5.2 describes the quadrature detection procedure (see sections 2.2 and 2.6.1).

For mathematical convenience (the computer does not understand this definition) these signals are usually referred as the real and imaginary parts of the FID signal, respectively, or as complex pairs. In some spectrometers rather than using two detectors one detector is used which alternates between the real and imaginary signals. In this case the number of data points in the transformed spectrum is halved. Phase sensitive detectors are used to monitor the absorptive signal, while suppressing the dispersive one.

The advantage of using quadrature detection is that the offset frequency (see below) can be placed in the middle of the spectrum halving the spectral width required.

5.1.1.4 Acquisition time (t_{aqc})

This is the time required for sampling (collecting DP) the whole FID once. This is not the time required for the entire spectrum to be acquired, inasmuch it does not include the relaxation delay and pulse width. t_{aqc} is dependent on the spectral width and/or the number of data points. The interconnection of these three parameters is described by the following fundamental NMR acquisition equation (eqn (5.2)):

$$t_{acq} = DP * DW = \frac{DP}{2 \times SW} \quad \text{or} \quad \frac{1}{t_{acq}} = \frac{SW}{DP/2} = DR \tag{5.2}$$

According to **eqn (5.2)**, changing one of the three parameters will force another to change in order to maintain the equality. For example, if *SW* is

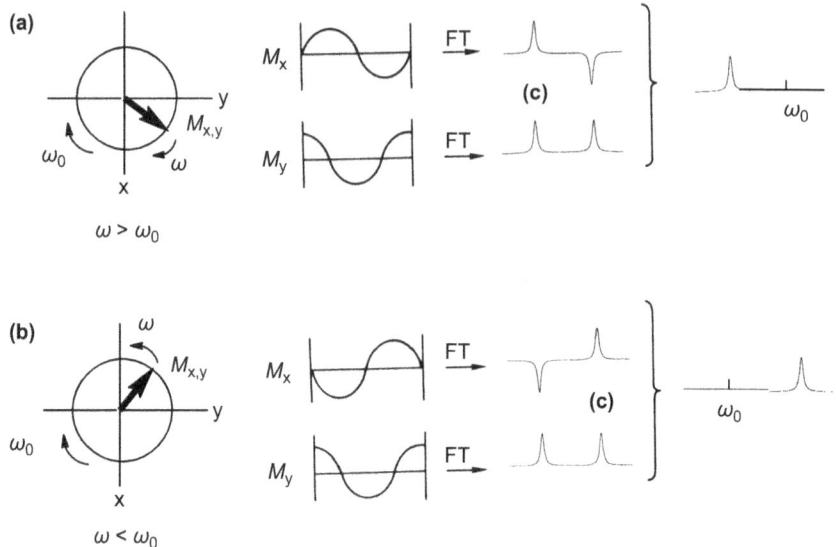

Figure 5.2   The principle of quadrature detection; (a) for the magnetisation vector
rotating clockwise on the (xy) plane, the x-component is detected as a
positive sine and the y-component as positive cosine. (b) For the magnetisation vector rotating counter-clockwise, the x-component is detected as
a negative sine, and the y-component again as positive cosine. (c) The
result of the FT is a dispersive signal for the x-component and an
absorptive signal for the y-component. Addition of the signals pairs
eliminates one signal, and therefore the sense of rotation and the relative
sign of the frequency can be determined.
(Reprinted from ref. 6. Copyright (1998), with permission from John
Wiley & Sons.)

doubled, either the number of points will double or the acquisition time will be
halved. This is explained by the fact that the larger spectral width requires a faster
sampling rate to assure that the highest frequency of the FID is sampled at least
twice in each cycle. Useful information extracted from this equation is the digital
resolution (DR), *i.e.* the width (in Hz) of a signal at half height. DR is equal to the
reciprocal of the acquisition time $(1/t_{aqc})$. For instance, for a spectral width
$SW = 4000\,Hz$, $DP = 16K$ data points are required to obtain resolution of
0.49 Hz. Digital resolution can be increased (DR decreases), either by increasing
the number of data points for a constant spectral width, or by decreasing the
spectral width for a fixed number of points. However, one must be cautious not
to exceed the limits, while changing the spectral width and causing aliasing.

5.1.1.5   *Relaxation Delay (D_1)*

The simple one pulse ($90°$) experiment is repeated a number of times in order to
average the individual FIDs for increasing the sensitivity of the experiment.

The relaxation delay is inserted in the acquisition parameters in order to guaranty that the spin population returns to the Boltzmann equilibrium before the application of the next pulse. The absence of delay results to partial or complete spin saturation and consequently there will be little or no signal in the spectrum. According to relaxation theory, full nuclear relaxation is achieved provided that the relaxation delay becomes five times the longest longitudinal relaxation time ($5 \times T_1$), prolonging thus significantly the duration of the NMR experiment. Although a long relaxation delay is a precondition for quantitative analysis, qualitative experiments can be conducted with shorter relaxation delay with a concomitant reduction of the signal intensity. This compromising value allows a larger number of scans and the accumulation of more data per unit time. The problem with shorter delay than $5 \times T_1$ is the fact that the second and subsequent scans will find the spin system not in equilibrium. Therefore, the data from the first 1–2 scans will not be the same. Nevertheless, after a few scans the system will reach a steady state, although not in equilibrium. In order to avoid collecting improper data points, dummy scans are used once at the start of the experiment. Dummy scans include all the parameters used in the particular pulse sequence, but the receiver is not operating to collect data. In this respect, the system is in the steady state before the start of the actual acquisition. 8 to 16 dummy scans are adequate for the system to reach the steady state. Another way to decrease the delay is to use a shorter pulse width that tips the magnetisation vector (M_0) at a flip angle, φ, smaller than 90° (eqn (2.15)). Then the return to equilibrium requires less time, since at the beginning of acquisition the M_z component has a value of $M_0 \cos \varphi$ (see section 2.2) instead of zero. Also, the smaller flip angle, φ, reduces the M_y components by a factor of $M_0 \sin \varphi$, being thus weaker. Therefore, one searches for an optimum value for the flip angle (the so called Ernst angle), φ_{opt} to detect the maximum M_y magnetisation. This optimum value has been theoretically expressed by the following formula, eqn (5.3):[7]

$$\cos \varphi_{opt} = \exp\left(\frac{D_1}{T_1}\right) \tag{5.3}$$

For protons with T_1 values 1–10 s and $D_1 = 5$ s, φ_{opt} has a value between 90° and 52°, whereas for ^{13}C, it ranges from 25° to 8°, for $T_1 = 10$–100 s and $D_1 = 51$ s.

5.1.1.6 Number of Scans

The number of scans, and hence the accumulated FIDs, depends on the sample concentration and the desired sensitivity. In section 3.2, a thorough discussion about the NMR sensitivity in terms of the S/N ratio has been given. Sensitive NMR spectrometers should have large S/N ratios. A ratio appreciably larger than unity ($S/N > 3$–10) is required to distinguish a signal from noise. As it will be discussed below much larger S/N ratios are required to ensure accurate and reproducible signal integrals for quantitative analysis.

5.1.1.7 *Offset Frequency*

The offset frequency (or transmitter offset) corrects the position of the reference frequency of a nucleus at the centre of a predetermined spectral width. In other words, the offset sets the position of the centre of the spectral width. Adding the offset frequency to the fundamental or basic resonance frequency for the nucleus does this. For instance, to move the reference frequency by 0.01 MHz (10 000 Hz) for a ^{1}H nucleus with a fundamental frequency of 300.13 MHz, an offset frequency of the same value should be added, yielding a reference frequency of 300.14 MHz. In addition, the spectral width can be moved down field or up field by adding an offset frequency of the appropriate size. To move the spectral width by 1 ppm (300 Hz) downfield, 300 Hz should be added in the offset frequency of 10 000 Hz, *i.e.* 10 300 Hz. The outcome of changing the offset frequency can be seen in the ^{1}H NMR spectrum of naringenin in DMSO-d_6 solution of Figure 5.3.

The reasonable question is why the spectral width needs to move down field or up field? The answer is given by considering the lock signal of the deuterated solvent. As the solvent changes, *e.g.* from CDCl$_3$ to acetone-d_6, the lock system changes slightly the spectrometer magnetic field strength striving to centre the deuterium frequency of the solvent at the null point of the lock feedback circuit. Needless to say, that a change in the field is accompanied by changes of all the

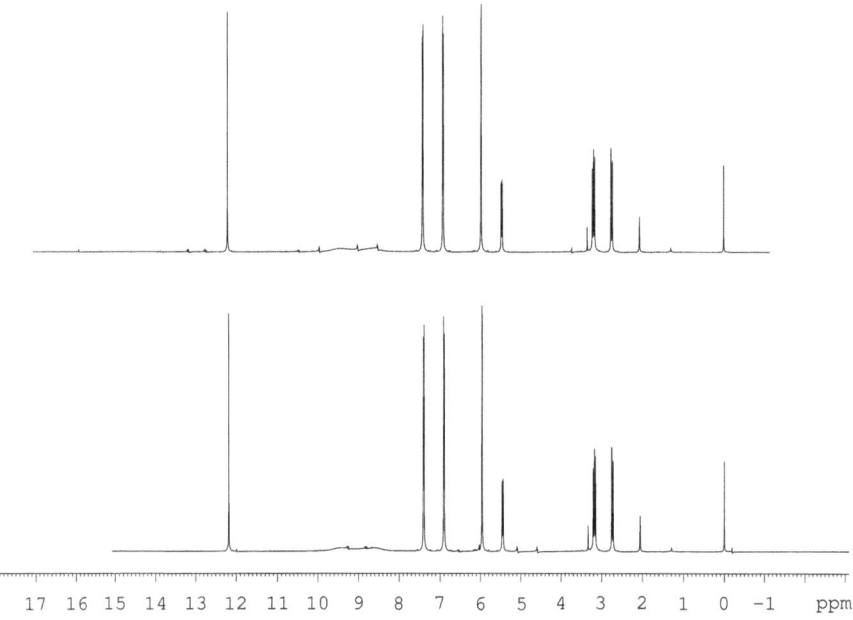

Figure 5.3   The effect of offset on the ^{1}H NMR spectrum of naringenin in DMSO-d_6. The spectral width of the normal spectrum (bottom) has moved by 2 ppm towards higher frequencies to the left (top) as the carrier frequency (offset) has increased by same value.

resonance frequencies of the spectrum by the same amount. To restore the initial position of the spectral width, an offset frequency of an equal and opposite amount should be added. If the field is not corrected, the spectrum will move out of the spectral width and some signals will be aliased.

Apart from the transmitter offset, there exists the decoupler offset. This is used whenever decoupling is carried out using the RF power of the decoupler. The decoupler offset represents the centre of the frequency range, which will be decoupled. For instance, broadband proton decoupling of ^{13}C NMR spectrum requires the decoupler offset to be at the centre of the proton spectrum. In selective decoupling, the offset is centred exactly on the frequency of the particular signal that will be irradiated.

5.1.1.8 Receiver Gain

The radio frequency FID signal coming from the probe is very week and needs amplification. The amount of amplification given to FID in the receiver is called the receiver gain. The amount of gain depends on the sample concentration and is adjusted before the start of the experiment for any new sample. This can be done either manually, or automatically by the spectrometer itself. In automated mode, a number of trial FIDs with varying receiver gain is performed until the optimum (maximum) value is found that does not cause overflow of the ADC. Concentrated solutions need less amplification than diluted samples. High receiver gain is also given to experiments with the less sensitive nuclei such as ^{13}C and ^{15}N. A gain setting too low will reduce sensitivity (S/N) and produce dynamic range problems, whereas too large a receiver gain will cut the most intense part of FID, and perhaps the manifestation of a distorted baseline in the transformed spectrum.

5.1.1.9 Dynamic Range

There is no such a parameter in the list to define, but it has to do with the digitisation (sampling) of FID by the ADC device (digitiser). A simple definition of dynamic range is the ability of ADC to detect weak signals in the presence of strong ones. How small are the signals to be detected, it depends on the ADC word length measured in bits (a binary digit). The word length defines the ADC resolution. Typical ADC resolutions on NMR spectrometers range from 12 to 16 bits. For a 12-bit word length, the maximum number of data points that can be collected and stored is $2^{11} - 1 = \pm 2047$; one bit is reserved for the sign of the signal. This means that the weakest (smallest) signal which can be digitised will have an intensity, H_w, (eqn (5.4)) such that:

$$\frac{H_s}{H_w} = 2^{11} - 1 = 2047 \tag{5.4}$$

H_s, is the intensity of the strongest (largest) signal. The ratio in eqn (5.4) provides the dynamic range of the spectrometer, indicating that signals with relative intensity less than 1 : 2047 is not digitised and is lost under the noise. Therefore, it is advisable to use the full dynamic range in order to detect the FID correctly.

5.1.1.10 *Solvent Suppression*

Dynamic range problems occur very often when strong signals appear in ^{1}H NMR spectra. In this case, the digitiser samples preferentially the strong solvent signal at the detriment of small signals, which are hardly visible in the spectrum. Dynamic range usually occurs in diluted aqueous solutions of biological samples, where the huge H_2O signal or the residual signal HOD of the deuterated D_2O solvent overwhelms any other signal in the spectrum. Another example is the strong ^{1}H signals of triacyglycerols of edible oils that obscure the appearance of the nearby signals of the minor mono and diacylglycerol compounds and other minor compounds in olive oil. Therefore, strong signals need to be removed from the spectrum. The easiest way to suppress a strong signal is the pre-saturation method applying a saturation pulse (a saturation pulse lasts several ms as opposed to an excitation pulse, which lasts a few μs) on solvent resonance transmitted by the decoupler of the spectrometer (Figure 5.4).[8]

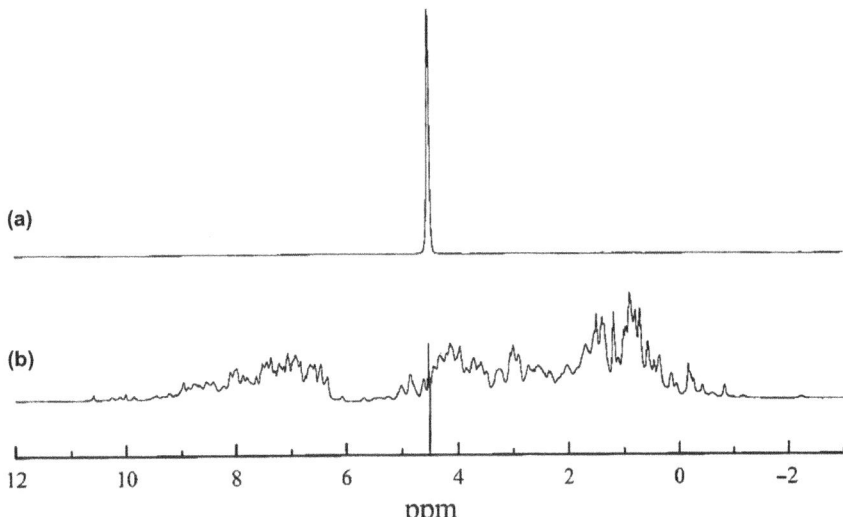

Figure 5.4 ^{1}H NMR spectra of a lysozyme solution in D_2O : H_2O solution. (a) Without water suppression and (b) with the water resonance suppressed using the Water-PRESS sequence. Both spectra were acquired at 36 °C at 300 MHz and in both cases a 90° observe pulse was used. To avoid saturation of the analogue-to-digital converter the receiver gain used in acquiring spectrum (a) was 28 dB less than that used in acquiring spectrum (b).
(Reprinted from ref. 8. Copyright (1997), with permission from Elsevier.)

Although it is simple, this method has the disadvantage in aqueous solutions that NH or OH protons that are exchanging with water also have their signals reduced or eliminated. Other disadvantages of this method are the large dispersive tail of the strong signal resulting in a tilted baseline, and the attenuation of the resonances of interest close to the strong signal. The degree of attenuation depends largely on the decoupler power and how close the small signals are from the signal to be suppressed. A multitude of techniques for solvent suppression have been invented and implemented in modern NMR spectrometers. The most effective of these techniques use pulsed field gradients yielding better suppression compared with other methods. Currently, the most satisfactory water suppression technique available is the so-called excitation sculpting,[9] which is based on an older suppression technique, the WATER-GATE[10] (water suppression by gradient tailored excitation) using pulsed field gradients. WATERGATE provides good water suppression, but gives spectra with baseline distortion and presents problems with signal phasing. The DPFGSE (double pulsed field gradient spin-echo) method, also called excitation sculpting solves these drawbacks by applying the WATERGATE sequence twice.

5.1.1.11   Memory Overflow

Another important aspect characterising the NMR experiment is the so-called memory overflow, which is related to the ADC resolution, the receiver gain, and the required S/N ratio. Computer memory overflow happens when the word length of ADC is not very long compared to that of the computer memory. Assuming that the receiver gain is at maximum, then for an ADC word length of $r = 10$ bits and a memory word length $p = 12$ bits, memory locations will be filled up after $n = 2^{(p-r)} = 2^2 = 4$ consecutive scans. Therefore a high S/N ratio requires a memory word length as long as possible and a minimum ADC resolution. Memory overflow will result in severe loss of information. Overflow can be prevented either by diminishing the signal amplification or more often by reduction of ADC word length done automatically.

5.1.1.12   VT Regulation

Several NMR experiments are conducted at low or high temperature. Therefore, temperature regulation and temperature stability over the whole experimental time is needed for high quality and reproducible results. This is especially important for long acquisitions in aqueous solution, where even a temperature drift by 0.1 °C with a D_2O lock would shift all resonance by 0.5 Hz on a 500 MHz spectrometer. The temperature control system should have the ability to set the temperature by 0.1 °C intervals, and approaching stability by 0.01 °C. For low temperature experiments liquid nitrogen or helium is needed. However, the use of refrigerated liquids could reduce the temperature as low as −20 °C. With the present NMR spectrometers variable temperature studies are feasible at temperatures ranging from $-150\,^\circ$C to $200\,^\circ$C. Before changing the temperature, however, one should know, (a) the solvent melting/boiling point for obvious reasons, (b), the sample solubility, since solubility can be reduced at lower temperature, (c) the durability of the NMR tube used, and (d) if an exact temperature is needed, *e.g.* while performing kinetic studies for exchange processes or conformational changes. In the latter case, temperature calibration should be made using methanol (low temperatures) or ethylene glycol (high temperatures).

5.1.2 Setting Up the 2D NMR Experiment

This section introduces some experimental aspects related to 2D NMR data sets. Since each 2D NMR experiment has its own practical set-up, the following discussion will focus on some general experimental characteristics, common to most 2D NMR experiments. Many of the discussions in this section are simply extensions of what has been said in the previous section for 1D NMR experiments.

Before starting a 2D experiment, a precise calibration of the 90° and 180° pulses is needed. Pulse calibration is more important for 2D NMR experiments. Pulse imperfections may contribute to a number of artefacts (*e.g.* ridges or t_1 noise) in the spectra that cannot be corrected with phase cycling or the use of field gradients. For more accurate results, composite pulses are used. Whenever possible the pulse calibration should be made with the sample to be measured. This is because the pulse angle may be sensitive to experimental conditions, such as solvent, sample temperature and the presence of salt. Also, calibration is necessary for pulsed field gradients, which constitute an indispensable part of recent versions of homonuclear and heteronuclear 2D NMR experiments.

Special care should be taken to properly adjust the receiver gain to avoid memory overflow. This is because the first FID with evolution time $t_1 = 0$ does not necessarily produce the largest signal. This can be realised by considering, for example, a COSY experiment with two consecutive 90° pulses. For $t_1 = 0$ the two 90° pulses are equivalent to 180° rotation, which generates no signal at all.

The spectral width in both dimensions is an important stage of parameters setting. Spectral widths too low may result in signal aliasing due to inappropriate sampling rate against the Nyquist theorem, whereas excessively large spectral widths may lead either to poor digital resolution for the final spectrum, or to the use of an unnecessary large number of data points. The t_1 increments, Δt_1, determine the Nyquist frequency in 2D NMR. For instance, a value of 1 ms for Δt_1 means that frequencies of 1 KHz can be sampled if quadrature detection is used. The spectral width in the ω_1 dimension (indirect dimension) is governed by the number and length of increments, Δt_1 (section 2.9.1). The number of increments used is dependent of the spectral width in the ω_2 dimension (direct dimension) after all. For most homonuclear 2D experiments (*e.g.* COSY, TOCSY, NOESY), $SW_1 = SW_2$ and the resulted spectrum is symmetrical. One exception is the homonuclear and heteronuclear 2D J spectroscopy, which reveals correlations between chemical shifts and coupling constants; here, the indirect dimension, ω_1, which displays coupling constants, extends to a few Hz. The smaller frequency range to be covered in ω_1 allows higher resolution along this axis, and when necessary saves acquisition time. The spectral width, SW_2, in the direct dimension should be chosen so that it will retain the spectral region that provides the desired information. For heteronuclear shift correlation experiments, the spectral width of X-nucleus, SW_1, is larger than the spectral width of protons, SW_2, reflecting the existing differences in the proton and X nucleus chemical shifts. These 2D spectra are lacking of a diagonal and hence diagonal symmetry.

Digitisation of the FID signal is performed in a high resolution ADC with a dynamic range longer than the maximum signal intensity to avoid degradation of the data. Digital resolution is equally important for both dimensions. For the direct dimension (ω_2), DR is given by the familiar SW_2/DP ratio (*i.e.* the reverse of the acquisition time), whereas that for the indirect dimension, ω_1, depends on the number of increments Δt_1, *i.e.* $DR = SW_1/n\Delta t_1$. Therefore, the resolution increases upon increasing the number of increments in the first dimension. In general for COSY and TOCSY experiments, n is equal to 32 or 64, whereas for heteronuclear experiments n ranges between 64 and 128, depending on the required resolution. It should be noted that higher resolution in the direct dimension costs little time, but higher resolution in the indirect dimension adds directly to the total time of the experiment; doubling the resolution in ω_1, means an experimental time twice as long. Thus, it is preferable to keep the number of t_1 increments to a minimum, but consistent with resolving the correlation of interest, and increasing the acquisition time, t_2, when higher resolution is required. Quadrature detection is also applied to both dimensions for frequency discrimination (positive and negative frequencies) by two detectors differing in phase by 90°. When resolution is not an issue, the absolute value mode is often used to display the data. In this approach, phase information for the absorption (real) and dispersion (imaginary) parts of the signal are mixed. When resolution is critical (*e.g.* observation of the fine structure of the cross-peaks), then the phase-sensitive method should be employed. In this mode, the two phase-sensitive detectors collect the real and imaginary data sets, either simultaneously (States' method), or sequentially (TPPI method). In the first method for each t_1 increment two data points are collected simultaneously, whereas in the second method only one data point is acquired for each t_1 period, but the sampling rate is faster and twice as many t_1 increments are collected. Thus, the total t_1 time and hence the digital resolution in the indirect dimension is equal for both methods.[11]

Another parameter that should be taken into consideration is the relaxation delay. For qualitative 2D experiments, relaxation delays of 1–2 s are adequate. Longer delays do not improve the quality of the spectrum.

The number of scans is an important issue in acquiring the FID signal in two dimensions. Before the introduction of pulsed field gradients in 2D pulse sequences, the minimum number of steps dictated the minimum number of scans in the phase cycles used to select the desired signals. Phase cycling, not used anymore at least in modern NMR spectrometers, involves repetition of pulse sequence(s) with identical timings and pulse tip angles, but with judicious changes (cycling) in the phases of pulse(s) and receiver phase, suppressing thus unwanted signals including artefacts (quad images, axial peaks, *etc.*) arising from instrumental imperfections. Signal averaging to increase the sensitivity of the experiment requires additional scans. Furthermore, as in 1D NMR, dummy scans are used to achieve a steady state magnetisation before the start of the experiment. Fortunately, dummy scans are collected only at the beginning of the experiment, and not in every increment as in older spectrometers, reducing considerably the total length of the experiment. Also, significant reduction of the experimental time was achieved using pulsed field gradient instead of phase cycling as mentioned in section 2.7.5.

5.1.3   Data Processing

Data processing aims at improving the quality of the spectrum with specific emphasis to sensitivity and resolution. Data manipulation is performed before and after the Fourier transform of the FID.

5.1.3.1   Window Functions

It has already been seen how sensitivity and resolution were benefited using the proper acquisition parameters and efficient digitisation. After the collection of data points and signal averaging in the computer memory, the resulting time domain signal (FID) is subjected to digital filtering for improving either the sensitivity or resolution or both if possible. The various digital or window or weighting functions used in 1D and 2D NMR experiments are depicted in Figure 5.5.[2]

Sensitivity is enhanced by multiplying (convoluting) each data point with a decaying exponential function of the form $\exp(-t/lb)$; $lb > 0$ is called the line broadening parameter and is measured in Hz. Since the initial part of the FID contains all the information and the tail mostly noise, multiplication with this window function results in reducing digitally the noise towards zero, increasing thus the sensitivity (Figure 5.6(a)).

Of course, cutting the FID tail is equivalent to decreasing the apparent decay rate (*i.e.* the acquisition time) of the NMR signal, and hence decreasing resolution. Therefore, the debt that will be paid upon increasing sensitivity is line broadening. If the parameter lb is a negative number ($lb < 0$) then the opposite effect will be observed, *i.e.* enhancement of resolution at the expense of sensitivity (Figure 5.6(b)). Resolution enhancement by multiplying the FID with the window function $\exp(-t/lb)$ increases not only the decay time, but also the noise amplitude in the tail of the FID, reducing sensitivity. Sensitivity enhancement is not well suited for proton NMR, where loss of resolution may prevent separation of closely spaced lines. The optimum balance between

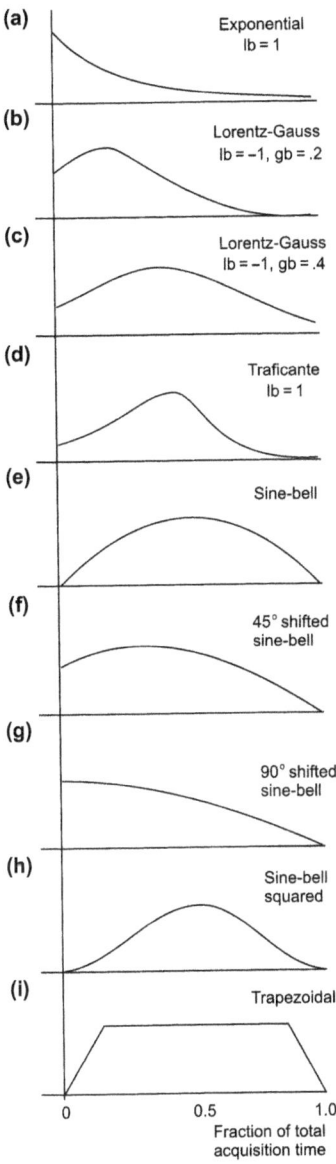

Figure 5.5 Some common employed window functions. These are used to modify the acquired FID to enhance sensitivity and/or resolution. (a) Decaying exponential function with $lb = 1$; (b) and (c) Lorentz-Gauss transformation with two different values of the Gaussian broadening parameter gb; (d) the TRAF function provides resolution enhancement with minimal loss of S/N; (e) sine-bell function; (f) and (g) shifted sine-bell by 45 and 90°, respectively; (h) sine-bell squared; (i) trapezoidal shaped function. Windows functions (e) to (i) are used in the processing of 2D data. (Reprinted from ref. 2. Copyright (1999), with permission from Elsevier.)

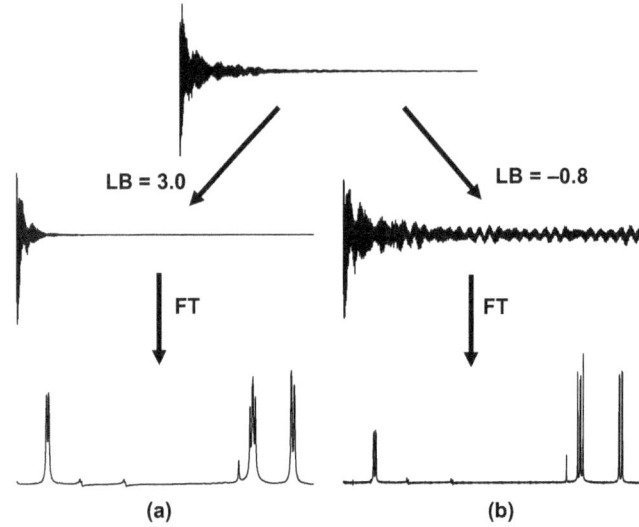

Figure 5.6   The use of window functions for (a) sensitivity (*lb*>0) and (b) resolution
(*lb*<0) enhancement for the spectrum of naringenin. Sensitivity enhancement
is accompanied by loss of resolution, whereas resolution enhancement goes
along with loss of sensitivity.
(With permission.)

reducing noise and increasing resolution can be obtained when the parameter *a*
matches the line width of the signal, which is proportional to the inverse of
the T_2 value. Practically, this matching condition rarely works for proton
spectra where signals of different line widths appear. The best way to increase
sensitivity with minimal loss of resolution is to broaden the lines by an amount
equal to digital resolution. For proton decoupled heteronuclear spectra
resolution is not an issue and line broadening of several Hz may be applied. A
better approach is to use a composite exponential function eqn (5.5):

$$G = \exp(-t/lb)\exp\left(-t^2/gb\right) \qquad (5.5)$$

The parameter *gb* is called the Gaussian broadening parameter. The first
term with the line broadening parameter *lb* equal to $-T_2$ corresponds to
resolution enhancement as discussed above, and the second term with parameter *gb* positive corresponds to a Lorentz–Gauss transformation of the signal.
Keeping the parameter *lb* fixed, the parameters *gb* is determined interactively by
trial and error.

Other functions for resolution enhancement widely used in the processing
of 2D data sets are the sine-bell window, $\sin(\pi t/t_{acq})$, the squared sine-bell,
$\sin^2(\pi t/t_{acq})$ functions with no adjustable parameters, and the phase-shifted
sine-bell, $\sin(\pi t + c)/(t_{acq} + c)$ with the adjustable parameter *c* (Figure 5.5).
Each function comprises one half a sine wave that brings the FID to a small

value or zero at the both ends of the acquisition time, while brings to a maximum the middle of the FID. These functions have similar properties, although the square sine-bell is preferred because it drops more gently at the end of the FID, avoiding thus undesirable line shape distortions. The use of the phase-shifted sine-bell allows the variation of the FID maximum.

5.1.3.2 Zero-Filling

Another means to increase the digital resolution of the spectrum is possible upon increasing the data points just before the FT of the FID. This is done by adding zeros to the end of the FID increasing artificially the size of the data points. How many zeros should be added depends on T_2, which characterises the natural decay of the transverse magnetisation component and the influence of the magnetic field inhomogeneity on nuclear relaxation (see section 2.4). Doubling the data points with zeros is recommended for routine experiments. Additional zeros may not improve the resolution of the spectrum, although they contribute to its better appearance. A better signals line shape can be very helpful in resolving very small couplings in multiple structures and/or in an accurate assignment of resonances, which is a prerequisite for decoupling. Figure 5.7 illustrates the amelioration of resolution upon doubling the data points by zeros.

However, one must be cautious and make sure that the FID has decayed to zero at the end of the acquisition time, otherwise zero-filling applied to a truncated FID produces baseline distortions after FT. However, this may be avoided by applying first a window function to enhance sensitivity, which by itself forces the FID to decay to zero and removes the truncation consequences. Zero-filling is extremely important in 2D experiments, where acquisition times are necessarily kept short. Lastly, it is recommended that zero filling should only be done after line broadening and baseline correction otherwise it introduces baseline artefacts.

5.1.3.3 Linear Prediction

The concept of linear prediction is similar to zero-filling. Rather than simply appending zeros, this method predicts values of the additional data points by using previous data points of the FID. Each new data point is a linear combination of the immediately preceding values, and is given by eqn (5.6):

$$r_n = b_1 r_{n-1} + b_2 r_{n-2} + b_3 r_{n-3} + \cdots \cdots \tag{5.6}$$

b_1, b_2, b_3, ... represent the linear prediction coefficients. Linear prediction appears to be superior comparing to zero-filling, since the missing data are not just zeros but values estimated from real data points. Linear prediction requires high sensitivity in FID, and demands that the number of data points to be

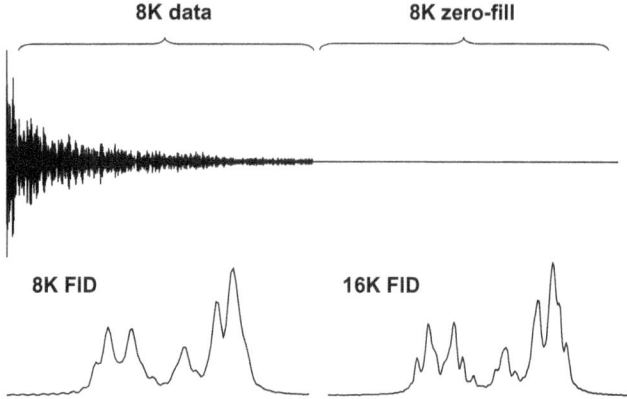

Figure 5.7 Digital resolution enhancement of the frequency spectrum by adding zeros in the time domain data, increasing artificially the data points. The number of zeros cannot be lower than the number of data points.

much greater than the signals (sine waves) contained in the FID. This is not always the case with 1D spectra with many component signals. Therefore linear prediction is rarely employed to increase resolution in 1D experiments. Linear prediction is very well suited for 2D NMR experiments, where the individual interferograms contain rather few signals, and in addition the short acquisition times lead to truncated FIDs. A detailed investigation of the advantages and limitations of linear prediction in processing 2D data sets has been reported.[12]

5.1.3.4 Fourier Transform (FT)

Fourier transform is a mathematical process that transforms any function from the time domain to the frequency domain. Actually, it is a complex transformation, which affords real and imaginary components after the transform. This mathematical treatment is applied to the time domain NMR signal, $F(t)$, the FID to produce the frequency domain signal, $F(\omega)$ The FID Fourier transformation can be written (eqn (5.7)) as:

$$F(\omega) = \int\limits_{-\infty}^{+\infty} F(t)e^{-i\omega t}dt \tag{5.7}$$

This integral can be split into two components (real and imaginary) with the aid of the Moivre's formula, $e^{-i\omega t} = \cos(\omega t) - i\sin(\omega t)$, *i.e.* the real part: $F_y(\omega) = \int_{-\infty}^{+\infty}\cos(\omega t)dt$ and the imaginary part: $F_x(\omega) = i\int_{-\infty}^{+\infty}\sin(\omega t)dt$. Taking into account the following factors: (a) the magnitude of the FID, $F_0 \propto M_0$; (b) the exponential decay of the transverse magnetisation with time constant T_2; and (c) the frequency offset, $\Omega = \pm(\omega - \omega_0)$, with respect to the reference frequency, ω_0, then the real and imaginary parts of the Fourier transform of FID can be written (eqn (5.8)) as:

$$F_y(\omega) = \int_{-\infty}^{+\infty} F_0 \cos(\Omega t) \exp(-t/T_2)dt \qquad (5.8a)$$

$$F_x(\omega) = i \int_{-\infty}^{+\infty} F_0 \sin(\Omega t) \exp(-t/T_2)dt \qquad (5.8b)$$

Summation of these two terms gives the correct absorption signal (Figure 5.8). Transposition of sine and cosine terms in the Fourier transform results in the dispersion signal as shown in Figure 5.8. The absorption signal has a Lorentzian shape, and is the signal of interest.

The trouble with the integration in eqn (5.7) is that it requires continuous functions for an infinite length of time. This condition cannot be met considering the finite data points acquired during the FID digitisation. This means that eqn (5.9) should be modified to apply to a discrete number of data points. The discrete FT for sampling at regular time intervals for N equally spaced data points is given by eqn (5.9):

$$F_i(\omega) = \sum_{k=0}^{N-1} X_k(t) \exp(-i\omega t/N) \quad (i=0, 1, ..., N-1) \qquad (5.9)$$

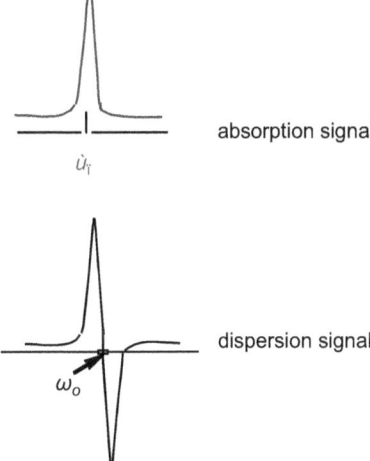

absorption signal

dispersion signal

Figure 5.8 The FID contains a real and an imaginary part. The real part after FT provides an absorptive signal and the imaginary part gives a dispersive signal. Both signals have a Lorentzian shape.

$F_i(\omega)$ is the intensity of the ith point in the frequency domain (the spectrum) and $X_k(t)$ is the corresponding value in the time domain signal (FID) at point k. FT is frequently performed using the Cooley–Tukey algorithm.[13] This algorithm works better for a number of points, N, which is a power of 2, *i.e.* 2^N, in agreement with the capacity of the computer memory expressed in multiples of K (see section 3.1.3.2). After the transformation, the spectrum consists of $N/2$ real points and $N/2$ imaginary points corresponding to the cosine and sine transformation of the data (eqn (5.7)).

5.1.3.5   Phase Correction

Phasing is the process that corrects inevitable instrumental imperfections involved, while acquiring the FID, which leads to significant distortions of the signal line shape from their normal Lorentzian line shape. Some signals appear upside down, while others may have a dispersive line shape (Figure 5.9).

There are two reasons for signals to be out of phase after FT. The first has to do with the so-called off-resonance effects. This term describes the inefficient pulse power to excite equally all the nuclear spins in the sample. In other words, the nuclei resonate outside the excitation frequency bandwidth of the pulse. For these nuclei, the 90° pulse, for instance, does not tip the magnetisation exactly on the (xy) plane as for nuclei on resonance (within the pulse excitation bandwidth). Pulse imperfection causes a phase difference between the nuclei on resonance and those driven away from the (xy) plane. For proton nuclei with relatively narrow frequency dispersion, phase errors of this kind are usually

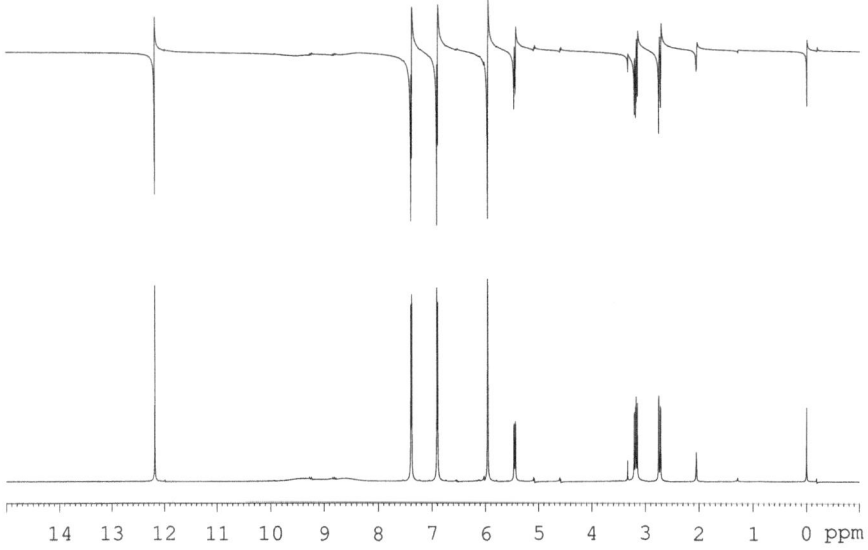

Figure 5.9   The ^{1}H NMR spectrum of naringenin in DMSO-d_6 before (top) and after (bottom) proper phasing.

small and almost linear function of frequency, so they can be easily removed by the so-called zero-order correction of the final spectrum. However, for nuclei (^{13}C or ^{31}P) with much larger range of chemical shifts, pulse imperfections may result in severe phase errors that cannot be corrected. One approach to overcome these limitations is to use composite pulses (see section 2.7.3). The second reason for phase errors is the short time period (tens of microseconds) imposed between the closure of the transmitter just after the pulse release and the opening of the receiver to acquire the time domain signal. This pre-acquisition delay is required to protect the receiver electronics from the disastrous effect of the pulse. During this short delay, after the $90°$ pulse, various nuclear vectors evolve a little on the (xy) plane with different precession frequencies, so they no longer have the same phase when digitisation begins. FT results in an absorptive and dispersive spectrum. In fact, the spectrum is a mixture of these two modes (Figure 5.9). The proportion of each mode in the spectrum varies with chemical shifts. If the pre-acquisition delay is small relative to chemical shifts, the phase errors are minor and can be corrected by the so-called first-order phase correction. It should be noted that if the pre-acquisition delay becomes large, the correction could not be performed without a rolling baseline. Most NMR spectrometers have pertinent software packages for zero and first-order phase correction. Correction is attainable either interactively or automatically. The former method is used for an effective phasing, which is a prerequisite for accurate signals integration. Automation in phase correction is vital for overnight runs of several samples using auto-samplers or flow-injection systems in metabonomic studies. Figure 5.9 shows the spectrum after the first and zero order correction.

The process of phasing the 2D spectra is not different than that applied to 1D spectra, although there are two phases to adjust. Phase correction is carried out on selected cross sections parallel to one of the axes, and then the correct phase is applied to the whole spectrum. The process is repeated for the other axis.

5.1.3.6 Baseline Correction

A flat baseline in 1D and 2D NMR spectra is an important prerequisite because it allows accurate integration and detection of small signals that may be hidden by the tails of large signals. An abnormal baseline could be critical for small cross-peaks close to large diagonal signals of certain 2D experiments. Moreover, these errors could be fatal in the study of metabonomics, which involves many small but statistically significant peaks that are sensitive to baseline distortions. Incorrect quantification of these peaks may result in failures in detection of important metabolites or identifying potential biomarkers in food matrices. Baseline distortions in NMR spectra are often caused by the corruption of the first few data points in FID. Corrupted data points resulted from high signal amplification or from incomplete recovering of the receiver electronics from the shock of the exciting RF pulse, add low frequency modulations (baseline variations) in the FT spectrum, and thus the formation of distorted baseline. Also, a distorted baseline may be the result of a truncated FID. These nuisances can be eliminated by careful adjustment of signal amplification and the proper choice of spectral width to allow full T_2 decay of the FID. Convolution of a truncated FID with a weighting function for sensitivity enhancement may prevent line distortion. When this function is used in that way the process is called apodisation.

Apart from the FID manipulation, contemporary NMR spectrometers afford baseline correction routines. These methods, which operate preferably in the frequency domain (real data points) have been implemented in commercial software and hand-written programs for NMR data processing. These baseline correction methods construct baseline curves in the spectra directly, and subtract these baseline curves to remove the distortion. There are several baseline corrections to perform based on polynomial, sine, and exponential functions. The selection of the appropriate function depends on the baseline shape to be corrected. The most popular method is to fit the baseline of the spectrum to a polynomial and then subtract the polynomial from the spectrum to produce a flat baseline. Some 1D spectra show a sloping baseline, which can be corrected with a polynomial of first degree (straight line). The construction of the polynomial function $f(x) = a_0 + a_1x + a_2x^2 + a_3x^3 + \cdots + a_nx^n$ requires the fitting of its order (n) and coefficients a_0 to a_n. Selecting baseline regions that contain only noise does this. The data points within each selected region are averaged to give one point for input into the polynomial fit. And this is going on until all baseline points are used to fit the polynomial. In 2D spectra, baseline correction can be applied to each dimension separately.

5.2   Quantitative NMR

Quantitative NMR spectroscopy offers a number of advantages for food analysis relative to other quantitative analytical methods, such as the traditional chromatographic methods: (a) no or little sample pre-treatment; (b) no prior separation of the analyte from the food matrix; (c) non-invasive and non-destructive techniques; (d) simultaneous analysis of multiple target analytes of a mixture in a single experiment; and (e) no need for distinct experimental setup for each component or family of components in the mixture. In this respect, NMR spectroscopy has been used intensively in recent years for the quantitative analysis of pharmaceuticals,[14] natural products,[15] and foods.[16]

5.2.1   1D NMR

The quantitative information obtained by NMR is contained in the area under the signal(s) of the spectrum, which is measured by integration using the integration subroutine implemented in the NMR spectrometer. The advantage of NMR relative to other spectroscopic and separation methods is the fact that the signal intensity (area under the signal) is proportional to the number of equivalent nuclei giving rise to that particular resonance. Peak height as an alternative and easiest measure for quantitative analysis is not recommended, since it is dependable on the signal line shape. The height of a broad signal may be less than a sharp signal, but its area may be greater. The accurate and reproducible measurement of integrals presupposes the fulfilment of a number of experimental and processing conditions (quantitative conditions) in addition to optimal shimming, tuning, and matching of the spectrometer.

Since most of quantification studies use 1D NMR spectra obtained by the single pulse experiment, it is wise to begin with the optimisation of the relevant acquisition parameters. There are three major acquisition parameters that should be optimised for a quantitative NMR experiment; the acquisition time, the relaxation delay, and the pulse width and pulse power.

5.2.1.1 Acquisition Time (t_{acq})

The acquisition time depends on the required digital resolution $(1/t_{acq})$ and the need to obtain normal line shapes. Therefore, t_{acq} must be long enough to avoid FID truncation that leads to intensity distortions, but short enough to allow the proper digitisation rate according to the Nyquist theorem in order to ensure the best possible digital resolution. In this respect, the spectral width and the number of data points, which are related to t_{acq} (eqn (5.2)) should be adjusted carefully. For ^1H NMR spectra, where resolution is an issue, t_{acq} is usually selected to correspond to a spectral resolution of about 0.25 Hz. A larger line width (0.5–0.8 Hz) is allowed for the X-nucleus in its proton-decoupled spectrum.

5.2.1.2 Repetition Time (t_{rep})

This term, often called recycling time, denotes the total time (acquisition time plus relaxation delay) during which the nuclear spins of the sample relax back to the state of equilibrium. For this to happen, the length of t_{rep} should be 5 times the longest spin-lattice relaxation time after a pulse 90° that flips the nuclear magnetisation onto the (xy) plane. Delay of this length guarantees that 99.3% of the magnetisation returns to the Boltzmann equilibrium between consecutive pulses. Therefore, the T_1 values of all nuclei, whose signals in the spectrum have been selected for quantitative analysis, should be measured. T_1 measurements must include the signal(s) of the internal standard, because many standards are small molecules characterised by very long T_1 values, often longer than that of the analyte.

However, the goal in a quantitative study is to perform the experiment in a reasonable time without degrading the quantitative information inherent in the NMR spectrum. This appears to be feasible with proton nuclei characterised with relatively short T_1 values (0.5–5 s). However, this is not the case with the less sensitive nuclei, *e.g.* ^{13}C, ^{31}P, ^{15}N, ^{29}Si known for their very long T_1, especially for the non-protonated nuclei. Here, long repetition times and large numbers of scans are required to obtain a reasonable S/N ratio, lengthening thus significantly the experimental time. Acquisition in the presence of paramagnetic relaxation reagents (PRR), such $Cr(acac)_3$ are used for a reduction of the relaxation time. Unpaired electrons on PRRs generate much stronger fluctuating local fields than the nuclear spins, owing to their greater magnetic moment, and therefore causing a more effective relaxation than the nuclear dipole–dipole interactions.

5.2.1.3   Pulse Power and Pulse Length

Another means to reduce the duration of the experiment is the adoption of a pulse width corresponding to an angle lower than 90°, despite the fact that the signal intensity will be decreased. The Ernst angle (eqn (5.3)) is recommended for optimising the pulse angle as a function of the relaxation delay and the longitudinal relaxation time, T_1. Of course, dummy scans should be used at the start of the experiment. The pulse power should be high enough and the pulse length sufficiently short for a uniform excitation of all nuclei within the selected spectral width. Efficient excitation precludes the observation of distorted signals in the spectrum or signals of smaller intensity due to off-resonance effects (see section 5.1.3.5). For ^{1}H NMR with typical spectral widths of 10–15 ppm this is not a problem, but for heavier nuclei such as ^{13}C, ^{19}F and ^{31}P with much larger chemical shift ranges, intensity distortions might occur particularly if measured in very high magnetic fields.

5.2.1.4   Additional Precautions

Extra care should be taken about NOE effects on signal intensities of proton-decoupled X-nuclei, such as ^{13}C, ^{31}P. Non-uniform distribution of NOE over the X-nuclei in the molecule and in particular amongst the non-protonated nuclei, or the operation of relaxation mechanisms other than the dipole–dipole relaxation (*e.g.* the chemical shift anisotropy relaxation mechanism for ^{31}P nucleus or the spin-rotation relaxation mechanism for ^{13}C in small sized molecules)[17] produce inherent signal distortions. Reduction of these distortions to 1% or less can be achieved by using the inverse gated decoupling technique.[18] According to this method; ^{1}H decoupling applies only during the acquisition period while the decoupler is off during the remainder of the experiment. This procedure requires more scans, because signal intensities are not benefited from NOE. A good practice is to avoid quantitative analysis based on non-protonated X nuclei, which are notorious for their very long longitudinal relaxation times and variable NOE effects.

Partially overlapped signals should not be considered for quantification. If this cannot be avoided, then signal areas can be measured by deconvolution or single value decomposition to extract approximate signal intensity. Besides the use of mathematical algorithms, optimal signal separation in 1D NMR spectra can be accomplished by modifying the experimental conditions including the solvent, temperature, sample concentration, chemical shift reagents, or the pH value when the analyte is basic or acidic. Details about these strategies can be found in the reference cited.[14]

Quantification studies based on hydroxyl proton signals that undergo chemical exchange should be avoided no matter how isolated these signals are in the spectrum. Hydroxyl protons involved in chemical exchange processes experience a faster decay relative to other non-exchangeable protons in the molecule, leading to line shape distortions.

Another source of interference is the presence of spinning sidebands (SSBs) and ^{13}C satellites in proton spectra that hinders the integration and in overcrowded spectral the spectral assignment. The origin of SSBs is the spinning of sample within the probe. They are immediately noticeable because they appear symmetrically around the main signal; their position depends on the spinning rate of the sample. In the case of a spin $\frac{1}{2}$ nucleus, SSBs do not significantly hinder interpretation of the spectrum. The only major difficulty comes in integration and identification of chemical shifts. Spinning sidebands can vanish by simply stop the sample spinning. Sample spinning was necessary in old instruments with relatively permanent magnets or electromagnets characterised by magnetic fields of poor homogeneity and stability. Contemporary NMR spectrometer operating with superconducting magnets produce very stable magnetic fields with superb homogeneity, and therefore there is no need for sample spinning.

The second problem is associated with the ^{13}C satellites, which are the result of ^{13}C–^{1}H coupling and appear as evenly spaced pairs around the main signals in the ^{1}H NMR spectrum. Each satellite is a miniature of the ^{1}H coupling, *e.g.* if the main ^{1}H signal is a doublet, then the carbon satellites will appear as a miniature of doublets, one doublet on either side of the main signal. These small signals, because of the low natural abundance of ^{13}C nucleus, create problems in integration and spectral assignment, although, contrary to SSBs, they can provide useful structural information if they are correctly identified. Satellites have been observed as well for other insensitive nuclei (^{15}N, ^{29}Si) in ^{1}H and ^{13}C NMR spectra. Suppression of ^{13}C satellites can be achieved by performing ^{13}C decoupling to collapse the ^{13}C satellites. Nevertheless, strong decoupling fields may cause sample overheating that could lead to degradation of sensitive samples. To avoid sample overheating during decoupling, the use of the composite pulses (Waltz-16 or GARP) is suggested (see section 2.7.3).

Once the NMR data has been acquired, it is necessary to process the data for obtaining high quality spectra. In this respect, post-acquisition parameters (window functions multiplication, phasing, zero-filling, *etc.*) described in the previous section will apply carefully to the time domain and frequency domain spectra in order to increase sensitivity and resolution of the signals. The final step is signal integration performed by the integration routine of the spectrometer. Provision should be made to choose the appropriate start and end of the integral trace around the signal. As a rule of thumb, each initial and final point from the signal should be about 64 times the signal line width. Another precaution concerns with the presence of satellites in ^{1}H spectra. If decoupling is not possible and they cannot be excluded, approximate areas should be calculated.

The disadvantage of ^{13}C NMR spectroscopy for quantitative analysis because of its low sensitivity leading to relatively long experimental times (see above) has been circumvented by introducing the slightly modified polarisation transfer sequences DEPT and refocused INEPT.[19,20] In particular, DEPT has been already used in the analysis of olive oil.[21]

5.2.1.5   *Validation*

Since NMR spectroscopy is not a standard method for quantitative analysis, it must be validated. This validation process includes several tests that confirm the experimental protocol used. Testing of linearity, robustness, specificity, selectivity, repeatability and reproducibility are required for a rigorous validation. Finally, inter-laboratory comparisons complete the validation tests and guarantee the precision and the accuracy of the applied NMR methodology. Validation of the quantitative NMR measurements has been made in several cases, including a recent study,[22] which comprises a detailed validation protocol for quantitative high-resolution ^{1}H NMR is described.

An alternative validation test for quantitative NMR is the comparison of the NMR method (the field method) with an official and well-established method of analysis (reference method). The common statistical approach to assess the degree of agreement between the field and reference methods is based on regression analysis of the results obtained by the two methods and the use of the slope of the regression equation and correlation coefficient as an indicator of agreement. Nevertheless, comparison of methods based on regression analysis appears to be inappropriate for several reasons.[23] One problem is the dependence of the correlation on the range of the results in the samples; a wider range would result in a better correlation, but not necessarily to a better agreement. An alternative approach to the use of linear regression and correlation was the difference or bias plot recommended by Bland and Altman.[24] On the abscissa they used the mean value of the methods to be compared and on the ordinate they plotted the calculated difference between measurements by the two methods. They further estimated the mean and standard deviation of differences and displayed horizontal lines for the mean and for the mean $\pm 2 \times$ the standard deviation. The two horizontal lines corresponding to the mean $\pm 2 \times$ the standard deviation constitute the limits of agreement, which represent the 95% confidence interval for individual differences between the field and reference method. In summary, this plot allows the assessment of how the differences differ systematically from zero (bias) and how much the difference varies (error). This statistical procedure, which is a common practice in medicine, has been applied to compare the amounts of certain constituents of olive oil samples determined by NMR spectroscopy with results obtained by independent laboratories using independent official and well recognised methods of analysis (GC, HPLC, titration).[25] As an example, Figure 5.10 illustrates the bias plots comparing the ^{1}H and ^{31}P NMR methodology with the conventional analytical methods of gas chromatography, and titration for the determination of the free acidity, total diacylglycerols, and fatty acids.

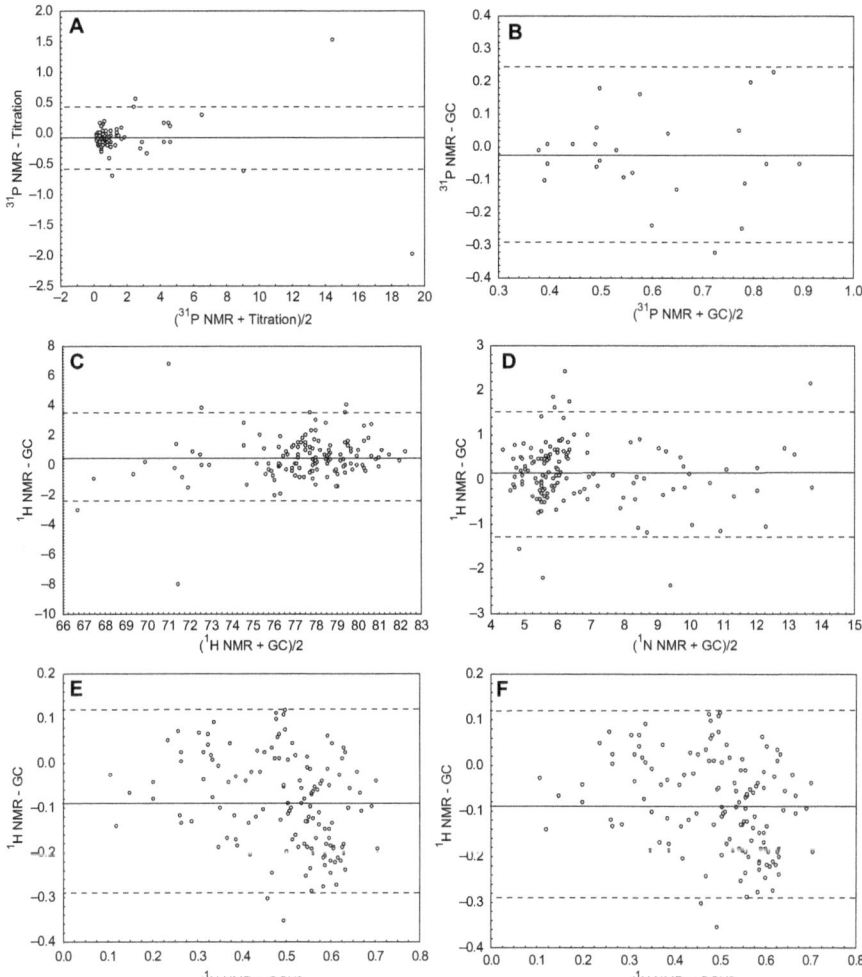

Figure 5.10 Overview of difference (bias) plots of olive oil samples measured by NMR spectroscopy and conventional methods against the average of measurements with the mean difference (solid lines) and limits of agreement (dotted lines) for (a) free acidity, (b) total diacylglycerols, (c) oleic acid, (d) linoleic acid, (e) linolenic acid, and (f) SFA.
(Reprinted from ref. 25. Copyright (2007), with permission from American Chemical Society.)

In Figure 5.10(a), which illustrates the distribution of the data points in the bias plot of free acidity, 132 measurements out of 137 (96.4%) are located within the limits of agreement, leaving only three measurements (2.2%) near or on the limits, while only two measurements (1.5%) corresponding to the highest free acidity values are well outside the limits. Apart from one measurement of total diacylglycerols (Figure 5.9(b)), which is close to the lower limit of agreement, the remaining 24 measurements for total diacylglycerols are within the limits of agreement in the respective bias plot. Five measurements (3.6%) for oleic acid, eight (5.8%) for linoleic acid, two (1.5%) for linolenic acid, and four (2.9%) for saturated fatty acids (SFA) out of a total of 137 measurements are outside the limits of agreement, as shown in Figure 5.10(c–f), respectively. Very good agreement was observed between NMR data and references methods for polyphenols and iodine value.[25]

5.2.2   2D NMR

In recent years several attempts have been made to introduce 2D NMR spectroscopy in quantitative analysis.[26] The key issue for using 2D NMR is the spread of chemical shifts in two dimensions that facilitates the separation of overlapping signals that hamper the reliable quantification with 1D NMR experiments. Figure 5.11 shows the application of 2D NMR to the quantitative analysis of an equimolar mixture of tropine–nortropine.[27]

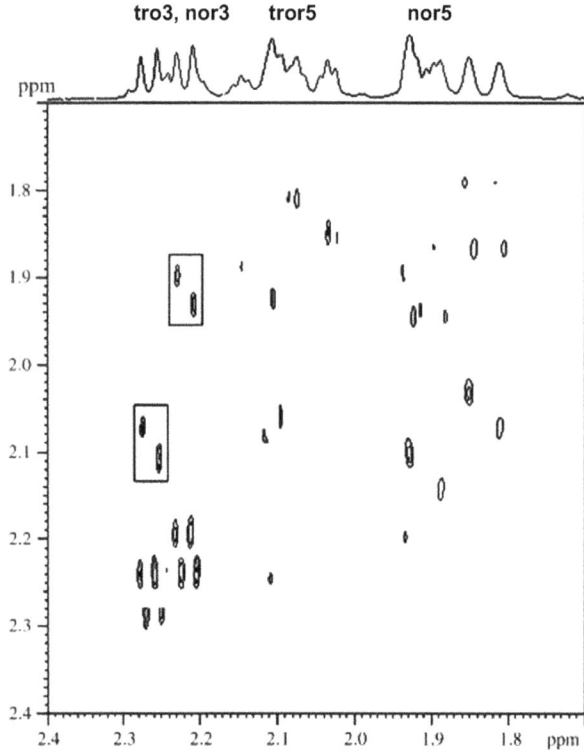

Figure 5.11   400 MHz 2D DQF-COSY spectrum of an equimolar mixture of tropine and nortropine, recorded after parameter optimisation in 12 min at 25 °C, with water signal pre-saturation. Framed areas correspond to integration zones. Only the positive components are plotted.
(Reprinted from ref. 27. Copyright (2007), with permission from Elsevier.)

The 2D DQF-COSY spectrum of the mixture separates the overlapping signals H3 of the two constituents, which correlate with the H5 protons at different chemical shifts. The usual practice in using 2D NMR experiments for quantitative analysis is the optimisation of the acquisition and processing parameters followed by construction of a calibration curve defined by several standard samples containing known amounts of the analyte(s) within the expected concentration range in the food sample.[26,27] This calibration curve can be used afterwards with real samples. This method presupposes that the observed cross-peak volumes have a linear relationship with respect to the concentration of the analyte. This is hardly true, since there are numerous factors that contribute to the volumes of the correlation peaks (*e.g.* relaxation rates, multiplicity, coupling constants). At any rate, any inconsistency in the measured peak volumes is absorbed in the calibration curve.

A rigorous approach for an accurate and self-sufficient quantification method is the theoretical and experimental examinations of the 2D NMR pulse sequence that intended to be used for quantitative analysis in order to understand the various factors affecting the polarisation transfer efficiencies. The final target, of course, is the achievement of correct volume integrals so that observed cross-peak volumes can be directly used to estimate the concentration

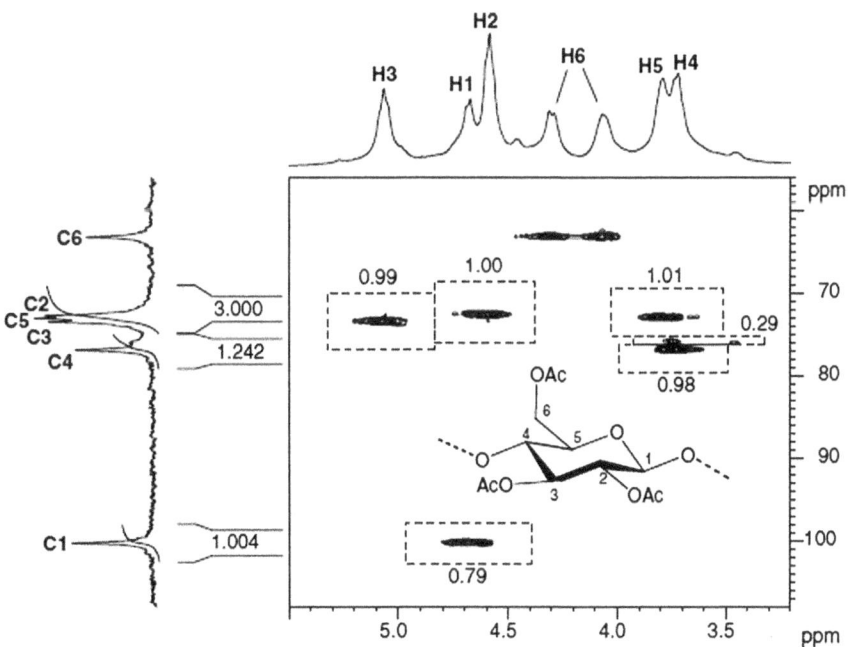

Figure 5.12 2D HSQC spectrum of cellulose triacetate with signal integration. Resonances of similar structural features and identical integrals (CH-2, CH-3, CH-4 and CH-5) are shown.
(Reprinted from ref. 28. Copyright (2007), with permission from Wiley Interscience.)

of the analyte. The pulse sequence that has gained interest in quantification studies is the heteronuclear single quantum coherence (HSQC), which correlates proton and ^{13}C chemical shifts. This experiment exploits the wide chemical shift range of carbon, offering therefore a significantly improved resolution for protons. Figure 5.12 depicts the HSQC spectrum of cellulose triacetate in the region where signals originating from the glucose ring protons resonate.[28] The well resolved C4 and C5 resonances allow a clear separation of the cross-peaks with their directly attached protons H4 and H5. Also, a much better separation is observed for protons H1 and H2 along the axis presenting the carbon chemical shifts.

Several modifications of the original HSQC pulse sequence have been proposed in recent years to make this sequence suitable for direct quantitative analysis, *i.e.* reducing or eliminating the influence of factors that contribute to the volumes of the correlation peaks as mentioned previously. A thorough discussion of these modifications is outside the scope of this book, and the interested reader may consult relevant references.[29–31] The best approach in this direction is the so-called Quick Quantitative HSQC (QQ-HSQC) experiment,[29] which improves earlier HSQC variants. The main advantage of QQ-HSQC is the significant reduction of the experimental time without a decrease in sensitivity as compared with the earlier versions Q-HSQC (Quantitative HSQC) and Q-CAHSQC (Quantitative-CPMG-adjusted HSQC),[30,31] that on the other hand are easier to implement.

References

1. A. D. Derome, *Modern NMR Techniques for Chemistry Research*, Pergamon Press, Oxford, 1997, ch. 4.
2. T. D. W. Claridge, *High-Resolution NMR Techniques in Organic Chemistry*, Pergamon Press, Oxford, 1999, ch. 3.
3. J. Keeler, *Understanding NMR Spectroscopy*, John Wiley & Sons Ltd., Chichester, 2005.
4. R. R. Ernst, G. Bodenhausen and A. Wokaum, *Principles of Nuclear Magnetic Resonance in One and Two Dimensions*, Oxford Science Publications, Oxford, 1987.
5. M. H. Levitt, *Spin Dynamics: Basics of Nuclear Magnetic Resonance*, 2nd edn, John Wiley & Sons Ltd., 2008.
6. H. Günther, *NMR Spectroscopy*. Basic Principles, Concepts, and Applications in Chemistry, John Wiley & Sons, New York, 1998.
7. R. R. Ernst and W. A. Anderson, *Rev. Sci. Instrum.*, 1966, **37**, 93.
8. W. S. Price, K. Hayamizu, Y. Avato, *J. Magn. Reson.* 1997, **126**, 256.
9. T.-L. Hwang and A. J. Shaka, *J. Magn. Reson. Ser. A*, 1995, **112**, 275; E. Prost, P. Sizum, M. Piotto and J.-M. Nuzillard, *J. Magn. Reson.*, 2002, **159**, 76.
10. M. Piotto, V. Saudek and V. Sklenár, *J. Biomol.*, 1992, **2**, 661; V. Sklenár, M. Piotto, R. Leppik and V. Saudek, *J. Magn. Reson. Ser. A*, 1993, **102**, 241.

11. J. Keeler and D. Neuhaus, *J. Magn. Reson.*, 1985, **63**, 454.
12. W. F. Reynolds, M. Yu, R. G. Enriquez and I. Leo, *Magn. Reson. Chem.*, 1997, **35**, 505; W. F. Reynolds and R. G. Enriquez, *Magn. Reson. Chem.*, 2003, **41**, 927.
13. T. C. Farrar and E. D. Becker, *Pulsed and Fourier transform*, Academic Press, New York, 1971, ch. 5, p. 72; J. W. Cooley, J. W. Tukey, *Math. Comput.*, 1965, **19**, 297.
14. U. Holzgrabe, *Progr. NMR Spectrosc.*, 2010, **57**, 229.
15. G. F. Pauli, B. U. Jaki and D. C. Lankin, *J. Nat. Prod.*, 2005, **68**, 133.
16. I. Berregi, J. I. Santos, G. Del Campo and J. I. Miranda, *Talanta*, 2003, **61**, 139.
17. J. R. Lyerla, Jr. and G. C. Levy, in *Topics in carbon-13 NMR spectroscopy*, ed. G. C. Levy, John Wiley & Sons, New York, vol. 1, ch. 3, p. 93.
18. R. Freeman, H. D. W. Hill and R. Kaptein, *J. Magn. Reson.*, 1972, **7**, 327.
19. T. J. Henderson, *J. Am. Chem. Soc.*, 2004, **126**, 3682.
20. A. V. Mäkela, I. Kilpeläinen and S. Heikkinen, *J. Magn. Reson.*, 2010, **204**, 124.
21. G. Vlahov, A. D. Shaw and D. B. Kell, *J. Am. Oil Chem. Soc.*, 1999, **76**, 1223.
22. F. Malz and H. Jancke, *J. Pharm. Biomed. Anal.*, 2005, **38**, 813.
23. M. A. Pollock, S. G. Jefferson, J. W. Kane, G. Mackinnon and C. B. Winnard, *Ann. Clin. Biochem.*, 1992, **29**, 556; S. Hollis, *Ann. Clin. Biochem.*, 1996, **33**, 1.
24. J. M. Bland and D. G. Altman, *Lancet*, 1986, **8**, 307.
25. P. Dais, A. Spyros, S. Christophoridou, E. Hatzakis, G. Fragaki, A. Agiomyrgianaki, E. Salivaras, G. Siragakis, M. Tasioula-Margari and M. Brenes, *J. Agric. Food Chem.*, 2007, **55**, 577.
26. H. Koskela, *Ann. Rep. NMR Spectrosc.*, 2009, **66**, 1.
27. P. Giraudeau, N. Guignard, E. Hillion, E. Baguet and S. Akoka, *J. Pharm. Biomed. Anal.*, 2007, **43**, 1243.
28. L. Zhang and G. Gellerstedt, *Magn. Reson. Chem.*, 2007, **45**, 37.
29. D. J. Peterson and N. M. Loening, *Magn. Reson. Chem.*, 2007, **45**, 937.
30. S. Heikkinen, M. M. Toikka, P. T. Karhunen and L. A. Kilpeläinen, *J. Am. Chem. Soc.*, 2003, **125**, 4362.
31. H. Koskela, L. A. Kilpeläinen and S. Heikkinen, *J. Magn. Reson.*, 2005, **174**, 237.

CHAPTER 6

Chemometrics in Food Analysis

6.1 Why Chemometrics?

The chemical details reflected on the NMR spectra would offer to the food scientist the opportunity to understand the molecular basis of the unique taste, texture, aroma and colour of a certain food. Moreover, the emergence of new signals and/or observed changes in signal intensities in the NMR spectra of certain food constituents are indicative of the occurrence of physicochemical alterations in food composition induced by endogenous or exogenous factors, *e.g.* climatic conditions, agronomical practices, ripening process, storage conditions and time, geographical origin. Also, food composition is the fundamental source of information to solve problems related to food processes, quality and authenticity. Due to the complexity of food composition as indicated by the multitude of signals in NMR spectra, univariate statistical analysis selecting one or two signals corresponding to certain food chemical constituents at a time is an oversimplification that puts the analysis at a risk. This is because enormous information spread across the entire NMR spectrum is lost. The selection of signals associated with crucial variables requires the *a priori* knowledge of the sample properties, which are not always available, while such a procedure, *i.e.* selecting already known variables reduces the chances for innovation. Finally, interferences in the spectrum due to over-lapping signals may preclude the accurate integration of the selected variable(s). Multivariate statistical analysis known better as chemometric data analysis deals with these problems, including complex spectral patterns such that the signal of a single variable cannot be easily identified in the spectrum.

A simplified definition of chemometrics could be the science that uses mathematics, statistics and informatics to extract the maximum useful information from a large number of data by reducing their dimensionality. The aim of chemometrics at analysing the NMR spectra is three-fold: (a) to classify and/or discriminate among groups of food samples, for instance as a

RSC Food Analysis Monographs No. 10
NMR Spectroscopy in Food Analysis
By Apostolos Spyros and Photis Dais

function of geographical origin or botanical origin (*e.g.* olive oil, honey, wine) or as a function of the mode of processing; (b) to study the relationship between composition and physiochemical properties, and (c) to build calibration–prediction models for identification of unknown samples and/or control food processes.

There are two ways by which chemometrics works with NMR data; *the targeted* or *comprehensive* or *metabolic profiling* and the *chemometric approach* or *metabolic fingerprinting*. In the first method, the focus is to identify and quantify the chemical compounds as much as possible and then performs multivariate statistical analysis either for classification purposes or to identify relevant biomarkers useful for the characterisation of the product detection of possible adulterants. The identification of chemical compounds (metabolites) is achieved by using model compounds, the NMR spectra of which could be assigned by using 1D and 2D techniques. In the second method, metabolites are not identified and quantified, but the NMR spectra of the food samples are statistically analysed to identify relevant spectral features or patterns that can differentiate samples. This mode of chemometrics is applied preferably to very complex spectra, where accurate signal integration is impossible. Both methods applied exclusively to NMR spectra are known with the generic name *metabonomics*. This term originated from the well-known methodology of *metabolomics*, which refers to the identification of small molecules that are constituents of cells, tissues, organs, or organisms.

6.2 Before the Application of Chemometrics

6.2.1 Sample Selection

The multivariate statistical analysis requires the recording NMR spectra for a large number samples in order to ensure variability in the data. The first objective is the collection of samples, the properties of which should be representative of the whole material to be studied. Selection of an appropriate fraction of samples (the so-called laboratory sample) is one of the most important stages of food analysis procedures. Inappropriate samples can lead to large errors in the analysis and application of chemometrics in erroneous data is useless. In this respect, it is necessary to develop a sampling plan that contains a precise description of the sample size, the date of sampling, the location from which the samples were originated, the method used to collect samples, and the method used to preserve samples prior to analysis. Also, the sampling document could include a short description of the gross samples properties (*e.g.* colour, taste, aroma), as well as the particular property or properties that will be measured. A few remarks about sampling have been given in section 4.1. Sampling methods are reported in several texts for the interested reader. As an example, the collection of virgin olive oil samples for their geographical classification requires the following: (a) the number of samples must be greater than the number of variables selected for chemometric analysis; (b) the harvesting period; (c) the geographical division and/or the site of collection; (d) the method of olive fruits collection from the olive trees and the extraction procedure; and (e) sample storage conditions (usually at $-20\,°C$).

6.2.2   Sample Preparation

The guidelines given in chapter 4 should be followed for sample preparation used directly for NMR experiments. Special care should be taken for samples that need extra physical treatment (extraction, separation). Some general clues on this direction are given in section 4.3.

6.2.3   Data Pre-Processing

This step ensures that the NMR data are processed equally well for all samples to conform to the prerequisites of chemometric analysis directly from the NMR spectrum. Pre-processing is applicable in the time domain (TD) data (use of window functions, zero-filling or linear prediction, FT) as well as in the frequency domain data (phasing, baseline correction). Relevant instructions for data pre-processing are furnished in section 5.1.3. Additional data processing is necessary for signals intensity, scales and chemical shifts that may change from sample to sample.[1–6]

6.2.3.1   *Quantification Conditions*

Identification and quantification of metabolites is carried out when the targeted profiling metabonomic method will be used. There are two quantification schemes: the absolute quantification, which provides the amount of each metabolite in a given sample with respect to an internal standard of known concentration, and the relative quantification, which affords the relative amount (%) of a given metabolite within the sample. Which quantification method will be used depends primarily on the number of samples to be analysed and the simplicity of the spectra to be analysed. In general, the quantification methods are applied to food samples that show relatively simple spectra, where accurate signal integration is affordable, and in addition when the statistically representative number of samples used is reasonable. The factor that influences mostly the choice of the proper quantification method is the NMR relaxation properties of the metabolite nuclei. In the absolute quantification method, all nuclei in the sample are allowed to return to their Boltzmann equilibrium before the application of the next $90°$ pulse. This means that the duration of the recycling time should be set to $5 \times T_{1,\mathrm{max}}$. However, this method places limitations on the experimental time. As a result it is not practical to allow full relaxation (*i.e.* five times $T_{1,\mathrm{max}}$) of the metabolite nuclei following each scan during data collection. Instead a compromise is made whereby an acceptable amount of time is allowed to sample nuclei to relax back to the Boltzmann equilibrium provided that an adequate S/N ratio is sufficient for signals of low intensity (see section 5.2.1). Although the lower repetition time does not allow complete relaxation for all nuclei, a comparison between samples is still possible because each nucleus essentially experiences the same extent of relaxation. To account for the partial relaxation and to improve the accuracy of the quantitative analysis, a correction factor has been introduced recently in the quantitative calculations that take into consideration the recycling time and differences in relaxation rates of the various nuclei (NMR signals) and particularly the signals of the internal standard.[7]

6.2.3.2 Scaling

This pre-processing technique, used exclusively in the chemometric approach of metabonomics, aims at reducing the influence of the dominant resonances in the spectrum that may not be necessary in the analysis as opposed to small resonances that show a much greater variability within the analysed samples. Amongst the various scaling techniques, mean centred scaling (removal of the mean spectrum) has been used to emphasise weaker spectral components. However, this scaling method produces positive and negative intensities that may not be usable with some statistical techniques utilising positive intensities only. For these situations, removal of the least common spectrum is suggested. This spectrum is constructed by using the minimum intensity of each frequency over the total number of spectra. Recently several methods for NMR data scaling (*e.g.* auto-scaling variable stability scaling, orthogonal signal correction) have been applied using scaling weights on the basis of prior knowledge of the significance of certain variables.[4]

The scaling methods mentioned above are not applicable for targeted profiling, where high-resolution line shapes are required for concentration determination. At any rate, the NMR spectra of all samples must have the same intensity scale in order to mark differences in signal intensities that reflect concentration changes of certain food components. In case that the intensity scale correction may not be possible due to instrumental imperfections and/or deficient experimental conditions (see chapters 3 and 5), normalisation with respect to an internal standard should be performed.

6.2.3.3 Spectral Editing

This pre-processing technique is another measure that should be taken into consideration before the application of the statistical analysis, *i.e.* a range of frequencies to the NMR data. Spectral editing means the removal or suppression of certain spectral regions that may hamper the statistical analysis. Spectral editing concerns with a range of frequencies which contain only noise, signals below a given threshold, or signals of other chemical species such as additives. Also, signals that are not of interest, but their presence may strongly bias or skew the subsequent chemometric analysis, should be removed. A sound example is the presence of the strong signal of water, the role of which in dynamic range problems and the suppression techniques used for its elimination have been discussed in section 5.1.1.10. However, it should be noted that by removing portions of the data set, the analyst might lose signals for species that are important markers for classification or outliers detection. Also, water suppression may cause significant attenuation and line shape variation not only on nearby signals, but also with more distant signals of compounds bearing OH and NH groups which participate in proton exchange with water.

6.2.3.4   Chemical Shift Variations

This pre-processing problem is manifested in the frequency domain NMR spectrum as a variation of chemical shifts from sample to sample or even from signal to signal. Instrumental small variations (*e.g.* in field stability and homogeneity, temperature fluctuations), variations in pH, solute–solute and/or solute–solvent interactions are the main reasons for such chemical shifts discrepancies. This problem is much more damaging for the chemometric approach, in which a large portion of the spectral width is used for statistical analysis. A common solution to this problem is the data reduction by binning the spectral region in question to a number of buckets or bins of equal width (0.01 to 0.05 ppm). In this spectral alignment, each bucket must contain the same frequency information. The major drawback of this solution is the loss of spectral resolution. When high-resolution is required, other alignment methods can be considered with satisfactory results. These methods are described in detail and compared with each other in references cited.[6–10]

6.3   Targeted or Metabolic Profiling

Targeted profiling works by defining and targeting all the metabolites of interest in the NMR spectra followed by the integration of the corresponding signals.[11] The spectral assignment is made either with the aid of data bases or on the basis of the chemical shifts of reference compounds. The signal intensities or the calculated concentrations of the food metabolites can then be used as input variables into multivariate statistical methods. This procedure requires the same experimental protocol for quantitative analysis and the necessary data pre-processing as discussed in section 5 and section 6.2 for all the samples under study. Nevertheless, not all the variables obtained from the spectra are suitable for statistical analysis. One should eliminate variables with poor predictive performance and/or poor discriminatory power. The first stage for data reduction (dimensionality reduction) is the data exploration, which gives information about the quality and suitability of each variable for subsequent statistical analysis.[4] The analysis of variance (ANOVA) is used extensively in targeted profiling.[12] ANOVA is a statistical method in which the total variability present in a dataset is split into its individual components in order to highlight the contribution of specific variables to the total variability of the dataset. ANOVA includes a group of appropriate statistical methods, such as one-, and two-ways ANOVA, capable of analysing data obtained by using experimental designs. One-way ANOVA checks the statistical significance of the mean parameter values of two or more groups of samples that are affected by a single independent variable. The statistical significance is evaluated by using the Fisher *F*-test, which tests the null hypothesis that two or more samples belong to the same group. A high value of the *F*-test for each variable suggests that the null hypothesis is wrong, the samples belong to different groups, and that particular variable may be used for classification. Two-way ANOVA improves the statistical analytical ability of one-way ANOVA by taking into account the effect of two independent variables on the response, in addition to the possible interactions between these two variables. It has been suggested that for large spectroscopic data ANOVA could be replaced with MANOVA (multivariate analysis of variance). MANOVA is an extension of ANOVA, which uses a linear combination of the dependant variables and performs ANOVA to distinguish amongst multiple groups.[13] Although MANOVA is a more complicated approach, it is superior to ANOVA because it reveals correlations that exist between variables and discloses differences not shown by ANOVA. It should be noted that variable selection and/or reduction of variables prior to chemometric analysis can be performed using other (global) statistical strategies,[4] including the automatic selection during chemometric manipulation of the NMR data, *e.g.* during the stepwise discriminant analysis and/or the PLS-DA statistical method (see below).

The targeted profiling methodology is a time-consuming metabonomic methodology, since it requires quantification procedures and the assignment of selected resonances in the spectrum. Another disadvantage of this metabonomic method is the fact that it is not amenable for automation. Nevertheless, the selection of signals and variables constitutes significant advantages for targeted profiling, since it prohibits the negative effects of data redundancy, and offers the knowledge of the individual metabolites contribution to the interpretation of statistical results.

6.4 The Chemometric Approach or Metabolic Fingerprinting

No chemical shift assignment is required for the chemometric approach. The spectra will be prepared for chemometric analysis upon applying the technique of buckets or bins. Each spectrum will be pre-processed and divided into N buckets of 0.01 to 0.05 ppm width. Each bucket will be integrated separately, and the number of buckets N and the S values of integrals will form an $N \times S$ matrix. The total data matrix of n spectra, $n \times N \times S$ will undergo direct chemometric analysis using a number of multivariate statistical techniques in an attempt to correlate the spectral data with constituents (variables) of the food system under study. The final step is the exploration of the ability of the statistical techniques to locate spectral regions (spectral fingerprints) that could be able to differentiate food samples with different physicochemical properties, *e.g.* originated from different geographical areas or varieties. These spectral patterns will be used subsequently to construct classification—prediction models for unknown samples. Manufacturers of NMR spectrometers have developed and implemented in the computers of the NMR spectrometer sophisticated software that contain all the pertinent routines for spectroscopic and statistical analyses involved in metabonomic studies. The main advantage of the chemometric approach is its susceptibility for automation. However, it is unable to identify and/or quantify useful metabolites and biomarkers.

6.5   Chemometric Techniques

The selected resonances of the NMR spectrum (targeted profiling) or the chemical fingerprint (chemometric approach) reflected on the NMR frequency spectrum are recognised as a pattern in statistical analysis. Given a pattern, its recognition leads to one of the following two tasks:[3,4,14] *supervised pattern recognition*, in which the input pattern (a single constituent or a spectral bucket) is identified as a member of a predefined class of samples with similar properties (class membership) (*e.g.* olive oil grade) and *unsupervised pattern recognition* in which the pattern is assigned to a hitherto unknown class. The unsupervised pattern recognition focuses on data exploration seeking groupings of the samples, and perhaps the presence of outliers. The goal of the unsupervised pattern recognition is to gain an understanding of the data set by examining the natural clustering of the samples and determining the contribution of each variable to the clustering (unsupervised learning). On the other hand, in supervised pattern recognition, class membership information is known. The goal of the supervised pattern recognition is to use a set of data with known classifications to train the computer to distinguish between classes (supervised learning), and then to build models to predict class membership of unknown samples. In the context of supervised learning, genetic algorithms and/or genetic programs can be considered, which consist a rapidly growing area of artificial intelligence.[4,15] They are mathematical or computational models that belong to the larger class of evolutionary algorithms inspired by the structure and functional aspects of biological systems. They are used to provide solutions in complex multi-dimensional optimisation and research problems. The artificial neural network will be presented in section 6.5.2.8.

Popular statistical methods for unsupervised pattern recognition are the Principal Component Analysis (PCA) and the Hierarchical Cluster Analysis (HCA). The most common multivariate statistical methods involved in supervised pattern recognition are various versions of Discriminant Analysis (DA), Soft Independent Modelling of Class Analogies (SIMCA), k-Nearest Neighbours (kNN), Classification and Regression Trees (CART), and Artificial Neural Networks (ANN).[3-6,14]

Apart from the pattern recognition methods, multivariate calibration and prediction models is another category of statistical methods. These methods aim at relating multiple data sets (spectra) to a property or properties of samples. For instance, the goal of such a statistical method is to predict the concentration of different chemical constituents in food sample from the NMR measurements. Partial Least Squares (PLS), Partial Components Regression (PCR), and Multiple Linear Regression (MLR) are the most commonly used calibration and regression methods. The most robust technique amongst the calibration methods is PLS, which is also used for supervised pattern recognition.[14]

Applications of the aforementioned multivariate statistical methods to food analysis will be described in chapters 7–11, Also, there exist in the literature several review articles[3–5,11,16–19] summarising recent NMR works in food science using targeting profiling and/or the chemometric approach for the build up of classification/prediction models to assess the quality and authentication of foodstuffs, and the use of PLS regression models for obtaining correlations between chemical composition and food properties or comparing results obtained by different analytical techniques including NMR. A brief presentation of the most important statistical methods in food analysis will follow.

6.5.1 Unsupervised Pattern Recognition Methods

6.5.1.1 PCA

PCA is one of the most common unsupervised methods used in analysis of NMR data, especially by the chemometric approach. It remains the primary statistical tool for initial investigations of large data sets to explore trends (similarities), classifications and detection of outliers. This method involves the reduction of the dimensionality (*i.e.* the number of variables) of the data set using as few axes or dimensions as possible. The new axes, the principal components (PCs), are linear combinations of the original variables, they are orthogonal one another and describe the variation within the data set. The first PC accounts for the maximum amount of the variability in the data set, and each succeeding component explains as much as possible the remaining variability. In the form of matrix notation, the data matrix X can be described as the product of a score matrix S and a loading matrix L, *i.e.* $X = SL^T$ (L^T being the transpose of the loading matrix). The S matrix represents the coordinates of the samples relative to PCs and contains information about the amount of variance each PC describes. The loading matrix describes the contribution of each variable to the construction of the PCs. Figure 6.1 shows the PCA score plot obtained from the chemical shifts of the ^{13}C NMR spectra of five different groups of fish.[20] The first two PCs describe the 39% and 9% of the variance in the data set. The northeast arctic cod (AC) samples are well separated from the other samples, and the pollack (P) samples are also grouped together. However, the coastal cod (CC) samples are positioned far away from each other, and overlap with the saithes (S) and the haddocks (H). Also, PC3, explaining the 9% of the variance, contains valuable information regarding the separation of the fish classes.

6.5.1.2 HCA

HCA is another unsupervised technique, which examines the natural classification (clustering) of samples or variables, or both. Samples (or variables) that

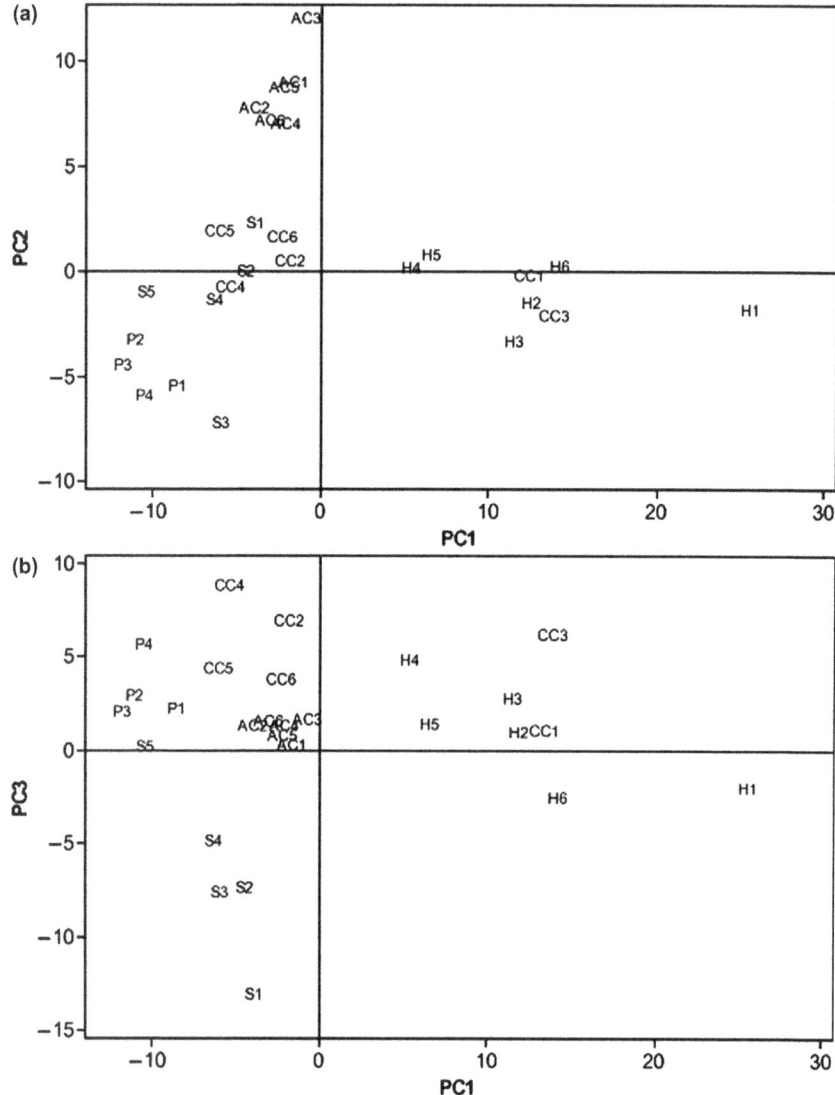

Figure 6.1 (a) PC1 *vs.* PC2 and (b) PC1 *vs.* PC3 score plots from NMR data for the five different groups of fish. AC: north-east arctic cod, CC: Norwegian coastal cod, H: haddock, S: saithe, P: pollack (254 chemical shifts as input variables).
(Reprinted from ref. 20. Copyright (2010), with permission from Elsevier.)

are similar or highly correlated with one another should be in the same cluster, whereas samples that are dissimilar or uncorrelated should be clustered in different groups. There are two main types of HCA: agglomerative or divisive. Agglomerating clustering starts with each sample forming its own solitary cluster, and iteratively merges pairs of clusters that are closest to one another

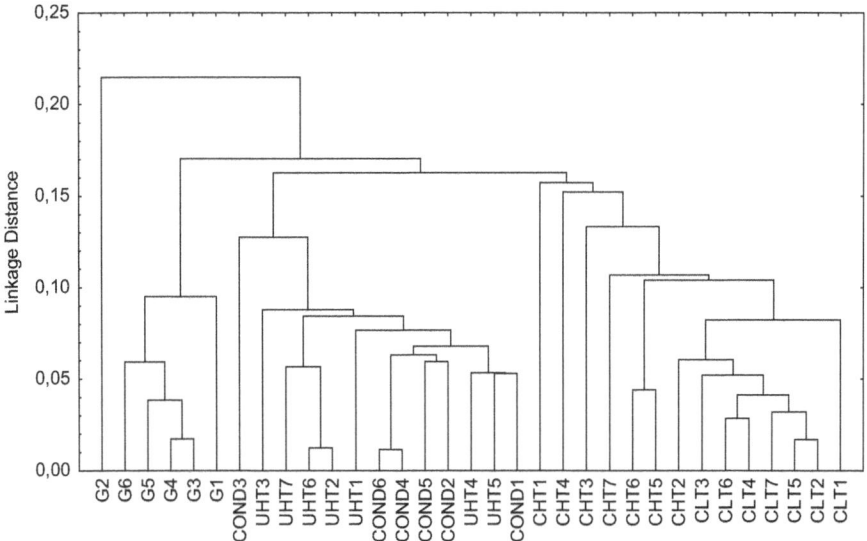

Figure 6.2 Clustering analysis to diglycerides content of milk fats. Euclidean distance metric and single linkage amalgamation method was used. Abbreviations: G = goat milk, COND = condensed milk, UHT = ultra-high temperature treatment, CHT = high temperature pasteurised cow milk, CLT = low temperature pasteurised cow milk.

until one cluster is formed. The divisive clustering follows the opposite direction. It starts with a single cluster with all data set in it, and iteratively divides it into clusters that are furthest apart from one another. Agglomerating clustering is the most popular HCA, and this technique will be described here.

HCA begins with the calculation of distances among all samples that form different clusters that are joined to larger clusters by amalgamation rules. Amongst the various distance measures and amalgamation rules, the simpler are the Euclidian distance and the single linkage, which joins two nearest members of two clusters. The end result is a 2D dendrogram (tree) regardless of the number of variables used for the clustering. Figure 6.2 depicts a vertical dendrogram constructed from ^{31}P NMR data of monoacylglycerols, diacylglycerols and sterols of various types of milk fats applying Euclidean distances and single linkage amalgamation between the variables forming the clusters.[21]

For the goat (G) and cow milk fats pasteurised at low (CLT) and high (CHT) temperature very good clustering was obtained, whereas the separation of the condensed (COND) and UHT milk fats was satisfactory.

6.5.2 Supervised Pattern Recognition Methods

Supervised pattern recognition methods have been applied to a wide range of chemical data obtained by NMR spectroscopy for a large number of foodstuffs with diverse objectives including profiling, fingerprinting, authentication, food quality control, data interpretation, *etc.*

6.5.2.1 PLS Regression

The idea behind this statistical method is to use a data matrix $\mathbf{X}$ that contains the independent or input variables extracted from the NMR spectra (*e.g.* signal intensities) of each sample to predict the data matrix $\mathbf{Y}$ that comprises quality attributes (dependent or output variables) in the sample (*e.g.* the concentration of certain constituents). In other words, this method seeks to find the components in the input matrix $\mathbf{X}$ that describe as much as possible the relevant variations in the input variables and at the same time have maximum correlation with the target value in $\mathbf{Y}$. This is achieved indirectly by extracting the latent (hidden) variables $\mathbf{T}$ and $\mathbf{U}$ from both $\mathbf{X}$ and $\mathbf{Y}$, respectively. The latent variable $\mathbf{T}$ is used to predict the latent variable $\mathbf{U}$, and then the predicted $\mathbf{U}$ is used to construct predictions for the input variable $\mathbf{Y}$. The model constructed as such links up the information from the NMR spectra to the properties of the foodstuff, and can be used to predict these properties. PLS works even for data sets with fewer data than variables and in cases where a high degree of correlation between the independent variables occurs. PLS has been applied to NMR data for the quality assessment of fruit juices[22] and beers[23] and for obtaining correlations between the composition and sensory attributes of wine.[24]

6.5.2.2 Linear Discriminant Analysis (LDA)

LDA is probably the most frequently used supervised pattern recognition method and the best studied. The basic idea underlying LDA is the construction of linear discriminant functions (L_i), also called canonical roots, which are linear combinations of the (independent) variables (x_n), also called input variables or predictors. Predictors with the highest discriminatory ability selected by ANOVA are used, *i.e.* eqn (6.1):

$$L_k = c + b_1 x_1 + b_2 x_2 + \cdots + b_n x_n \tag{6.1}$$

where b_1, b_2, *etc.* are the discriminant coefficients and c is a constant. Each discriminant coefficient indicates the relative importance of the respective variable in the construction of the discriminant function. In total, there are $k-1$ discriminant functions for k classes of samples. LDA selects the proper discriminant functions that achieve the maximum separation among the given classes. The first discriminant function L_1 distinguishes the first group from $2, 3, \ldots, n$ groups; the second discriminant function L_2 distinguishes the second group from $3, 4, \ldots, n$ groups, and so on. An example of LDA is reported in Figure 6.3. In this targeted profiling study,[12] LDA has been applied to the compositional data of 192 samples of 13 types of edible oils determined by ^{1}H and ^{13}P NMR spectroscopy to create a classification model for the detection of seed oils in olive oil.

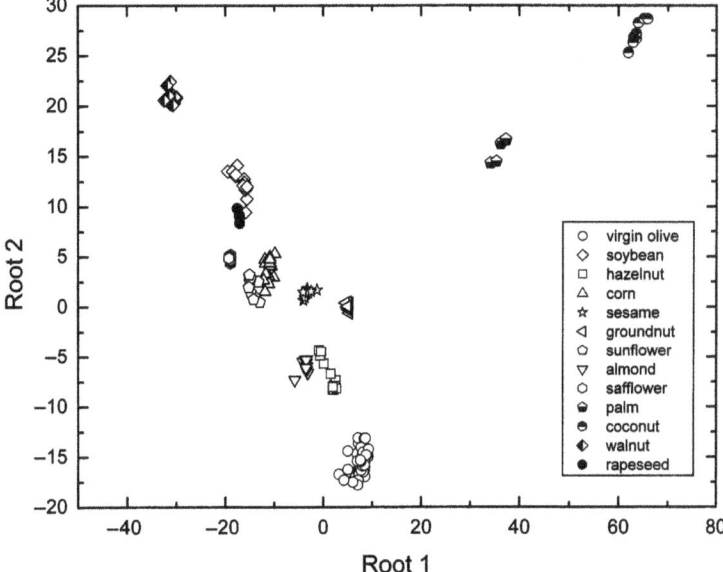

Figure 6.3 Plot of discriminant functions roots 1 and 2 for 192 samples of 13 types of vegetable oils.
(Reprinted from ref. 12. Copyright (2003), with permission from American Chemical Society.)

LDA analysis suffers from two drawbacks: (a) the presence of large differences in the number of samples in each class, which may twist the classification in favour of the most populated class(es); and (b) the number of samples must be larger than the number of variables. When the number of samples is comparable or lower than that of variables, then the classification model fails to predict new data. This phenomenon is called over-fitting. Several other variants of discriminant analysis are used in the literature. However, two of them, the stepwise DA and the PLS-DA, have been used more often in food analysis.

6.5.2.3 Stepwise Discriminant Analysis (SDA)

The forward stepwise methodology in DA aims at building a classification–prediction model stepwise. At each step all variables are reviewed and evaluated to determine which one will contribute most to the discrimination between classes. This particular variable is included in the model and the process starts again, until all variables have been undergone the selection criteria. An alternative methodology is the backward stepwise DA. In which all variables are incorporated in the model and then, at each step, the variable that contributes the least to the prediction of class membership is removed. Finally, the model is built up by variables with a greater discriminatory power. The so-called F to enter and F to remove values, respectively, govern the forward and backward stepwise procedures, which characterise the statistical significance of a variable in the discrimination between classes. A large F to enter value and a low F to remove value for a particular variable signifies the importance of this variable in the model.

6.5.2.4 PLS—DA

The main objective upon combining PLS with DA is to exploit the advantages of two robust statistical methods in order to maximise the separation amongst classes of food samples. In the context of PLS-DA, PLS regression model is calculated, which relates the independent variables (the original matrix **X**) to an artificial (dummy) **Y** matrix, which is constructed with zeros and ones, and has as many columns as there are classes. For example a vector [0,1,0,0] indicates that of the four possible classes, the particular sample belongs to class 2. Classification of an unknown sample is derived from the value predicted by the PLS model, which computes a bilinear decomposition of both X and Y matrices, under the assumption that a relationship between these two matrices exist. The result is a linear classification that has proven to be equivalent to LDA. The contribution of each variable to the classification model is evaluated by examining the VIP scores (Variable Importance in Projection) and the regression coefficients. PLS-DA appears to be superior to LDA because it is applicable in all cases, *e.g.* when the number of samples is lower than the number of variables. The chemometric approach was used for the Amarone dry red wines produced in Verona (Italy) in order to find possible correlations between metabolic content and vintage/ageing.[25]

6.5.2.5 kNN

The kNN statistical method is simple; it classifies unknown samples based on their similarity with samples (the training data set or k-subset) of known membership. The procedure is to find for a given unknown sample the k nearest sample in the training data set and assign this sample to the class that appears most frequently in k-subset. For this to occur, kNN requires (a) an integer k; (b) a set of known samples (training set or k-subset); and (c) a metric to measure the nearness between samples. The most commonly used distance between samples is the Euclidean distance (**eqn (6.2)**), although other distance measures can be used.

$$\text{Distance} = \sqrt{(x_1 - y_1)^2 + (x_2 - y_2)^2 + \cdots + (x_n - y_n)^2} \qquad (6.2)$$

This equation calculates the distance between samples x and y, which have coordinates x_i and y_i over the n-dimensional space (the index i ranges from 1 to n variables). kNN presents several advantages: (a) analytically tractable because of its simple mathematics and implementation; (b) free from statistical assumptions, *e.g.* the requirement for normal distribution of the variables;

(c) nearly optimal for the large sample limit ($n \to \infty$); and (d) it lends itself very easily to parallel implementations. As for its disadvantages, (a) it has large storage requirement; (b) it cannot work well if large differences are present in the number of samples in each class; and (c) the use of the Euclidean distance makes kNN very sensitive to noisy features. Solution of the latter problem is to modify the Euclidean metric by a set of weights that stress the importance of a known sample as a neighbour to an unknown sample. This metric system ensures the nearest neighbour influences more the classification than the farthest ones.

6.5.2.6 Soft Independent Modelling of Class Analogies (SIMCA)

SIMCA uses a training set of samples with known identity, which is divided in separate sets one for each class, and PCAs are calculated separately for each of the classes. The number of relevant principal components is determined for each class and defining boundary regions for each of the PCA models completes the SIMCA model. Each region may contain samples on a line (one PC), on a plane (two PCs), on a 3D space (three PCs), or it can be extended to higher dimensional regions (n PCs) (Figure 6.4).

The number of PCs in the training set is determined by cross-validation. Unknown samples are then compared to the training samples in the class models and assigned to classes according to their analogy with the training samples. A new sample will be recognised as a member of a class if it is enough similar to the other members of the class, or else, it will be rejected (possibly as an outlier). To determine whether an unknown sample belongs to a certain class, SIMCA uses a combination of the distance of the unknown from the PCA model and the distance from the PC component(s). SIMCA was used to discriminate olive oil samples originated the Italian region Liguria from non-Ligurian olive oils.[26] The SIMCA model depicted in the form of a leverage *versus* orthogonal distance plot (Figure 6.5), resulted in a 84.3% sensitivity (118 out of 140 Ligurian samples accepted by the model) and in a 70.6% specificity (92 of the 132 samples from the other regions rejected by the model of Liguria).

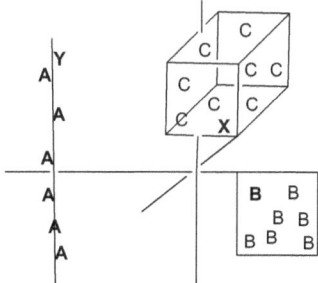

Figure 6.4   3D example of SIMCA. A, B, C represent samples from three different classes, and X and Y are unknowns.

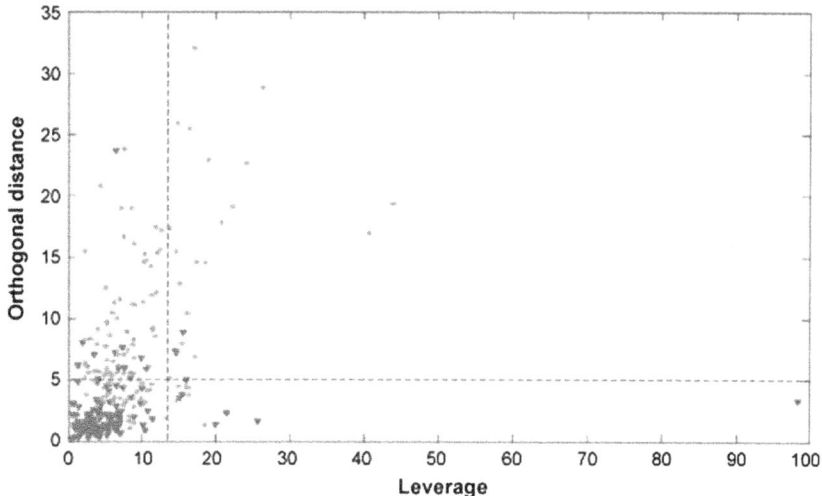

Figure 6.5   SIMCA model relative to three harvesting years. ▼ = Ligurian olive oils,
                  * = non-Ligurian olive oils.
                  (Reprinted from ref. 26. Copyright (2010), with permission from
                  Elsevier.)

Similar results were obtained also in cross-validation. The model was also used for prediction of unknown samples showing comparable specificity (70.6%), and a significantly higher sensitivity (92.8%, with only 5 Ligurian samples misclassified.[26] SIMCA has the ability to detect automatically whether an unknown sample belongs to any class in the training set. Moreover, it uses the PCA model for defining the class and, therefore, it detects outliers. As disadvantages, the following may be considered: (a) the statistical assumptions made for classification are often not obeyed by chemical data; and (b) many samples should be analysed to obtain a good representation of the size and dimensionality (number of PCs) for each class. Another limitation of SIMCA is the possibility of class overlap in the training set, which has an impact in the confidence of making future predictions. Using proper diagnostic tests can identify class overlap.

6.5.2.7   Classification and Regression Trees (CART)

The Classification and Regression Trees (CART) are used in order to produce a set of simple rules, which will predict the origin of any new sample. When the predicted outcome is the class membership, a classification tree is constructed, whereas a regression tree is formed when the predicted outcome is a real number (*e.g.* the value of a variable).[4,27] CART has three major tasks: (a) the build-up of a complete decision tree; (b) the 'pruning' of the tree; and (c) selection of the optimum tree and the prediction of new samples. The first task comprises a recursive binary splitting of the training samples into groups on the

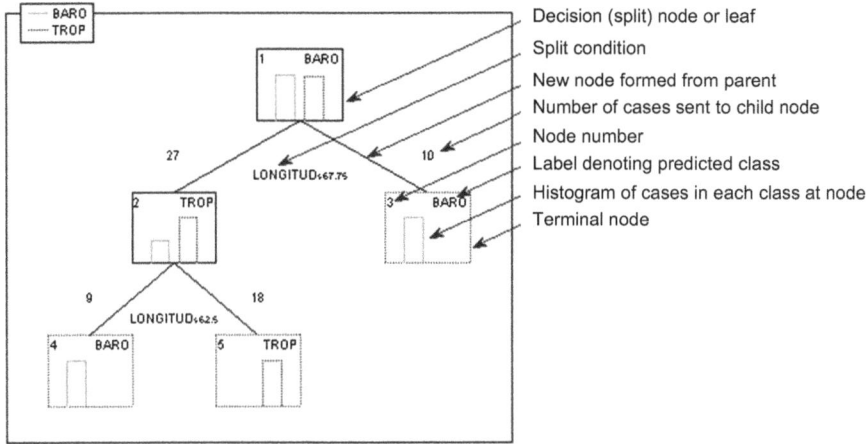

Figure 6.6 Schematic classification tree for 37 Atlantic hurricanes formed from either tropical (TROP) or baroclinical-influenced (BARO) sources. The six independent variables were the dates, longitudes, and latitudes of where the storms first achieved tropical depression and then when the storm achieved hurricane status. A portion of the data is shown. Number of splits 2, number of terminal nodes 3, tree impurity 0.0.
(Reproduced from ref. 28. Copyright (2004), Stasoft Inc.)

basis of a criterion (splitting condition, or classifier) involving the best diagnostic variables that can predict the group affiliation of samples from the entire set of measured variables. Samples are classified repeatedly into the 'leaves' (nodes) of the tree. Each node, which is further divided into two new nodes, is called a parent node, and the two new ones are called as child nodes (the term binary describes exactly the splitting into two groups). Nodes with no child nodes are called terminal nodes, whereas the initial node is called the root node and contains the training samples. Figure 6.6 shows[28] schematically a classification tree with relevant notations for grouping of the Atlantic tropical (TROP) and baroclinical-influenced (BARO) hurricanes studied by Elsner et al.[29]

Successful classification is reflected on the purity or homogeneity of the terminal nodes, i.e. each terminal contains samples of the same class. As can be seen in Figure 6.6, the splitting conditions is given in the form of questions (the type of question depend on the particular software used); is $X_i \leq d_i$? Where X_i is some dependent variable and d_i is a numerical constant within the range of that variable. Samples whose variable(s) conform to splitting conditions are sent to the left node, whereas the remaining samples are classified in the right node. This procedure continues repeatedly through all possible variables that fulfil the splitting conditions until the tree is completed. 'Pruning' is the second task in constructing a tree. This step is necessary for large-sized (complex) trees, i.e. trees with a large number of terminal nodes that are difficult to understand and interpret, and/or resulting in poor prediction for unknown samples. Pruning a branch from a tree consists of stopping or deleting all descendant nodes from that particular branch leading to an optimal sub-tree, namely three with a smaller number of terminal nodes, but without reducing predictive accuracy. There are several algorithms for pruning. The most commonly used criterion for 'pruning' is the minimal cost-complexity pruning.[27] In the third task, the best sub-tree is selected from the optimal sub-trees on the basis of the quality of the prediction for unknown samples by carrying out a cross-validation test or using the test set (samples not used for the construction of the tree). The sub-tree with the lowest estimated misclassification rate must be selected. Figure 6.7 shows a real classification tree for extra virgin olive oils from four different divisions of Greece.[30]

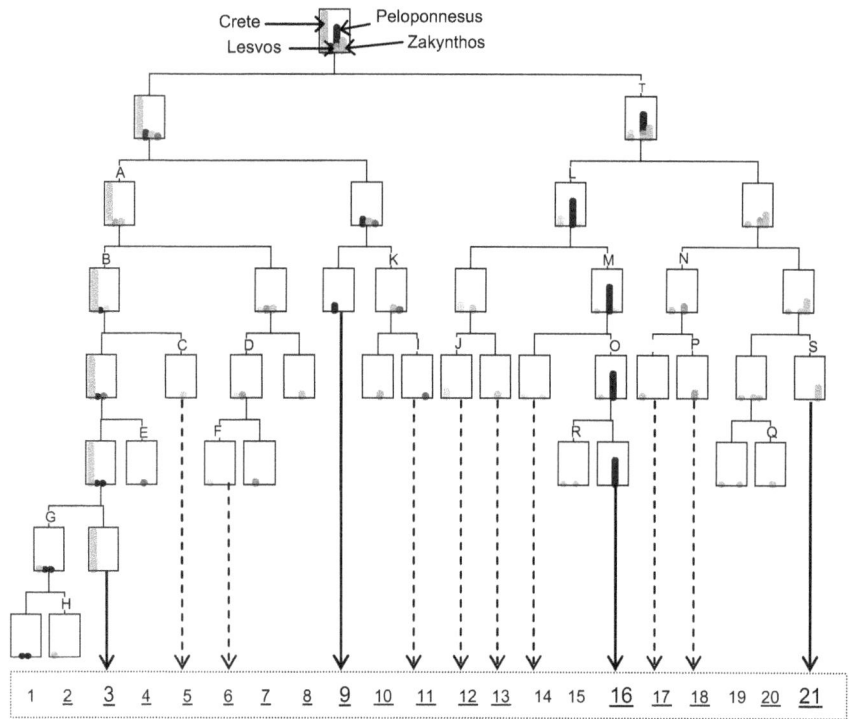

Figure 6.7 Classification tree (expanded version) of EVOO samples characterised by standardised data of the compound content and grouped according to the four geographical divisions (Crete, Lesvos, Peloponnesus, and Zakynthos) indicated by thick arrows. The overall reduction in error is high (0.97), while 21 terminal leaves are produced and numbered in the bottom box. Underlined leaves are pure and those among them written in higher font size and pointed with thick arrows bear the bulk of EVOO oils in a division, that is node 3 [Crete], node 9 [Lesvos], node 16 [Peloponnesus] and node 21 [Zakynthos]. Arrows with broken lines indicate terminal leaf numbers.
(Reprinted from ref. 30. Copyright (2012), with permission from Elsevier.)

CART as a classification method presents a number of advantages: (a) it does not require the variables selection in advance; (b) it is invariant to variable transformation (logarithms or square roots), although the splitting values will be different; (c) it is unaffected by outliers; and (d) it deals with missing variables and it is easy to interpret.

6.5.2.8 Artificial Neural Network (ANN)

This genetic algorithm imitates the operation of the biological neural network in the brain.[4,31] The operation of ANNs is as follows: as natural neurons receive signals through synapses, so the artificial networks receive input data that subsequently are multiplied by weights that mimic the strength of the signals, and then are computed by a mathematical function, which determines the activation of the neurons. Another mathematical function computes the output of the artificial neurons. Figure 6.8 depicts a simple artificial neuron introduced by Warren McGulloch and Walter Pitts as early as 1943.

The higher a weight of an artificial neuron is, the more influential is the input variable multiplied by it will be. Negative weights restrain the variables from being part of the solution. By adjusting the weights and the variables that fed the artificial neural network specific outputs that explain the particular complex problem are obtained. In other words the ANN learns the training data and after that is able to make predictions with minimal error (difference between actual and expected results).

The number of ANN types and their uses is very high. Also, there are many ANN hybrids. The ANN variants differ in the input and/or output functions, the accepted values, the learning algorithms, *etc.* An ANN presents several advantages, such as (a) it completes tasks that a linear program cannot; (b) its parallel nature guarantees the continuous function even when an element of the neural network fails; (c) it learns and does not need reprogramming; and (d) it

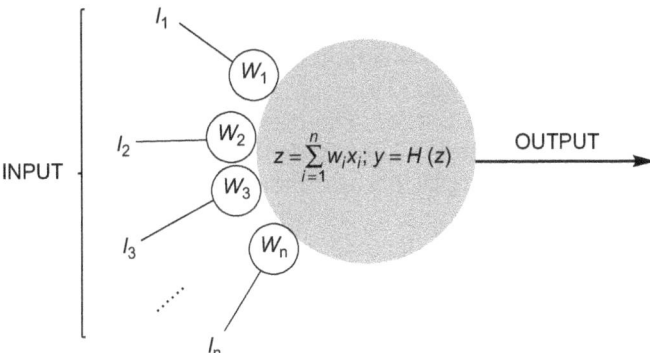

Figure 6.8 Schematic representation of the McGulloch–Pitts artificial neuron. Ws = the weights.

can be implemented in application without serious problem. The need for training of the model, the high processing time, and the difficulty to interpret the original variables consist the main disadvantages of ANN. The use of ANN in food analysis is scarce[31] and other chemometric techniques, such as PCA, LDA and PLS-DA are preferred.

6.6 Validation of Chemometric Methods

Undoubtedly, one of the most important aspects of pattern recognition is the validation of chemometric models. Model validation implies first the evaluation of the significant variables needed to construct the model, and second the recognition and prediction ability of the model. For instance, it is possible to obtain as close a fit as desired by using more and more PCA components until the data set is fitted exactly. However, these PCA components are unlikely to be physically meaningful. Therefore, a validation procedure is needed to limit the number of PCs to those that are sufficient to describe the data.

There are two ways to perform validation of the supervised statistical methods: the *external validation* and the *cross-validation*, or *internal validation*. Basically both procedures are similar and the choice of one or the other depends on the number of samples that are available. When there are enough samples, the use of the external validation is suggested, whereas for a limited number of samples, which is the usual situation in food analysis, cross-validation is applied. Both methods use training and test sets, *i.e.* a number of samples (training set) to build the model, and a smaller number of samples (test set) to validate the model. The two methods differ in the arrangement of the test set. In external validation the test set is completely independent from the model building process employing the training set in the cross-validation, the model is determined with part of the data set (training set) and using another part of the data (test set) for testing the model. Both, training and test set should contain samples representative of each class. This procedure, comprising the model development and the model testing is repeated several times assigning samples randomly, so that the same samples have the probability to be used as training and test set samples. Cross-validation is frequently called as *k-fold cross-validation* or *jack-knife method*. There are several approaches of splitting the data into the training and test set. The most popular is to form the test set containing $1/k$ (k takes values of 3, 4 or 5, depending on the number of samples) of the total number of samples (*e.g.* for data set of 30 samples and $k = 3$, 10 samples constitute the test set). There is a special case of the *k-fold cross-validation*,[4] in which k equals the total number of samples. In this validation procedure the so-called *leave-one-out cross-validation*, one sample is removed at a time from the training set and is considered as a test set. This method is suitable for small sized data set, where the split of the data into training and test set is problematical. However, the use of this method may lead to data over fit and prediction errors, when the size of data set is not large enough. The *leave-multiple-out cross-validation* has been suggested to overcome this deficiency.[32]

A few remarks concerning the validation procedure applied to the unsupervised statistical methods.[14] The validation of a PCA model is associated with the number of PCs (inherent dimensionality) that fit the data set. This is performed by several ways, including the percent variance explained by each PC, and the plot of residuals as a function of the number of variables for each PC. Cross-validation for PCA is performed on the calculated residuals, which represent the portion of the original data set that is not described by a given number of PCs. The result of the applied cross-validation is a set of residuals for a given number of PCs. The cross-validation residuals summarised in eqn (6.3) are plotted as a function of the number of PCs.

$$RMSECV\_PCA_k = \sqrt{\frac{\sum\limits_{i=1}^{nsamp}\sum\limits^{nvar} [residuals_k(i, j)]}{nsamp}} \qquad (6.3)$$

RMSECV_PCA stands for *root mean squared error of cross-validation of PCA*, and it is a measure of the magnitude of the residuals using k PCs for the ith sample and the jth measured variable. After the inherent dimensionality is reached, additional PCs are unnecessary (describing noise), and the plot is levelling off at a particular number of PCs, which represents the smallest number of PCs required by the model. Figure 6.9 shows such a plot (the socalled scree plot, indicating that the maximum inherent dimensionality of the PCA model is three.

The process of validating the results of HCA includes[14] (a) evaluation of how well a clustering solution fits the given data set, when this data set is the only information available (internal validation); and (b) the evaluation of how well a clustering solution agrees with partitions obtained based on other data source (external validation).

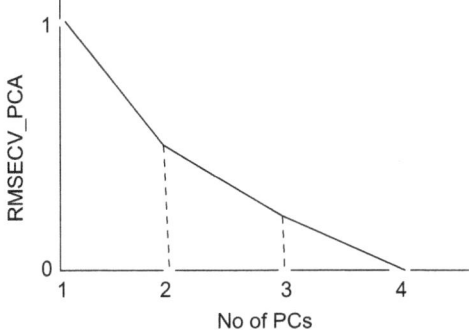

Figure 6.9 Validation plot of RMSECV_PCA *versus* number of PCs. For this particular data set, the plot is levelling off at 4 suggesting that the inherent dimensionality (number of PCs) of the system is 4.

6.7   Which Statistical Method to Choose for Food Analysis?

The choice of the proper statistical method for food analysis depends on a number of factors, such as the number of variables, the number of samples, possible noise, and the metabonomic procedure used. The decision tree in Figure 6.10 can be used to choose the appropriate pattern recognition method(s) for a given problem. This scheme proposed by some authors[14] and extended by us to include additional multivariate statistical methods starts with the first question asking about the purpose of constructing the model. If the model will be used to explore similarities and/or differences amongst samples, the unsupervised pattern recognition techniques, HCA and PCA should be used. Also, these statistical methods are more suitable for the chemometric approach, since their implementation in the form of computational tool-kits is much easier. If the goal is to predict the class membership of future samples then supervised pattern recognition techniques are appropriate. However, before the selection of the proper supervised statistical method, one should consider the question regarding the structure of the data set. For discrete measurements (signal intensities) of variables, supervised techniques are recommended.

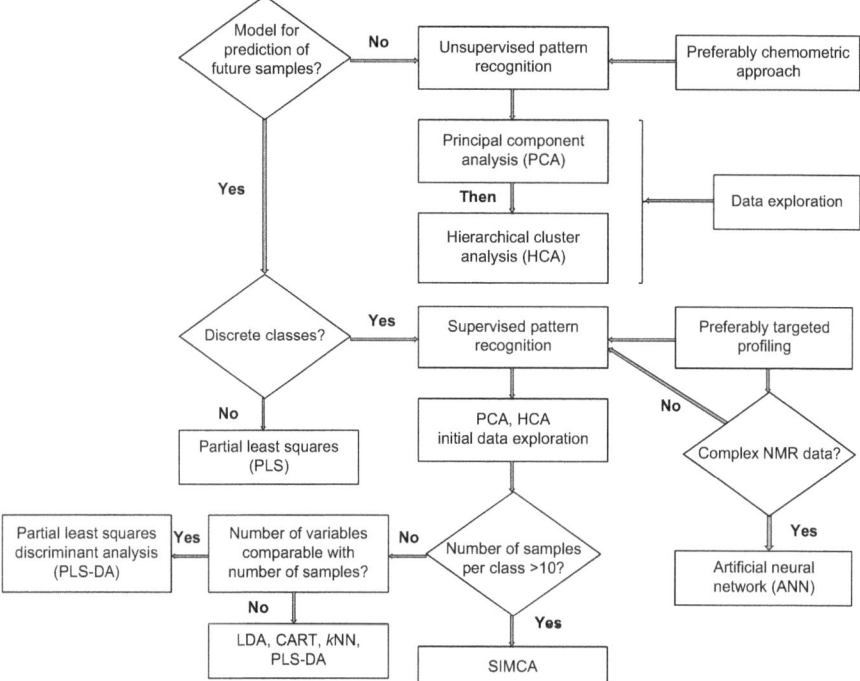

Figure 6.10   Pattern recognition decision tree.

As discussed in previous sections, the supervised methods are frequently used in targeted profiling. Before the application of the selected supervised statistical method, it is recommended to start examining the data using the unsupervised methods HCA and/or PCA. The use of these methods allows an overview of the data set that reveals data homogeneity, data co-linearity, any groupings, outliers and trends. Strong outliers observed as deviating points in the PCA scatter plot, or isolated points in the HCA dendrogram can adversely affect the prediction ability of the model among the supervised techniques, SIMCA requires a greater number of samples because a PCA model is constructed for each class, and therefore it is necessary that there are sufficient samples to build-up a sufficient model. Comparable or even lower number of samples with respect to the number of the measured variables may result in data over fit, a fact that engenders the prediction ability of the model. In this case, application the PLS-DA supervised technique is advised. Selection of important variables for the classification of samples through the VIP scores and the regression coefficients of PLS moderates the low number of samples to number of variables ratio. In cases when NMR data demonstrate unrecognisable, complicated, vague or incomplete patterns that cannot be managed by any of the available supervised and unsupervised statistical methods, the use of ANN is recommended. The complex data are decomposed into simpler elements (artificial neurons) and analysed by the proper ANN algorithm in order to understand their structure. Subsequently, these elements are processed as interconnected computational units to obtain a global output.

References

1. N. Viereck, L. Nergaard, R. Bro and S. B. Engelsen, in *Modern Magnetic Resonance*, ed. G. A. Webb, Springer, 2008, p. 1833.
2. N. Aranibar, K.-H. Ott, V. Roongta and L. Mueller, *Anal. Biochem.*, 2006, **355**, 62.
3. T. M. Alam and M. K. Alam, *Ann. Rep. NMR Spectrosc.*, 2005, **54**, 41.
4. L. A. Berrueta, R. M. Alonso-salces and K. Héberger, *J. Chromatogr. A*, 2007, **1158**, 196.
5. J. S. McKenzie, J. A. Donarski, J. C. Wilson and J. Charlton, *Progr. NMR Spectrosc.*, 2011, **59**, 336.
6. J. L. Izquierdo-Garcva, P. Villa, A. Kyriazis, L. del Puerto-Nevado, S. Pιrez-Rial, I. Rodriguez, N. Hernandez and J. Ruiz-Cabello, *Progr. NMR Spectrosc.*, 2011, **59**, 263.
7. E. J. Saude, C. M. Slupsky and B. D. Sykes, *Metabolomics*, 2006, **2**, 113.
8. R. Stoyanova, A. W. Nicholls, J. K. Nicholson, J. C. Lindon and T. R. Brown, *J. Magn. Reson.*, 2004, **170**, 329.
9. F. Savorani, G. Tomasi and S. B. Engelsen, *J. Magn. Reson.*, 2010, **202**, 190.
10. J. Forshed, R. J. O. Torgrip, K. M. Åberg, B. Karlberg, J. Lindberg and S. P. Jacobsson, *J. Pharm. Biomed. Anal.*, 2005, **38**, 824.

11. A. M. Weljie, J. Newton, P. Mercier, E. Carlson and C. M. Slupsky, *Anal. Chem.*, 2006, **78**, 4430.
12. G. Vigli, A. Philippidis, A. Spyros and P. Dais, *J. Agric. Food Chem.*, 2003, **51**, 5715.
13. W. W. Cooley and P. R. Lohnes, *Multivariate Data Analysis*, John Wiley & Sons Inc., New York, 1971.
14. K. R. Beebe, R. J. Pell and M. B. Seasholtz, *Chemometyrics a Practical Guide*, John Wiley & Sons Inc., New York, 1998.
15. T. M. D. Ebbels and R. Cavill, *Progr. NMR Spectrosc.*, 2009, **55**, 361.
16. L. Mannina, A. P. Sobolev and S. Viel, *Progr. NMR Spectrosc.*, 2012, in press.
17. S. Moco, R. J. Bino, R. C. H. De Vos and J. Vervoort, *Trends Anal. Chem.*, 2007, **26**, 855.
18. J. M. Cevallos-Cevallos, J. I. Reyes-De-Corcueraa, E. Etxeberriaa, M. D. Danyluka and G. E. Rodrick, *Trends Food Sci. Technol.*, 2009, **20**, 557.
19. D. S. Wishart, *Trends Food Sci. Technol.*, 2008, **19**, 482.
20. I. B. Standal, D. E. Axelson and M. Aursand, *Food Chem.*, 2010, **121**, 608.
21. F. Michalopoulou, A. Spyros and P. Dais, unpublished results.
22. M. Spraul, B. Schütz, E. Humpfer, M. Mörtter, H. Schäfer, S. Koswig and P. Rinke, *Magn. Reson. Chem.*, 2009, **47**, S130.
23. L. L. Nord, P. Vaag and J. Ø. Duus, *Anal. Chem.*, 2004, **76**, 4790.
24. K. Skogerson, R. Runnebaum, G. Wohlgemuth, J. De Ropp, H. Heymann and O. Fiehn, *J. Agric. Food Chem.*, 2009, **57**, 6899.
25. R. Consonni. L. R. Gagliani, V. Guantierg and B. Simonato, *Food Chem.*, 2010, **129**, 693.
26. L. Mannina, F. Marini, M. Gobbino, A. P. Sobolev and D. Capitani, *Talanta*, 2010, **80**, 2141.
27. I. Breiman, J. H. Friedman, R. A. Olshen and C. J. Stone, *Classification and Regression Trees*, Wadsworth International Group, Belmont, CA, 19884.
28. Statsoft Inc., STATISTICA 7, Version 6, Tulsa, OK, 2004.
29. J. B. Elsner, G. S. Lehmiller and T. B. Kimberlain, *J. Clim.*, 1996, **9**, 2880.
30. A. Agiomyrgianaki, P. V. Petrakis and P. Dais, *Food Chem.*, 2012, **135**, 2561.
31. F. Marini, *Anal. Chim. Acta*, 2009, **635**, 121.
32. K. Baumann, *Trends Anal. Chem.*, 2003, 22, 395.

CHAPTER 7
Fats and Oils

7.1 Vegetable Fats and Oils

7.1.1 Triacylglycerols

Chemically, both vegetable fats and oils are composed of triacylglycerols (TAG), which constitute approximately 98% of the total lipids. They are esters of glycerol with fatty acids of different chain lengths, degree of unsaturation and glycerol positional distribution. The naturally occurring fatty acids in TAG are aliphatic monocarbonic acids, mainly saturated, mono- and bi-unsaturated acids with small amounts of tri-unsaturated acids. The stereochemistry of the double bonds is *cis*. A small amount (2–5%) of *trans* fatty acids were found in tissues of ruminant animals (cattle and sheep) along with the predominant *cis*-isomers. Instead of the formal chemical names, trivial names are used for fatty acids (*e.g.* oleic acid stands for *cis*-9-octadecenoic acid). For saturated and monoenoic straight chain fatty chains, we adopt the shorthand nomenclature that uses two numbers separated by colon. The number before the colon indicates the number of carbon atoms of the chain, whereas the number after the colon designates the number of double bonds. For example, the symbol 18 : 1 (or C18 : 1) means a straight fatty acid chain containing 18 carbons and one double bond (oleic acid). The position of the bond(s) within the chain is denoted in the form (n-x), where n is the chain-length of the fatty acid and x is the number of carbon atoms from the double bond to the terminal carbon atom (usually a methyl carbon) of the molecule, *i.e.* oleic acid is denoted as 18 : 1(n-9). Figure 7.1 shows the chemical structures of fatty acids with linear chains consisted of 18 carbon atoms and different degree of unsaturation, along with their formal and trivial names. Table 7.1 lists the trivial name of most common saturated and monoenoic fatty acids and their occurrence along with their shorthand designation.

The number of TAG isomers depends on the nature of the fatty acids and their glycerol positional distribution; this number increases as the cube of the

RSC Food Analysis Monographs No. 10
NMR Spectroscopy in Food Analysis
By Apostolos Spyros and Photis Dais
© Apostolos Spyros and Photis Dais 2013
Published by the Royal Society of Chemistry, www.rsc.org

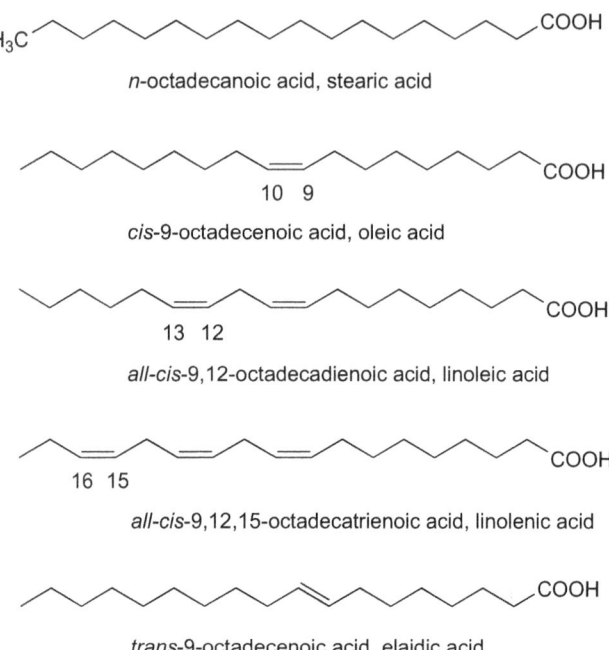

Figure 7.1   Chemical structures and formal names of fatty acids with linear chains of
18 carbon atoms and different degree of unsaturation. The position of the
double bonds is indicated by the numbering system, which begins from the
carbonyl carbon.

number of fatty acids. With only two fatty acids, *e.g.* A and B, a total of eight
triacylglycerols isomers are possible, namely **BBB, BAB, BBA, ABB, BAA,
AAB, ABA** and **AAA**. Since mixed TAG molecules may not have an element
of symmetry, the molecule is chiral and exists in enantiomers. The triacylglycerols **BAA** and **AAB** of the previous example have two enantiomers each.
Figure 7.2 illustrates the chemical structures and stereospecific numbering
of acylglycerols.

The carbons are numbered 1 to 3 from the top. The prefix *sn-* (for stereospecific numbering) denotes a particular enantiomer. Simplified structures are
often used; *e.g.* 1-palmitoyl-2-linoleoyl-3-oleoyl-*sn*-glycerol is abbreviated to
PLO. Detailed information about the structure and composition of lipids and
their constituents is given by the references cited.[1,2] Apart from TAG, mono-
acylglycerols (mono esters of glycerol with fatty acids) (MAG) and diacylglycerols (diesters of glycerol with fatty acids) (DAG) have been found in vegetable oils. In general, these molecules are not considered as significant
components of good quality oils. For instance, fresh virgin olive oil contains
less than 3% of **DAG** and less than 0.3% of **MAG**. Figure 7.2 illustrates the
chemical structure of **MAG** and **DAG** isomers.

Figure 7.3 shows the ^{1}H NMR spectra of olive, sunflower and soybean oils at
300 MHz in CDCl$_3$ solvent.[3]

Table 7.1 Major saturated and monoenic straight-chain fatty acids and their occurrence in plant and/or animal lipids.

Trivial name	Occurrence in common oils and fats	Shorthand designation
Saturated fatty acids		
Butyric acid	milk fat	4:0
Caproic acid	milk fat, coconut oil	6:0
Caprylic acid	milk fat, coconut oil, palm kernel	8:0
Capric acid	milk fat, coconut oil	10:0
Lauric acid	coconut oil, palm kernel	12:0
Myristic acid	coconut oil, palm kernel	14:0
Palmitic acid	coconut oil, palm kernel, sesame oil, milk fat, lard, tallow	16:0
Margaric acid	milk fat of ruminants, lard, tallow	17:0
Stearic acid	lard, tallow	18:0
Arachic acid	peanut oil, milk fat	20:0
Behenic acid	peanut oil	22:0
Lignoceric acid	peanut oil	24:0
Mono-unsaturated fatty acids		
Myristoleic acid	milk fat	14:1 (*n*-9)
Palmitoleic acid	milk fat, fish oil	16:1 (*n*-7)
Margaroleic acid	lard, tallow	17:1 (*n*-8)
Oleic acid	olive oil, rapeseed oil, peanut oil, sesame oil	18:1 (*n*-9)
Elaidic acid[a]	ruminants	18:1 (*n*-9)
Vaccenic acid[a]	butter fat, beef tallow	18:1 (*n*-11)
Eicosenic acid	cod oil	20:1 (*n*-9)
Erucic acid	rapeseed oil	22:1 (*n*-9)
Nervonic acid		24:1 (*n*-9)

[a]*Trans* fatty acids

Their appearance is similar, with differences only apparent in the intensities of the signals due to differences in the composition of fatty acyl chains. Olive oil is rich in oleic acid, sunflower oil in linoleic acid, and soybean oil in linolenic acid. Each spectrum in Figure 7.3 is composed of 10 main signals attributed to TAG constituents. Eight signals belong to the protons of the various fatty acids esterified by the glycerol molecule and two signals correspond to the glyceridic protons. Detailed chemical shift assignment of ¹H signals in vegetable oils has been performed by several authors in the past.[3–8] These values are summarised in Table 7.2 together with the protons of the different functional groups which give rise to these signals.

Integration of these signals under quantitative experimental conditions can provide the relative concentration of fatty acid constituents of TAG. Since most of these signals are assigned to non-equivalent groups of protons, which are common to fatty acyl chains (Table 7.2), the concentration of the individual acids are calculated only by combination of the various signal intensities in the ¹H spectrum through relevant mathematical relationships.[4–7,9] An exception is the signal at δ 0.98 ppm, which corresponds to the methyl protons of linolenic

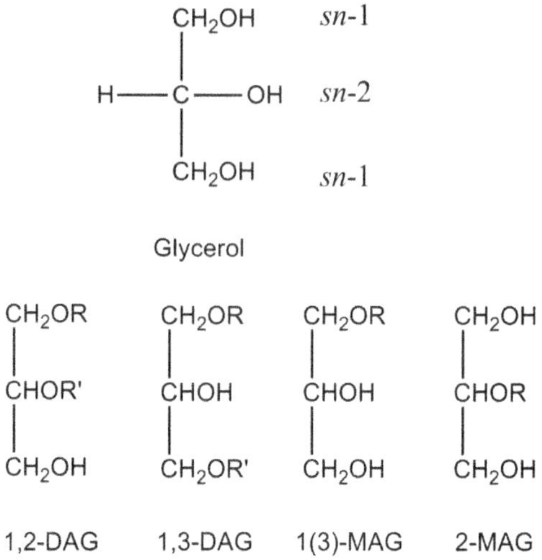

CH₂OH   sn-1

H——C——OH   sn-2

CH₂OH   sn-1

Glycerol

CH₂OR   CH₂OR   CH₂OR   CH₂OH

CHOR'   CHOH   CHOH   CHOR

CH₂OH   CH₂OR'   CH₂OH   CH₂OH

1,2-DAG   1,3-DAG   1(3)-MAG   2-MAG

Figure 7.2 Chemical structures and stereospecific numbering of acylglycerols.

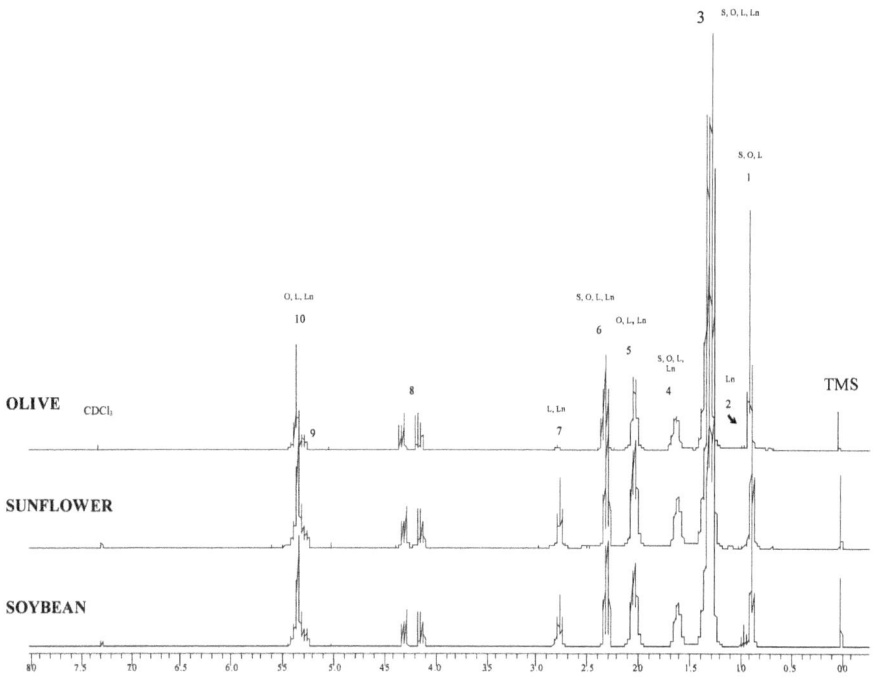

Figure 7.3 ¹H NMR spectra of olive, sunflower and soybean oils. S, O, L and Ln refer to saturated, oleic, linoleic and linolenic acyl groups, respectively. Signal numbering corresponds with that in Table 7.2.
(Reprinted from ref. 3. Copyright (2001), with permission from Elsevier.)

Table 7.2 Chemical shift assignments of the ^{1}H NMR signals of the main components of edible oils and fats.[3]

Signal	Chemical shift/ppm	Functional group
1	0.90–0.80	–CH_3 (acyl group)
1a	0.823	Saturated and oleic (or ω-9)
1b	0.839	Linoleic (or ω-6)
2	1.00–0.90	–CH_3 (acyl group)
2a	0.925	Linolenic (or ω-3)
3	1.40–1.15	–$(CH_2)_n$ (acyl group)
3a	1.194	Saturated
3b	1.230	Oleic
3c	1.280	Linoleic and linolenic
4	1.70 – 1.50	–OCO–CH_2–CH_2– (acyl group)
4a	1.553	Saturated
4b	1.557	Oleic
4c	1.567	Linoleic and linolenic
5	2.70–1.90	–CH_2–CH=CH– (acyl group)
5a	1.948	Oleic
5b	1.996	Linoleic
5c	1.994 and 2.030	Linolenic
6	2.35–2.20	–OCO–CH_2– (acyl group)
6a	2.219	Saturated
6b	2.226	Oleic
6c	2.238	Linoleic and linolenic
–	2.38[a]	–OCO–CH_2–CH_2– (docosahehanoic acyl group)
7	2.80–2.70	=CH–CH_2–CH= (acyl group)
7a	2.718	Linoleic
7b	2.754	Linolenic
8	4.32–4.10	–CH_2OCOR (glyceryl group)
9	5.26–5.20	>CHOCOR (glyceryl group)
10	5.40–5.26	–CH=CH– (acyl group)

[a]Signal only present in fish oils.
(Reprinted from ref. 3. Copyright (2001), with permission from Elsevier.)

acid. The relative compositions of saturated and unsaturated fatty acids of several vegetable oils were determined by ^{1}H NMR in a recent publication.[10] The NMR data for fatty acids in olive oil are well correlated with those obtained with the official method of gas chromatography.[11] An inspection of Table 7.2 reveals a number of ^{1}H signals corresponding to a particular proton of the acyl chains. For instance, the terminal methyl protons are split into three components with chemical shifts δ 0.823, δ 0.839, and δ 0.925 ppm, which correspond to the saturated and oleic chains, linoleic and linolenic chains, respectively. Slight differences in chemical shifts and shapes of ^{1}H signals allowed a satisfactory discrimination of vegetable oils with different proportions of saturated and unsaturated acyl groups.[10]

The ^{13}C NMR spectra of vegetable oils contain the resonances of carbons from the TAG fraction, *i.e.* resonances of fatty acids and the glycerol moiety. Four spectral regions can be distinguished in the spectra owing to the four

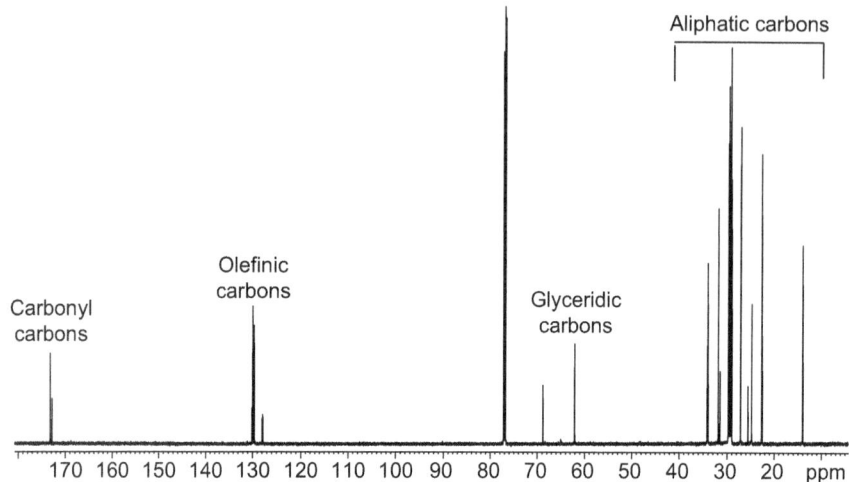

Figure 7.4 ^{13}C NMR spectrum of a virgin olive oil sample.

different groups of carbon resonances. The first region 172 to 174 ppm contains the carbonyl resonances of fatty acids, the second region 124–134 ppm involves resonances of the unsaturated (olefinic) carbons, the signals of glycerol backbone carbons appear in the third region between 60 and 72 ppm, and finally, the fourth region from 10 to 35 ppm contains signals of the aliphatic carbons. These four regions can be seen in Figure 7.4, which shows the ^{13}C spectrum of a virgin olive oil sample.

The assignment of the spectrum has been the objective of many laborious studies using standard TAG molecules, relaxation time measurements and chemical shift substituent parameters.[12–16] However, no agreement in the literature exists for the assignment of signals within the overcrowded region of 29–30 ppm. The situation becomes worse by the fact that ^{13}C chemical shifts are slightly concentration dependent especially for the carbonyl carbons.[16] An alternative NMR approach to assign the ^{13}C signals of standard TAG molecules was proposed by Simova *et al.*[17] using the hybrid 2D NMR pulse sequence HSQC–TOCSY. This technique, which exploits the resolving power of two powerful pulse sequences (HSQC and TOCSY), is based on magnetisation transfer between a carbon atom and all remote hydrogen atoms, belonging to a common coupling pathway with its directly bonded hydrogens. Figure 7.5 shows the HSQC–TOCSY spectrum of tripalmitin in the region of the methylene carbons C4–C6.

Carbons C4 through C6 are unambiguously assigned *via* the observed correlations with the methylene protons H2 and H3 of the palmitoyl chains. The ^{13}C NMR chemical shifts of TAG molecules contained in vegetable oils have been tabulated in several reports.[12–17] Edible vegetable oils usually exhibit some quantitative differences in acyl chain composition rather than in qualitative acid profile, exhibiting thus proportional signal intensities in their ^{13}C NMR

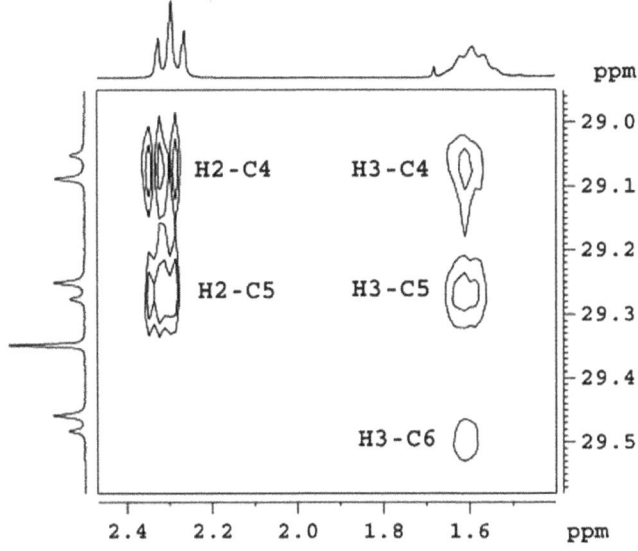

Figure 7.5 C4–C6 region of the HSQC-TOCSY spectrum of PPP triacylglycerols (P = palmitic acid); mixing time 50 ms.
(Reprinted from ref. 17. Copyright (2003), with permission from Elsevier.)

spectra.[15] ^{13}C NMR is an easy method to test the hypothesis that in vegetable oils unsaturated acids predominate at *sn*-2 position, with more saturated acids at *sn*-1 and *sn*-3 positions.[18] The distribution of fatty acids at *sn*-1 and *sn*-3 positions is often similar, although not identical, in accord with the widely accepted 1,3-random, 2-random model.[18] The carbonyl region of olive oil shown in Figure 7.6 reveals two groups of signals; the first al high frequencies, centred at about $\sim \delta$ 173.1 ppm includes the fatty acids esterified at *sn*-1(3) glycerol position, while the second group centred at $\sim \delta$ 172.6 ppm is ascribed to fatty acids bonded to the *sn*-2 glycerol position.[19]

Solid proof of the identity of these two groups of signals was given by performing an HMBC experiment, which correlates the carbonyl carbons with protons three bonds away, namely the allylic and bis-allylic protons of oleyl, linoleyl, and linolenyl chains in the ^{1}H spectrum.[20] The signals in Figure 7.6 belong to saturated (S), oleic (O), and linoleic (L) fatty acids; linolenic acid exists at very low concentration (<0.9%) in olive oil and the relevant carbonyl signal cannot be resolved in the spectrum. The saturated chains are restricted to *sn*-1(3) position corroborating thus the validity of the 1,3-random, 2-random model. Oleyl and linoleic chains are distributed between the *sn*-1(3) and *sn*-2 glycerol positions. The presence of the saturated chain at the *sn*-2 glycerol position renders the olive oil suspicious of adulteration with synthetic esterified oils. Similar results can also be obtained when analysing the olefinic carbon region.[19] This ^{13}C NMR methodology should be contrasted with the EU official method, which requires several steps including enzymatic hydrolysis of

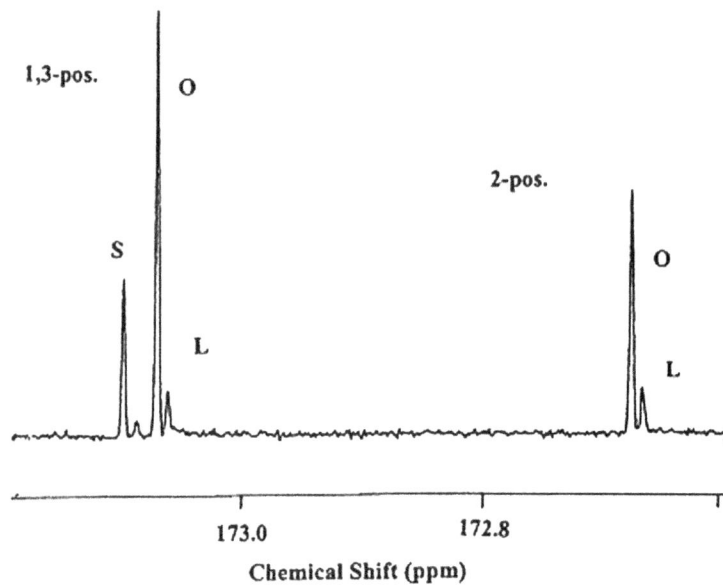

Figure 7.6 The carbonyl carbon region 172–174 ppm of the 300 MHz ^{13}C spectrum of an olive oil sample. The resonances of saturated (S), oleyl (O) and linoleyl (L) chains esterified at 1,3- and 2-glycerol positions are indicated. (Reprinted from ref. 19. Copyright (1999), with permission from Elsevier.)

TAGs at *sn*-2 position and GC analyses. Wollenberg[21] using the carbonyl spectroscopic pattern succeeded in determining the composition and positional distribution of acyl chains of non-edible vegetable oils containing eicosenoic acid and erucic acid. Two other studies reported the compositional and positional distribution of fatty acids of palm oil,[22] avocado oil, mango kernel oil, and grape seed oil.[23]

7.1.2   Minor Compounds

The unsaponifiable matter of vegetable oils contains a number of minor compounds, most of them not detected in ^{1}H NMR spectra of the bulk oils, either because they are overlapped by the huge TAG signals, or because their signals are lost under the noise due to dynamic range problems. The signals of minor compounds are of great interest, since they provide valuable information for the quality of vegetable oils (see below). Some signals of minor compounds have been found in the ^{1}H NMR spectra of the bulk olive oil,[3,6,24] Guillén *et al.*[3] provided a detailed list of the ^{1}H chemical shifts of minor compounds of the bulk edible oils and fats collected from various sources. ^{1}H NMR analysis of the unsaponifiable fraction of olive oil verified previous assignments and permitted the assignment of additional signals of minor compounds.[25] The chemical shifts of 86 signals of minor compounds observed in the unsaponifiable fraction of olive oil are tabulated in the reference cited.[25]

No mono-acylglycerols (MAG) chemical shifts were reported in the afore-mentioned ^{1}H NMR studies. The ^{1}H chemical shifts of the 1-MAG and 2-MAG isomers are known from model compounds and from a recent NMR study on the DAG oil.[26]

7.1.2.1 MAG and DAG

^{1}H NMR spectra of bulk oils obtained at higher magnetic field strengths allowed the detection of DAG. Sacchi *et al.*[6] conducting ^{1}H NMR experiments on a 600 MHz spectrometer observed in the region 4.4–3.5 ppm a number of low intensity signals around the huge signals of the glyceridic protons of TAG (Figure 7.7).

These signals were assigned to *sn*-1,2 and *sn*-1,3 DAG. DAG components in olive oil are considered as useful indices for quality control and authentication of olive oil. No MAG were detected presumably due to their presence in olive oil at very low concentration (<0.3%). Detection of these compounds using ^{1}H NMR was possible in the DAG oil, which contains significant quantities (~3%) of both *sn*-1(3) and *sn*-2 isomers.[26] The composition of DAG constituents have been, also, determined by ^{13}C NMR.[26,27–30] on the basis of the glyceryl ^{13}C resonances.

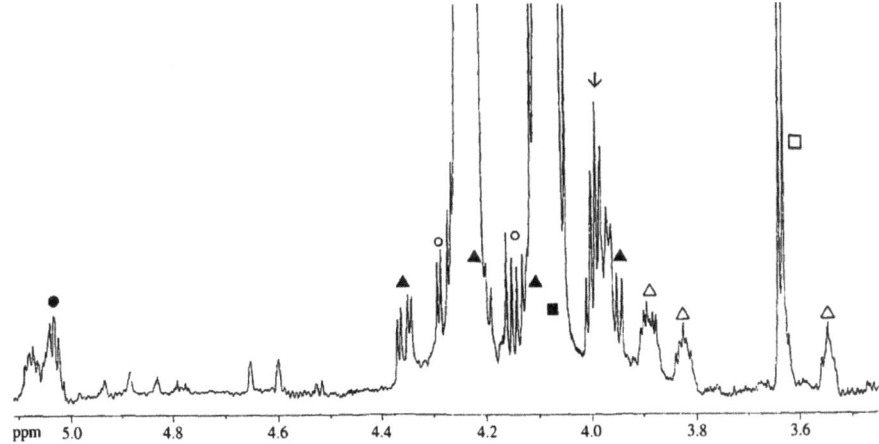

Figure 7.7   Expansion of the 3.5–4.5 ppm region in the 600.13 MHz ^{1}H NMR spectrum of virgin olive oil. The strong resonances, out of scale, are due to α, α′ protons of triacylglycerols; their ^{13}C satellites are marked with a filled triangle. At 5.07 ppm, the CH of *sn*-1,2 diacylglycerols is marked with a filled circle. At 4.29 and 4.17 ppm, the α-CH$_2$ of *sn*-1,2 diacylglycerols are marked with an open circle. At 4.07 ppm, the CH of *sn*-1,3 diglycerols is marked with a filled square. At 3.99 ppm, the CH$_2$ of *sn*-1,3 diacylglycerols are marked with an arrow. At 3.66 ppm, the α′-CH$_2$ of *sn*-1,2 diacylglycerols are marked with an open square. Resonances marked with an open triangle are due to saturated alcohols present as minor components.
(Reprinted from ref. 6. Copyright (1996), with permission from AOCS Press.)

However, due to the low concentration of these compounds in virgin olive oil, the low natural abundance of ^{13}C, and the long carbon spin–lattice relaxation times quantitative analysis of DAG constituents requires long repetition times, lengthening thus significantly the duration of the experiment.

Spyros and Dais[31] proposed an alternative methodology to detect and quantify minor compounds in olive oils, which can be applied to any vegetable oil. This method is based on the derivatisation of the labile hydrogen atoms of the hydroxyl and carboxyl groups with 2-chloro-4, 4, 5, 5-tetramethyldioxaphospholane (**I**) according to the reaction in Scheme 7.1, and the use of the ^{31}P chemical shifts to identify the phosphitylated products (**II**).

Compound **I** reacts rapidly ($\sim$15 min) and quantitatively under mild conditions (within the NMR tube) with functional groups bearing labile protons. An experimental protocol of this NMR methodology was developed for the phosphitylation reaction and the quantitative analysis of minor compounds upon integration of selected signals in combination with the phosphitylated cyclohexanol used as an internal standards.[31] A typical ^{31}P NMR spectrum of a virgin olive oil sample recorded at 202.2 MHz for the phosphorus nucleus is shown in Figure 7.8a.

The excellent resolution between the ^{31}P NMR chemical shifts of MAG constituents of olive oil allows the reliable quantification of these molecules. This spectrum can be compared with the 500 MHz ^1H NMR spectrum (Figure 7.8(b)) of the same olive oil sample in the region where the MAG and DAG resonate. No MAG is discernible in the ^1H spectrum, whereas the extensive overlap of the DAG signals with those of the *sn*-1(3) glyceryl protons and the ^{13}C satellites prohibits an accurate quantitative analysis. The signal overlap is more severe at lower magnetic field strength.[32] In addition to MAG and DAG determination, this technique is able to specify in a quantitative manner, total free sterols, and free acidity from the signals of the phosphitylated hydroxyl groups of sterols (δ 145.0 ppm) and the phosphitylated carboxyl groups of free fatty acids (δ 134.8 ppm), respectively, as shown in Figure 7.8(a). Dayrit *et al.*[33] used the same ^{13}P NMR methodology to differentiate virgin coconut oil from refined, bleached and deodorised coconut oil on the basis of MAG and DAG components and other minor components. Application of PCA on the ^{13}P

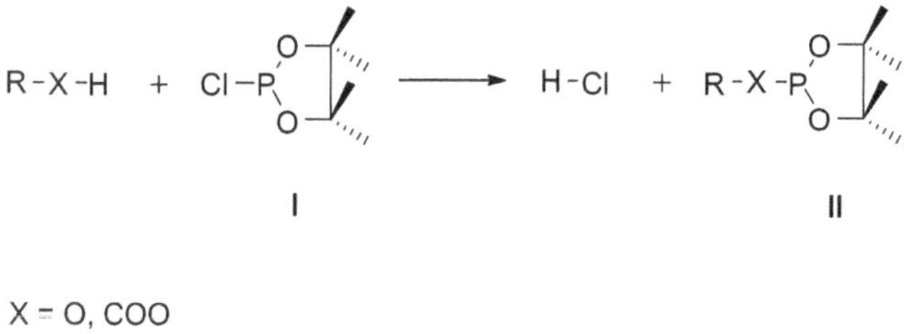

I   II

X = O, COO

Scheme 7.1   Reaction of hydroxyl and carboxyl groups with the phosphorus reagent 2-chloro-4,4,5,5-tetramethyldioxaphospholane (**I**) and the phosphitylated product (**II**).

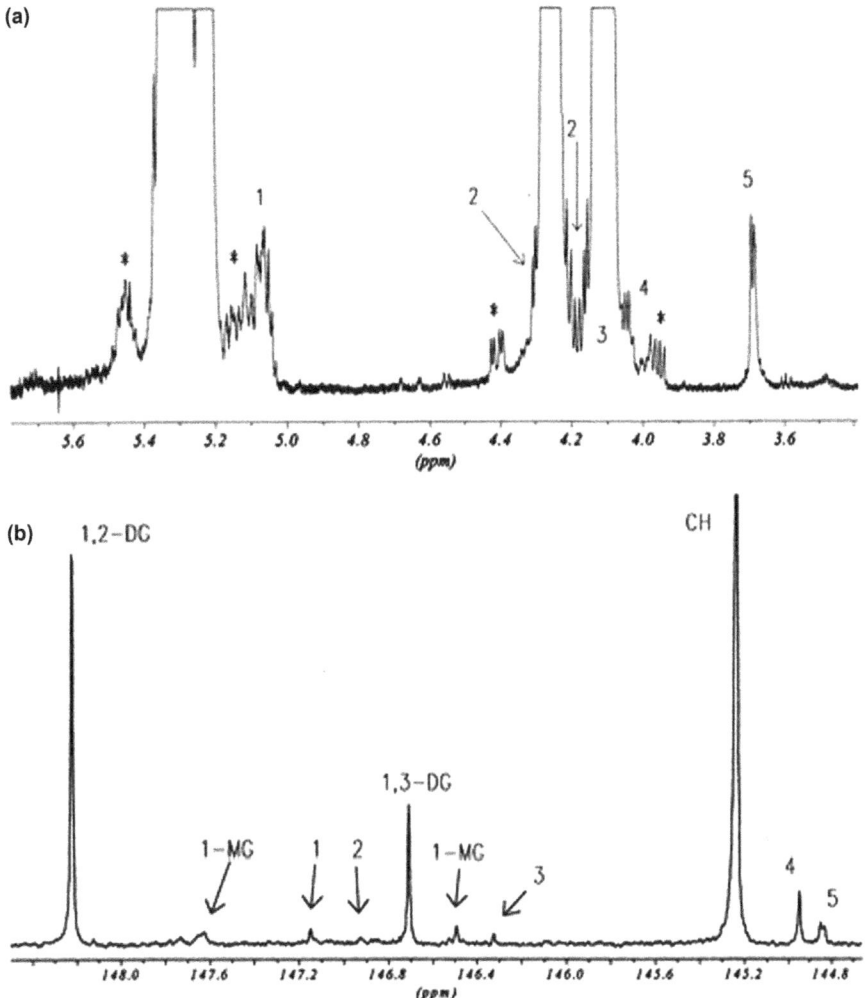

Figure 7.8 (a) ^{31}P NMR and (b) ^{1}H NMR of an olive oil sample. The region where mono- and diglycerides resonate is illustrated in both spectra. In the ^{31}P spectrum the peaks with Arabic numbers are assigned as follows: 1, aliphatic alcohols; **2**, tyrosol (aliphatic hydroxyl); **4**, sterolic hydroxyl; **3** and **5**, unknown. The peak denoted CH belongs to the internal standard cyclohexanol. In the proton spectrum the peaks are assigned as follows: **1**, CH; **2**, α-CH$_2$; **5**, α, α'-CH$_2$ of 1,2-diglycerides; **3**, CH; **4**, CH$_2$ of 1,3-diglycerides; *, ^{13}C satellites of triglyceride signals.
(Reprinted from ref. 31. Copyright (2000), with permission from American Chemical Society.)

NMR data resulted in different grouping for the virgin and refined coconut oil samples. The loading plot indicated that 1,2-DAG, 1,3-DAG and free acidity are the variables that contribute the most to the discrimination between virgin and refined coconut oil.

7.1.2.2 Sterols

Important sterols in vegetable oils include β-sitosterol, stigmasterol, and campesterol. The major phytosterol is β-sitosterol, the concentration of which could reach 90% of the sterolic faction. Sterols exist in their free and esterified forms. The amount of esterified sterols ranges between 30 and 50% of the sterolic fraction. Virgin olive oil, rapeseed, cottonseed, peanut and sunflower oil contain significant amounts of sterols ranging from 200 to 2000 mg kg^{-1}.

Free and esterified sterols as a whole can be detected by the ^1H signal of the methyl protons at position 18 of the steroid skeleton of sterols. This signal resonates in a narrow spectral zone 0.60–0.70 ppm. For virgin olive oils, this signal appears at δ 0.62 on a 500 MHz spectrometer and as mentioned before is attributed to β-sitosterol, which is the dominant sterol (75–90%) in this oil. This signal can be used for the quantification of total sterols in any vegetable oil. Free sterols can be detected in vegetable oils by using the ^{31}P NMR spectrum in Figure 7.8(a) and the signal of the phosphitylated hydroxyl group of sterols at δ 145.00. In a recent study, Hatzakis *et al.*[34] suggested a facile analytical method to quantify separately free and esterified sterols by combining ^1H NMR (determination of total sterols) and ^{31}P NMR (free sterols) spectroscopy. This methodology speeds up the time-consuming approach, involving three analytical steps; prior separation of free and esterified sterols followed by hydrolysis/transesterification and analysis by gas chromatography. However, the NMR methodology cannot give detailed information about individual phytosterols.

7.1.2.3 Phenolic Compounds

Phenolic compounds usually named as polyphenols play an important role in the nutritional and organoleptic characteristics of edible vegetable oils. Moreover, phenolic compounds act as natural antioxidants contributing to the oxidation stability of oils. The majority of seed oils contain traces of these polar compounds, whereas virgin olive oil is exceptionally rich (150–350 mg kg^{-1}) in these natural antioxidants; it contains more than 20 phenolic compounds, which are obtained from olive oil by extraction with methanol–water mixture (80 : 20 v/v). The more polar part of the extract contains free phenols and phenolic acids, whereas the less polar part contains derivatives of tyrosol and hydroxytyrosol (secoiridoids), flavonoids and lignans. Earlier attempts to study phenolic compounds in olive oil involved the use of preparative or semi-preparative TLC, liquid chromatography, and column chromatography to isolate individual phenols, the structure of which is determined subsequently by ^1H NMR off-line.[35–40] The isolation procedure is often a very laborious and time-consuming work, although it offers phenolic derivatives, especially those produced by the hydrolysis of oleuropein and ligstroside, not available in the market. Christophoridou *et al.*[41] using the LC-SPE-NMR technique, *i.e.* the powerful coupling of the NMR spectrometer with the LC chromatograph and adding a post-column solid-phase extraction (SPE) unit, succeeded to identify 27 constituents in the phenolic fraction of olive oil, including the detection and structure elucidation of new phenolic compounds, such as homovanillyl

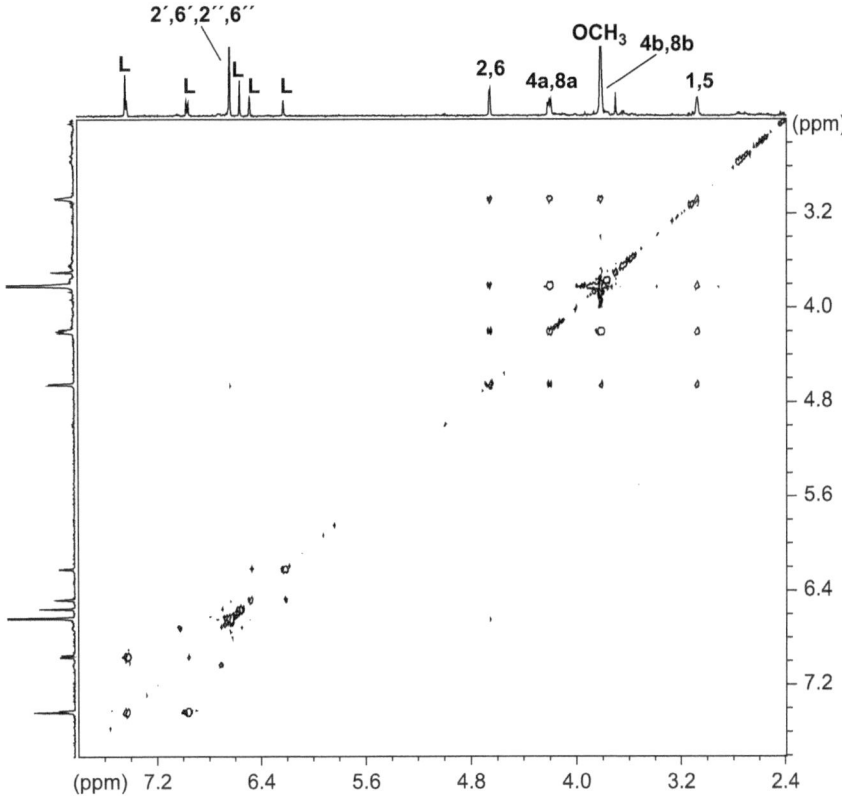

Figure 7.9 600 MHz TOCSY spectrum of the chromatographic fraction of the fla-
vonol luteolin detected by LC-SPE-NMR, indicating the presence of the
lignan syringaresinol. The signals denoted by L belong to luteolin.
(Reprinted from ref. 41. Copyright (2005), with permission from American
Chemical Society.)

alcohol, syringaresinol, the 5S, 8S, 9S isomer of the aldehyde form of oleuropein, the two isomers of the aldehyde form of ligstroside, the ligstroside
aglycon, the dialdehydic form of elenolic acid lacking a carboxymethyl group,
and, finally, maslinic acid, not detected in the past. The potential of the LC-
SPE-NMR is shown in Figure 7.9, which depicts the 600 MHz TOCSY spectrum of the LC fraction of olive oil that contains luteolin.

Apart for the luteolin (L) signals, the spectrum shows several additional signals
attributed to the newly discovered lignan syringaresinol in olive oil, which was co-
eluted with luteolin.[41] A subsequent NMR study managed to identify and
quantify phenolic compounds in olive oil by employing 1D high-resolution [1]H
NMR spectroscopy.[42] The ^{1}H signals in the spectrum of the polar extract of olive
oil were identified on the basis of chemical shifts of a large number of model
compounds assigned using 2D NMR techniques. This simple and non-destructive

NMR methodology disclosed the presence of nineteen phenolic compounds in olive oil including hydrolysis products of oleuropein and ligstroside. It is suggested[42] that the use of high throughput NMR probes in combination with solid-phase extraction and automation will be an important future step for the on line screening of a large number of samples in a short period of time, compensating thus the high cost of the analysis. A third methodology for the determination of phenolic compounds in olive oil was developed by Christophoridou and Dais[43] based on ^{31}P NMR spectroscopy after derivatisation of phenolic hydroxyls with the phosphorus reagent I (Scheme 1). Assignment of the signals in the ^{31}P NMR spectrum was based on the chemical shifts of a large number of model compounds assigned by 2D NMR techniques.[43,44] Figure 7.10 depicts the aliphatic and aromatic regions of the 202.2 MHz ^{13}P NMR spectrum of the polar extract of a virgin olive oil sample, where the phosphitylated aliphatic and aromatic hydroxyl groups of phenolic compounds resonate.

In addition to phenolic compounds, the spectrum in Figure 7.10 shows signals attributed to the MAG components, carbohydrate molecules and those of the triterpenic maslinic acid, indicating that this methodology is able to detect and quantify, in a single spectrum, a number of different compounds.

7.1.2.4 Phospholipids

This term phospholipids implies any derivative of sn-glycero-3-phosphoric acid that contains at least one O-acyl, or O-alkyl or O-alk-1'-enyl residue attached to the glycerol moiety and a polar head made of a nitrogenous base, a glycerol, or an inositol unit All the glycerophospholipids are derived from the simplest member of this family, i.e. phosphatidic acid (PA), which consists of a glycerol segment bearing a phosphoric group in sn-3 position and acyl chains in sn-1 and sn-2 positions. Figure 7.11 depicts the chemical structure of phosphatidic acid its major derivatives with a nitrogenous base or glycerol molecules. Lysophosphatidic acid (LPA) or 1-acyl-sn-glycerol-3-phosphate differs from phosphatidic acid in having only one fatty acid chain attached. Lyso- derivatives exist for all the other phospholipids depicted in Figure 7.11.

^{31}P NMR spectroscopy has proven to be a powerful technique for the quantitative determination of phospholipids in lipids. The benefits of using ^{31}P NMR spectroscopy have already described (see section 2.1). Various factors (pH, temperature, and the presence of paramagnetic metals) that affect the chemical shifts of phospholipids have been investigated thoroughly.[45–47] Most of these ^{31}P NMR studies have been made with biological fluids and tissues. Little has been done for vegetable oils, despite the fact that phospholipids are major polar lipids in several oils (e.g. soybean oil, sunflower oil, and rapeseed oil).[2] In a recent study,[48] the phospholipids extracted from virgin olive oil with a mixture of ethanol/water (2:1 v/v) were identified and quantified by high-resolution ^{31}P NMR spectroscopy. The main phospholipids found in olive oil were phosphatidic acid, lysophosphatidic acid, and phosphatidylinositol. Sensitivity was satisfactory within the detection limits of 0.25–1.24 μmol mL^{-1}. In addition, the composition of the fatty acids in phospholipids in olive oil

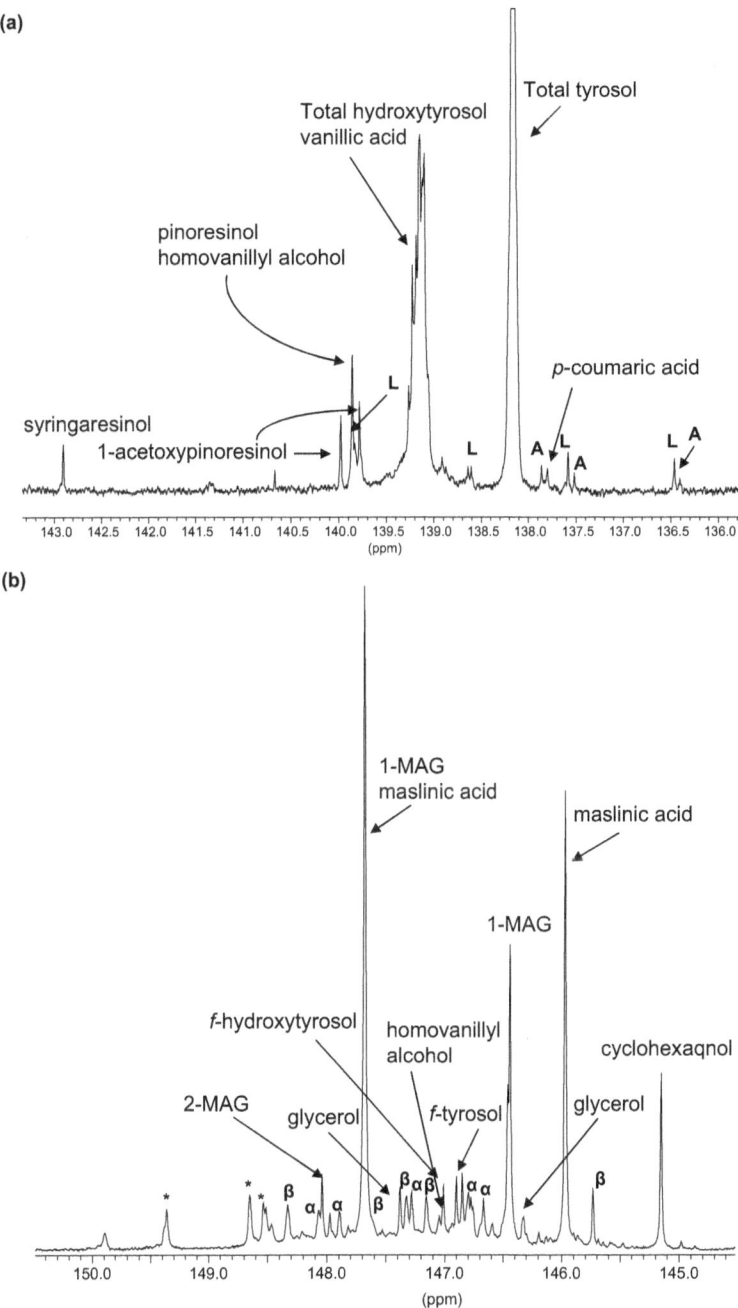

Figure 7.10 202.2 MHz ^{31}P NMR spectrum of the phosphitylated polar fraction of a virgin olive oil sample in chloroform/pyridine solution: (a) aromatic region; (b) aliphatic region. A, apigenin; L, luteolin; 1-MAG, 1-monoacylglycerols; 2-MAG, 2-monoacylglycerols; *f*-hydroxytyrosol, free hydroxytyrosol; *f*-tyrosol, free tyrosol; α, α-D-glucopyranose; β, β-D-glucopyranose. Asterisks denote unidentified signals.
(Reprinted from ref. 43. Copyright (2006), with permission from American Chemical Society.)

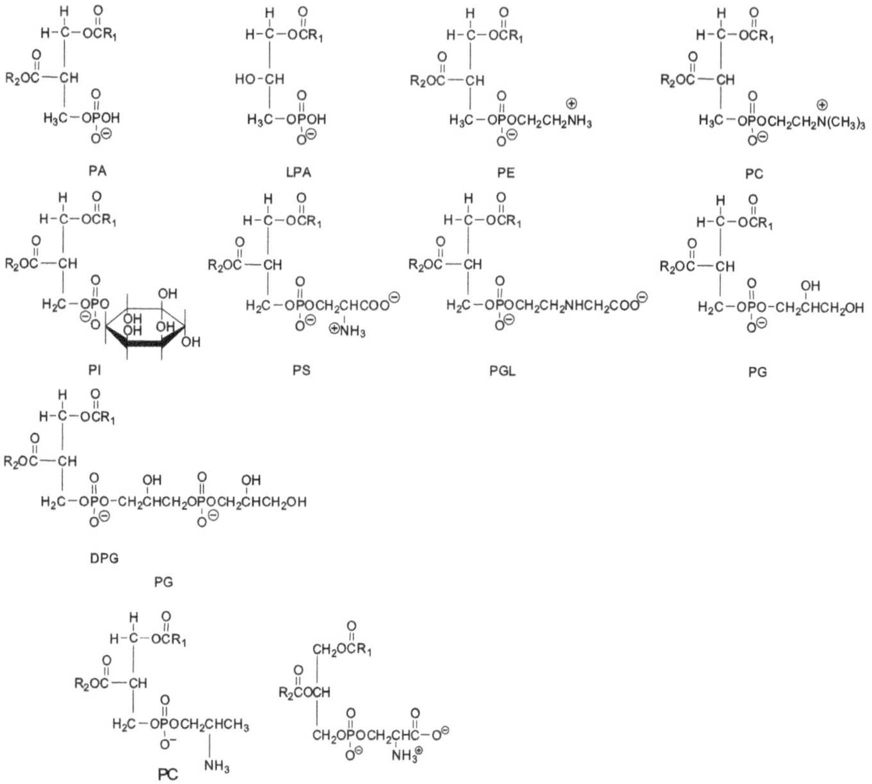

Figure 7.11 Chemical structures of the most important phospholipids detected in fats and oils; phosphatidic acid (PA), lyso-phosphatidic acid (LPA), phosphatidyl ethanolamine (PE), phosphatidylcholine (PC), phosphatidylinositol (PI), phosphatidylserine (PS), phosphatidyl glycine (PGL), phosphatidylglycerol (PG), diphosphatidyl glycerol (cardiolipin) (DPG). R_1 and R_2 are alike or different acyl chains. Derivatives of PG contain mono- and fully acylated hydroxyls on the second glycerol moiety.

samples was estimated by employing 1D and 2D ^{1}H NMR. The results indicated that the fatty acid composition in phospholipids and triacylglycerols of olive oil was similar. Another research paper[49] compared the efficiency of three analytical techniques (TLC, HPLC and NMR) in the determination of phospholipids in soybean and sunflower lecithins. Although the results obtained by the three methods were comparable, ^{31}P NMR was recommended as the preferred technique for phospholipid determination due to being rapid, easy to perform, and showing the highest selectivity.

7.1.3   Quality Assessment

Numerous NMR methodologies have been developed in recent years to examine the quality of vegetable oils. Most of these studies have been devoted to olive oil, which represent a valuable commodity of the countries located in the Mediterranean basin. Edible oil quality can be related to their oxidation stability, sensory characteristics and nutritional aspects. Multi-nuclear (^{1}H, ^{13}C, and ^{31}P) NMR spectroscopy offers complimentary information on edible oil quality.

7.1.3.1 Oxidation Stability

Quality deterioration has been observed during the storage of vegetable oils due to the oxidation of unsaturated fatty acids by atmospheric oxygen. ^{1}H NMR, and to a lesser extent, ^{13}C NMR spectroscopy has been proposed as an efficient methodology in the assessment of the oxidative state and oxidative stability of vegetable oils. The first attempts concerned with the determination of the ratio between the olefinic and aliphatic protons and the ratio between the aliphatic and bis-allylic protons in fatty acids by employing ^{1}H NMR.[50] A good correlation between these ratios and the classical peroxide value (PV) has been observed. This methodology, however, has not found a wide application in food analysis probably because it is not sufficiently sensitive. An alternative to PV for the assessment of the oxidative state in edible oils was proposed recently.[51] According to this method the peroxide value was expressed as the total amount of hydroperoxides detected in the ^{1}H NMR spectra of the oxidised oils. This value was determined by integration of the OOH signals resonating as singlets at 10–11 ppm and expressed in mmol kg^{-1}. The NMR results were in good agreement with the classical peroxide values for most of the oils examined. In the case of olive oils, PVs were too low when compared with the values obtained by ^{1}H NMR, which was attributed to the presence of specific phenolic compounds. An indirect measure of lipid oxidation in vegetable oils is the iodine value (IV). Both ^{1}H and ^{13}C NMR spectroscopy can easily measure the IV of edible oils.[52,53] Through calculation based on the unsaturated proton[52] or carbon[53] signals and the average molecular weight derived directly from the spectra, the IV determined by NMR compares favourably with the conventional titration method.[11] A more rigorous approach to monitor and understand the oxidation process occurring in edible vegetable oils was offered by ^{1}H NMR spectroscopy through direct detection and quantification of specific products arising from the oxidative degradation of oils.[54–58] Since oxidation is a rather slow process, oils were thermally stressed at high temperatures (70 – 180 °C). From these studies, it was concluded that the oxidation process occurs in two stages. In the first stage, fatty acids are oxidised to hydroxyperoxydienes, which in the second oxidation state degrade to generate secondary oxidation products. Among these secondary products, saturated aldehydes are responsible of the undesirable rancid flavour of the oxidised oils. The ^{1}H chemical shifts of the primary and secondary oxidation products for vegetable oils are tabulated in a recent review article.[3] The rate of formation of oxidation products determines the oxidation stability of the edible oils. This trend is greater for oils characterised by a higher degree of unsaturation.[54] Virgin olive oil, with its low composition of polyunsaturated fatty acids and its high content in natural antioxidants (phenolic compounds and tocopherols), has shown the highest oxidation stability amongst the vegetable oils. The use of ^{13}C NMR spectra provides additional information about the oxidation process and oxidation products of lipids.[59] The diagnostic ^{13}C chemical shifts of oxo and hydroxy acids have been reviewed by Gunstone and Knothe.[60] The oxidation stability of vegetable oils, such as sunflower, corn, grapeseed, soybean oil, and olive oils of various grades (virgin, lampante, and refined olive oil) was predicted; taking into account the TAG and fatty acid composition, as well as the phenolic and tocopherol content as determined by ^{13}C NMR spectroscopy.[61] The signals of the minor compounds were observed in the ^{13}C NMR spectra after fractionation of the oils by column chromatography.[62] The prediction stability was determined by a stepwise linear regression applied to the oils compositional variables, the latter being selected as those most contributed to oil stability. The ^{13}C NMR results were validated by the Rancimat method.[61] The olive oil stability was evaluated as well by adopting the chemometric approach, which was applied to the whole ^{1}H NMR spectra of the oils.[63] Small changes of the ^{1}H signals, in particular spectral regions and the appearance of some new, low intensity signals, indicated that some oxidation started after a time period of one year.

7.1.3.2 Free Acidity and Water Content

The determination of free acidity is of prime interest and the first measurement made in order to classify oils according to their quality. For instance, extra virgin olive oil must have the lowest free acidity ($<0.8\%$ expressed in oleic acid) amongst the various olive oil grades. Free acidity can be determined, either by comparing the carbonyl signal of free acids at 176–178 ppm with those of TAG at 172–174 ppm, or by using the signal of the phosphitylated carboxyl groups of free acids in the ^{31}P spectrum at δ 134.8 in combination with the cyclohexanol internal standard (see Figure 7.8). The latter method is faster, but destructive. Free acidities determined by the more sensitive ^{31}P NMR methodology are in very good agreement with the classical titration method.[11] This method can be used for any vegetable oil.

Water content has long been recognised as an important factor determining the quality of olive oil. Small quantities of water in olive oil are responsible for the creation, and persistence, of the suspended and dispersed material that constitutes the so-called "veiling" of virgin olive oil. ^{31}P NMR spectroscopy proved to be an accurate method for the determination of the water content (moisture) in vegetable oils.[64] This technique has been applied to olive oil and is consisted of the derivatisation of the water molecule by the phosphorus reagent **I**. However, reagent **I** gave three separate products with water.[64] Therefore, this phosphitylating reagent was replaced later by the less reactive diphenylphosphinic chloride, which produced a single reaction product with water without detectable formation of any by-products. The sole reaction product gives a signal at δ 28.05 in the ^{31}P NMR spectrum.[64]

7.1.3.3 Storage History

Prolonged storage of vegetable oils deteriorates their quality mainly due to the oxidation of the TAG molecules. The situation becomes even worst when storage conditions are inappropriate favouring the oxidative and lipophylic attacks to oil lipids. The DAG profile has been used to assess, qualitatively, the freshness of virgin olive oil.[32,65] Fresh olive oils contain mostly *sn*-1,2-DAG, the concentration of which decreases upon storage due to their isomerisation to the more stable *sn*-1,3-DAG. Therefore, the ratio of *sn*-1,2-DAG over *sn*-1,3-DAG or the ratio between *sn*-1,2-DAG and total DAG content (*sn*-1,2-DAG plus *sn*-1,3-DAG) (*D* ratio) as measured by NMR spectroscopy can be used as a useful index for the quality of olive oil. Figure 7.12 shows the plot of *D* ratio *versus* total DAG content for 96 samples of extra virgin olive oils (EVOO) from various regions of Greece, 15 commercial oils labelled as EVOO and pure or blended olive oils (POO), 3 samples of refined olive oils and 3 samples of olive-pomace oils.[32]

In this plot, the majority of virgin oils from the various regions of Greece are characterised by high *D* values and low total DAGs, and tend to gather at the

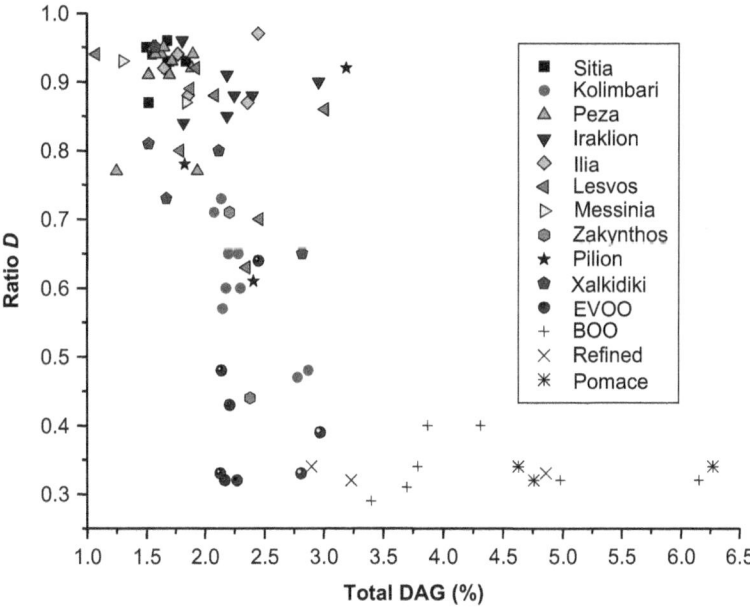

Figure 7.12   Plot of the ratio *D versus* the total DAG for virgin olive oils of various regions of Greece, commercial extra virgin (EVOO), commercial pure olive oils (POO), refined olive oils, and pomace oils. For the four regions of Crete (Sitia, Kolymbari, Peza, and Iraklion) the solid symbols correspond to crops of the period 2000–2001, whereas the empty symbols correspond to crops of the period 1999–2000.
(Reprinted from ref. 32. Copyright (2002), with permission from American Chemical Society.)

upper left corner of the plot, close to $D = 1$, indicating that they are fresh oils. The commercial EVOOs have low D values, and depending on the total DAG content, tend to the lower left or right corner of the plot. At any rate, the commercial oils do not fall at the upper left corner of the plot, indicating that they are of inferior quality, probably because of their long shelf life. The tendency of the commercial blended olive oils to gather at the lower right corner of the plot is pronounced. This was expected, since blended olive oils are legal mixtures of virgin olive oil and refined olive oil. Finally, the refined and the olive-pomace oils, which are not virgin, tend clearly to the lower right corner of the graph as expected. In a subsequent study by the same group,[66] the evolution of the two DAG isomers and free acidity of EVOO samples for a total storage time of 18 months was monitored by ^{31}P NMR spectroscopy. Through the kinetic treatment of the TAG hydrolysis and the isomerisation of *sn*-1,2 DAG to *sn*-2 1,3-DAG in olive oil at room temperature in the light and the dark, a mathematical expression was derived connecting the storage time or age of olive oil with the DAG concentration in terms of the ratio D and free acidity.

7.1.3.4   Oil Content in Oilseeds

The oil content in oilseeds is important analytical information for the edible oil industry, marketing and preservation studies, and plant breeding programs. Moreover, oil content may be correlated with the quality factors of the produced oil. Pulsed low-resolution ^{1}H NMR (LR NMR), also called Time Domain (TD) NMR has been recognised as an international standard method for the determination of oil content in oilseeds.[67] These experiments are performed at low magnetic field strengths (10–20 MHz) using low-cost spectrometers that can be used in industrial applications for on-line automated quality assessments.[68,69] A review article has been published for the applications of TD NMR to oilseeds.[70] The free induction decay (FID) method (see section 2.12.1.1) has been employed to measure the oil contents of mustard, sunflower and peanut seeds.[68] The TD signal was measured 110 µs after the end of the 90° pulse and converted using a calibration curve into the percentage of oil present in oilseeds. The delay time was necessary in order to eliminate the time domain signal (FID) from the water protons, which relax faster (shorter T_1s) than the TAG protons of the oil. The introduction of the spin–echo method (see section 2.8) and in particular the multi-echo, CPMG method (see section 2.8 and section 2.12.1) made the simultaneous determination of the oil and water contents possible, although calibration curves are still needed.[70,71] The CPMG method appears to be more accurate, since it eliminates the effects from diffusion. The application of this methodology to olives and/or olive paste requires pre-drying for the removal of free water. Contrary to the bound water in oilseeds, the free water in the flesh (mesocarp) of the olives is free and the relaxation times of protons in free water and oil are rather similar. In this case, a distinction of the signals from water and oil can be achieved by combining a relaxation and diffusion experiment, *i.e.* by combining the CPMG pulse sequence with the pulse field gradient spin–echo sequence (PFGSE).[72] The

CPMG signal contains information about both oil and water, whereas the diffusometric PFGSE and/or PFGSTE experiment can be controlled in such a way that the resulting signal contains a contribution from the oil only. Subtraction of the PFGSE signal from the CPMG signal, the water content can be determined (see section 2.12.1.1). Two recent reports[73,74] introduce chemometrics in the analysis of the CPMG relaxation spectra of oilseeds in an attempt to develop classification–prediction models for treated and untreated seeds.

Recently, MRI (see section 2.11) was applied to measure the oil content in oilseeds and provide information about the water and oil distribution during olive and palm fruit ripening.[75–78] Kotyk *et al.*[75] were able to measure the oil content of intact corn kernels using a high-throughput MRI method. Measurements were performed simultaneously on a large number of corn kernels arranged in layers stacked to form a rectangular sample cube using a 1.5 T MRI scanner. The MRI results were found to be in very good agreement with those obtained with TD NMR, NIR and the traditional method of solvent extraction. The MRI studies in olive fruits[77,78] provided information about the oil growth and mobilisation during the ripening of the fruit. Additional information about the water and oil content of processed olives was given,[77] whereas application of CPMAS solid-state ^{13}C NMR experiments confirmed the MRI results regarding the presence of mobile oil adsorbed into the solid matrix of the kernel in ripe olives.[78] The distribution of water and oil in olives at three different phases of ripening are shown in the chemical shift selective images of Figure 7.13.[77] Samples 2, 3, and 5 were collected on 11/11/2002, 27/11/2002, and 13/01/2003, respectively.

7.1.3.5   Harvest

There are several exogenous and endogenous factors that affect the quality of the olive fruit and hence the extracted olive oil. These factors are primarily genetic (cultivar) and pedoclimatic (environmental and/or climatic conditions) factors. Additional factors that may affect the fruit physiology and the composition of the extracted olive oil include agronomic practices, such as irrigation, fertilisation, and harvest method, and technological procedures such as processing and storage. NMR spectroscopy has made a significant contribution to the clarification of the influence of several of these factors on olive oil quality. One important factor is the harvest period. Studies with mono-cultivar virgin olive oils have shown that alteration of the climatic conditions (*e.g.* temperature, rainfall) from one crop season to the next are significant factors affecting oil composition, especially that of minor compounds (*e.g.* phenolic compounds).[79,80] This was verified by NMR spectroscopy in a recent study.[81] 221 extra virgin olive oil samples extracted from four olive mono-cultivars, originated from four divisions of Greece and collected in five harvesting periods (2002–2006 and 2007–2008) were analysed by ^{1}H and ^{31}P NMR spectroscopy.[81] Application of the forward stepwise canonical discriminant analysis (CDA) to the compositional data (fatty acids, phenolic compounds, diacylglycerols, total free sterols, free acidity, and iodine number) grouped the 221 oil samples into

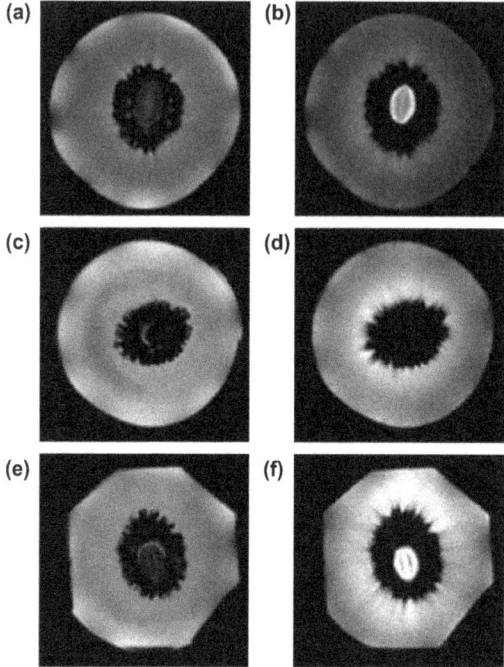

Figure 7.13   Axial images of olives. (a) Water selective image and (b) oil selective image of sample 2; (c) water selective image and (d) oil selective image of sample 3; (e) water selective image and (f) oil selective image of sample 5. (Reprinted from ref. 77. Copyright (2007), with permission from Elsevier.)

five well separated harvesting periods. The variation of the data is distributed into three significant CDA axes accounting for by 49.2, 29.3 and 16.9% (total 96.4%) of the data variation, respectively, with a very good classification rate (94%). The 3D CDA score-plot is shown in Figure 7.14.

The very good discrimination of oils harvested in different periods confirms the differences in the influence of the various factors on the oil composition. The influence of the time period and method of harvest on the olive oil composition and their interactions with other primary (genetic and pedoclimatic) and secondary (agronomic practices and technological procedures) factors was examined by NMR spectroscopy and classical analyses in combination with univariate and multivariate statistical analysis.[82] The most interesting results of this study are summarised as follows:[82] (a) the minor compounds hexanal, *trans*-2-hexenal, *sn*-1,3-diglycerides, and squalene significantly decreased during ripening; and (b) the relative values of the ΔK parameter (oxidation index) and hexanal are higher in the olive oils obtained from olives harvested by shakers, whereas the unsaturated fatty chains are higher when the comb hand-held machines are used. These results were confirmed by performing multivariate statistical analyses with PCA and LDA.[82]

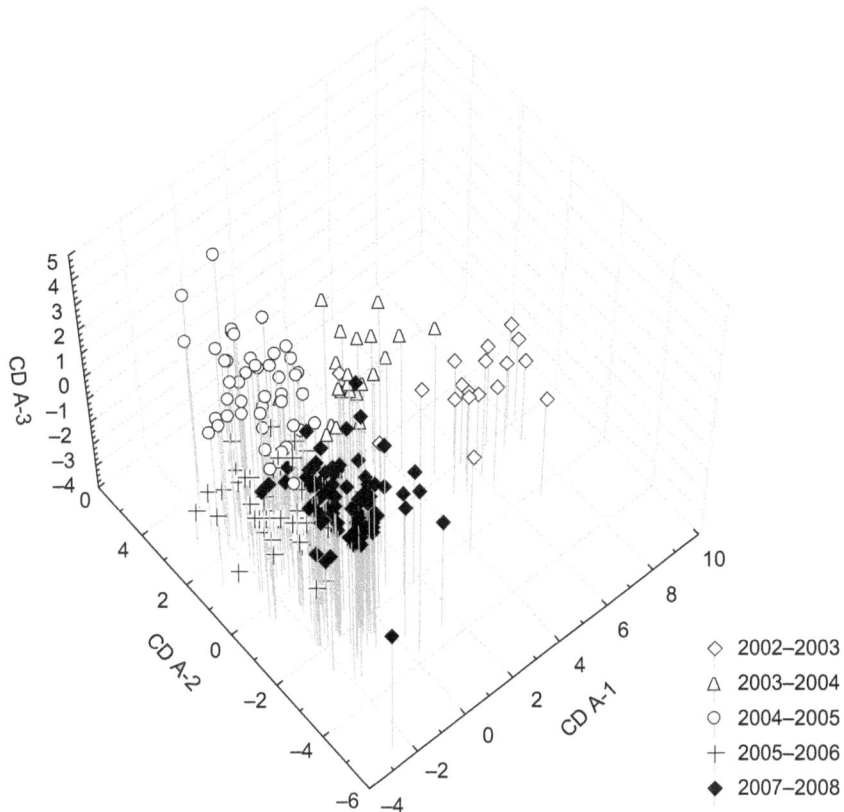

Figure 7.14 3D CDA score-plot for the 221 EVOO samples separated according to five harvesting periods (2002–2006 and 2007–2008). The axes CDA-1 = 49.2%, CDA-2 = 29.3%, and CDA-3 = 16.9% account for by the 96.4% of the total variability.
(Reprinted from ref. 81. Copyright (2012), with permission from Elsevier.)

7.1.4 Authentication

The vast majority of authentication studies by NMR concern extra virgin or virgin olive oil. Therefore, this section will be devoted to aspects of authentication associated with this oil. Authentication involves the detection of adulteration caused either by the addition of a foreign oil to olive oil, or from false information about its geographical and/or botanical origin. Addition of an adulterant to virgin olive oil is expected to induce alterations in the concentration of its constituents or at least to indicate an anomaly in its chemical composition. Clues of these changes can be observed in the NMR spectra. For instance, careful inspection of particular regions in the ^{1}H NMR spectra of vegetable oils reveals slight differences in the chemical shifts and signal intensities of the saturated and unsaturated proton signals, which might provide discrimination of these oils and possible detection of adulteration.[10] Addition of seed oil to virgin olive oil can be detected by the appearance of a minor signal in the carbonyl region of the ^{13}C spectrum (Figure 7.6) corresponding to saturated acyl chain at the *sn*-2 glycerol position. The signal of the methyl protons of linolenic acid (Figure 7.2 and Table 7.2) can be used to detect adulteration of virgin olive oil with oils, which far exceed the permissible limit (0.9%) of linolenic acid of virgin olive oil (*e.g.* soya bean, rapeseed, linseed oil). However, slight differences in the NMR spectral parameters could become less reliable when a real adulteration problem is confronted, especially when the foreign oil has very similar chemical composition to that of virgin olive oil (*e.g.* hazelnut oil).

New developments in the detection of fraud in virgin olive oil apply multi-variate statistical methods to the NMR data. The metabonomic method of analysis involves the targeted profiling an/or the chemometric approach (see chapter 6). A good example of the first method is depicted in Figure 7.15, which summarises the targeted profiling methodology.[7]

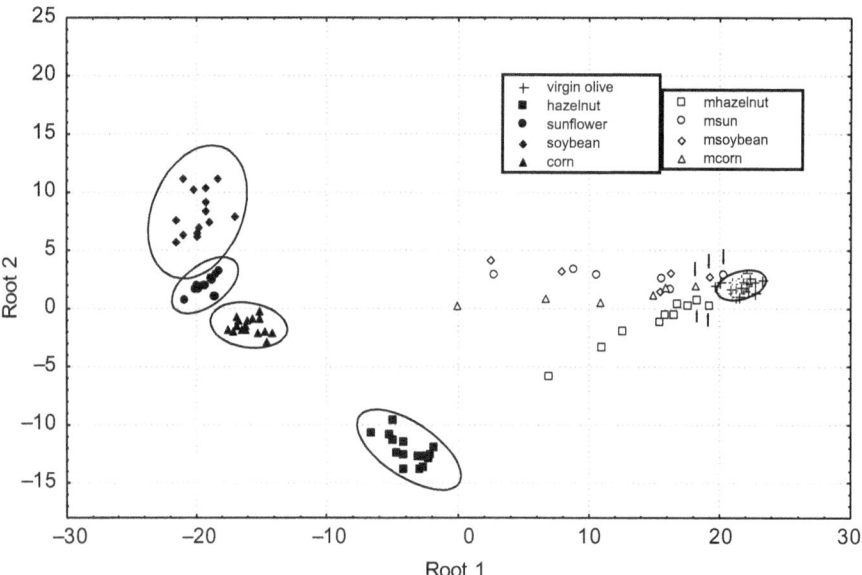

Figure 7.15   Plot of discriminant functions roots 1 and root 2 for five types of edible oils. Virgin olive oil, hazelnut oil, corn oil, and sunflower oil are shown by crosses and solid symbols. Four sets of mixtures of 5, 10, 15, 20, 35, and 50% w/w of virgin olive oils with hazelnut (mhazelnut), sunflower (msun), soybean (msoybean), and corn (mcorn) oils and one set of mixtures of 5, 10, 15, and 20% of virgin olive oils with hazelnut oils (mhazelnut) are denoted by open symbols. Arrows indicate mixtures of 5% w/w of seed oils in virgin olive oils.
(Reprinted from ref. 7. Copyright (2007), with permission from American Chemical Society.)

The chemical composition of a large number of virgin olive oil and seed oil samples (corn, soybean, sunflower, and hazelnut oil) is determined by ^{1}H and ^{31}P NMR. The oil constituents with the highest discriminatory power selected by one-way ANOVA are used as input variables in a subsequent linear discriminant analysis. LDA was successful in discriminating the five oils and build up a classification/prediction model.[7] The next step of the analysis involves the preparation of a series of mixtures of virgin olive oils with different adulterant concentrations, and the use of the previous prediction model to place the mixtures considered as unknown samples. The adulterated olive oil samples lie between the group of genuine olive oils and the respective groups of seed oils. Their position in the DA score-plot depends on the amount of foreign oils in the mixtures. The lowest detectable concentration of the foreign oil in the mixture defines the detection limit of the method. In this particular study the detection limit was 5% with the instrumentation used (500 MHz spectrometer). Several studies along these lines were reported in the literature including fraud of virgin olive oil with refined hazelnut oil.[7,83–89] This oil is difficult to detect, because it has very similar chemical composition with olive oil. Other studies following the chemometric approach proceed with binning of interesting regions of the ^{1}H NMR spectrum with buckets or selecting specific resonance intensities, and used the data matrix (number of samples × number of buckets or signal intensities) for pattern recognition analysis.[15,25,62,90,91] A few studies did not use metabonomic analysis, but detection of adulteration was based on comparison of specific signal intensities before and after the addition of the adulterants at different levels.[10,92] Recently, Šmejkalová and Piccolo[93] examined the potential of high pulsed field gradient (PFG) for screening extra virgin olive oil adulteration with seed oils. Application of a spin–echo pulse sequence allowed the determination of the diffusion coefficients for each signal in the ^{1}H NMR spectra of all vegetable oils and admixtures of extra virgin olive oil with seed oils (sunflower, soybean, and hazelnut oil). Subsequent multivariate statistical analysis of the NMR diffusion data allowed the build-up of a classification/prediction model to detect adulteration of extra virgin oil with these seed oils. Table 7.3 displays a list of olive oil adulteration studies performed so far by using NMR spectroscopy.

7.1.5 Geographical Origin and Cultivar

The characterisation of geographical and/or varietal origin of EVOOs is becoming increasingly important. This is because false labelling of the origin of olive oil is considered as another facet of fraud. With regard to the geographical origin, EVOO is permitted to be marketed under a Protected Designation of Origin (PDO), or Protected Geographical Indication (PGI) label on the basis of the area, cultivar and methods of production. This labelling protects the reputation of the regional food and eliminates the unfair competition and misleading of consumers by non-genuine products, which may be of inferior quality. Moreover, olive oils made from a certain cultivar (mono-cultivar olive oils) are being increasingly introduced in markets and their quality control

Table 7.3 Olive oil adulteration[a] detected by NMR.

Biomarker	Adulterant	Exprimental conditions	Nucleus/Frequency	Detection limit	Statistical methods[c]	Ref.
Targeted profiling						
Unsaturated FA	Vegetable oils	CDCl₃ solutions	^{13}C/75.5 MHz	5%	DA	83
Volatiles and FA	Deodorised hazelnut oil	NMR for volatiles, GC for fatty acids	^{1}H/600 MHz	10%	LDA	88
^{1}H NMR fingerprint	Sunflower, hazelnut oil	CDCl₃ solutions	^{1}H/500 MHz	10%	ANOVA, LDA	84
Di-acylglycerols, fatty acids	Vegetable oils	Phosphitylation pyridine/CDCl₃ solutions	^{1}H/500 MHz, ^{31}P/202 MHz	5%	ANOVA, LDA	7
Di-acylglycerols	Refined olive oil and lampante olive oils	Phosphitylation, pyridine/CDCl₃ solutions	^{31}P/202.2 MHz	5%	ANOVA, LDA	85
Phenolic compounds, fatty acids[b]	Refined hazelnut oil	Phosphitylation, pyridine/CDCl₃ solutions	^{1}H/500 MHz, ^{31}P/202.2 MHz	1%	SCDA, CBT	86
^{13}C NMR fingerprint	Classification of Vegetable oils	CDCl₃ solutions	^{13}C/75.5 MHz	—	SDA, HCA	87
^{13}C NMR fingerprints	Classification of olive oil grades	CDCl₃ solutions	^{13}C/125.7 MHz	—	ANOVA, LDA	89
Chemometric approach						
^{13}C NMR fingerprint[c]	Vegetable oils	CDCl₃ solutions	^{13}C/75.5 MHz	5%	SDA	15
^{1}H and ^{13}C NMR fingerprint	Hazelnut oil	CDCl₃ solutions	^{1}H/600, ^{13}C/150.7 MHz	8%	ANN	90
^{1}H NMR fingerprint[b]	Refined hazelnut oil	CDCl₃ solutions	^{1}H/400, 500, and 600 MHz	≤10%	ANOVA, PCA, LDA, LMR	91
^{1}H NMR diffusion spectra	Vegetable oils	Neat samples	^{1}H/300 MHz	10% sunflower, soybean oils, 30% hazelnut oil	ANOVA, DA	93
^{13}C NMR fingerprint	Classification of Vegetable oils	Fractionation of polar and non-polar part of oils, CDCl₃ solutions	^{13}C/75.5 MHz	—	SDA	62
Non-chemometric analysis						
Fatty acids	Discrimination of vegetable oils	CDCl₃ solutions	^{1}H/300 MHz	—	—	10
Unsaturated FA	Soybean oil	DEPT sequence, CDCl₃ solution	^{13}C/75.5 MHz	7–20% depending on fatty acids	Regrassion Analysis	92

[a]Extra virgin olive oils, unless otherwise specified.
[b]Refined olive oil.
[c]ANN = artificial neural networks; ANOVA = analysis of variance; CBT = classification binary trees; DA = discriminant analysis; LDA = linear DA; SCDA = stepwise canonical DA; SDA = stepwise DA; LMR = linear multiple regression; PCA = principle component analysis.

requires the development of new and effective analytical methodologies to detect fraud. Mono-cultivar olive oils have certain specific characteristics ascribed to the olive cultivar, and thereby easier to elaborate. Several authors in different countries, especially from countries located in the Mediterranean basin, have investigated the possibility of discriminating olive oils originated from different areas in the same country or from different countries by employing high-field NMR spectroscopy in combination with multivariate statistical analysis. Their efforts have focused on selecting those biomarkers that are capable of discriminating the oils. The examined biomarkers ranged from the major olive oil constituents fatty acids and triacylglycerols to minor components sterols, phenols, volatiles, hydrocarbons, *etc.*, or combinations of both. Minor components provide more useful information than major constituents, and they have been more often used to discriminate olive oils according to their geographical or varietal origin. It appears that minor components are prone to greater concentration changes under the influence of various exogenous and endogenous factors (cultivar, ripening conditions, storage, climatic conditions, agricultural practices, and extraction technology).

In almost all of these applications a statistical approach is employed implying that many samples are required for the statistical elaboration. In addition, the sampling method and the quality of the collected samples appear to be one of the most important prerequisite for this analysis (see section 4.1). Another criterion is the use of statistical methods. Sophisticated statistical approaches, such as artificial neural network, should be used sparingly and only when other much simpler statistical methods (see chapter 6) fail to give the required information. Table 7.4 depicts the capability of NMR in classifying olive oils according to geographical origin and cultivar. This Table contains the biomarkers used, the chemometric treatment, the cultivars geographical origin and the most important information obtained from each study. The listing in Table 7.4 follows the same pattern as in Table 7.3, *i.e.* the various studies are divided according to the statistical methodology applied to the NMR data sets.

7.2 Animal Oils and Fats

7.2.1 Milk Fat

Milk fat or butter is the lipid fraction of milk derived from cattle species; it contains about 80–82% of fat, 16–17% water and 1–2% solids other than fat. About 98% of its content is consisted of TAG. TAG molecules of milk fat are characterised by short-chain ($4:0$–$8:0$, 8.3%), medium-chain ($10:0$–$12:0$, 6.6%) and long-chain ($14:0$–$18:0$, 81.9%) length fatty acids. Moreover, milk fat is a relatively high saturated fat comprising about 65% saturated fatty acids (mainly $16:0$, $18:0$ and $14:0$) and about 35% unsaturated fatty acids (mainly $18:1$). The TAG composition of the milk fat in fatty acids has been determined mainly by gas chromatography after hydrolysis of the triacylglycerols. TAG composition is greatly influenced by physiological (*e.g.* cattle species, stage of lactation) and dietary factors. This situation has been exploited to discriminate

Table 7.4 Geographical and varietal classification of extra virgin olive oil by NMR.

Biomarker	Nucleus/ frequency	Cultivars	Geographical area of origin	Chemometric treatment[a]	Observations	Ref.
Targeted profiling						
¹H NMR (minor compounds (volatiles, sterols, *etc.*)	¹H/600 MHz	36 mono-cultivars	Liguria, Puglia, Sicily	ANOVA, TCA, LDA	Good geographical classification was achieved regardless of the cultivar.	94
¹H NMR (minor compound.) ¹³C NMR (TAG)	¹H/600 MHz ¹³C/150.9 MHz	17 mono-cultivars	Italy, Argentina	TCA, LDA	FA composition was determined by GC. NMR and GC data treated separately permitted 100% correct geographical classification of the 17 cultivars.	95
¹H NMR (minor compounds)	¹H/600 MHz	Frantoio, Leccino, Moraiolo, Merino, Quercetana (mono- and multi-cultivars)	Tuscany (Seggiano, Arezzo, Firenze Lucca micro-areas)	HCA, K-means, DA	Very good geographical discrimination of oils from four different micro-zones of Toscany.	96
¹³C NMR (DAG, TAG)	¹³C/150.9 MHz	Nocellara del Belice, Cerasuola, Biancolila, Tonda Iblea	Castel Vetrano, Trapani, Paceco, Caltabellota, Delia, Ragusa	MANOVA, PCA, TCA, MDS, LDA	Fatty acids composition was determined by GC. ¹³C NMR and GC data treated separately permitted 100% correct classification of cultivars.	97
¹³C NMR (FA)	¹³C/75.5 MHz	Coratina, Dritta, Grossa di Cassano Moraiolo, Picholine	Abruzzo, Calabria, Lazio, Lombardia, Marche, Molise, Puglia, Toscana	PCA, PLS, PCR, MLR	70–100% correct varietal classification and 93–100% correct geographical classification	14
¹H NMR (volatiles)	¹H/600 MHz	Different mono-cultivars	Campania, Lazio, Sicily, Umbria	PCA, HCA	96% geographical classification regardless of the cultivar.	98

Technique	Field	Cultivar/samples	Geographical area	Statistics	Results	Ref.
^{13}C NMR (TAG)	^{13}C/125.7 MHz	Coratina, Ogliarola, Peranzana	Terra di Barri, Colline di Brindisi, Dauno	MANOVA, LDA	Correct geographical classification ranged from 72% (Dauno) to 90% (Terra di Bari).	99
^{1}H NMR (minor compound)	^{1}H/600 MHz	Casaliva, multi-cultivar	Garda, Veneto (Valpolicella, Euganei-Berici, Grappa micro-areas)	ANOVA, PCA	Satisfactory geographical classification of oils even for different micro-areas of the same region	100
^{1}H and ^{13}C NMR (volatiles, sterols, terpenes, squalene)	^{1}H/600 MHz, ^{13}C/62.9 MHz	22 mono-cultivars	Lazio area	ANOVA, PCA, LDA	Satisfactory geographical separation was observed according to irrigation practice and the altitude in the Lazio area	101
^{1}H NMR (FA, DAG), ^{31}P NMR (phenolic compounds)	^{1}H/500 MHz, ^{31}P/202 MHz	koroneiki	Crete (Heraklion, Sitia, Chania), Peloponnesus (Messinia, Lakonia), Zakynthos.	SCDA, CBT	87% correct prediction for the three areas and 74% for the six sites. The former becomes 92% when harvest year is included in the statistical treatment.	102
^{1}H NMR (minor compound)	^{1}H/600 MHz	DO or EVOO from different Mediterranean areas; three harvest periods (2004–2006) (896 samples)	Italy, Liguria, Spain, France, Greece, Turkey	SIMCA, PLS-DA	17 selected ^{1}H resonances allowed a good discrimination between Ligurian and non-ligurian olive oils.	103
^{13}C NMR (DAG, TAG)	^{13}C/75.5 MHz	Leccino, Moraiolo, Dritta	Abruzzo, Puglia, Tuscany	PLS, PCR	100% correct varietal prediction and ~100% geographical prediction	106
^{1}H NMR (FA, Phenolic compounds)	^{1}H/400, 500 MHz	Coratina, Leccino, Ogliarola, Olivastro, Simone	Apulia region	PCA, HCA, DA	Fatty acids composition allowed 100% correct varietal classification. Phenolic compounds permitted discrimination only for samples originated from northern Apulia	107

Table 7.4 (*Continued*)

Biomarker	Nucleus/ frequency	Cultivars	Geographical area of origin	Chemometric treatment[a]	Observations	Ref.
^{13}C NMR (FA, TAG)	^{13}C/125.7 MHz	Coratina, Leccino, Peranzana, Ogliarola	Apulia area	ANOVA, PCA, HCA, DA	FA, TAG, sterols composition was determined by GC. 92% correct prediction with GC data and 95% upon application of DA to the NMR data.	109
^{1}H NMR (minor compound)	^{1}H/600 MHz	Casaliva, multicultivar	Garda lake (Veneto and Lombardia banks)	PCA	Satisfactory discrimination of oils originated from two different micro-areas of the same region (Garda lake)	110
^{1}H NMR (Fatty acids, iodine number) ^{31}P NMR (sterols, DAG, phenolic compounds)	^{1}H/500 MHz ^{31}P/202 MHz	koroneiki, tsounati, throubolia and adramitini	221 samples from Crete (Sitia, Heraklion, Chania, Rethymnon), Peloponnesus (Messinia, Lakonia), Lesvos, Zakynthos; four harvest periods (2002–2005, 2007)	SCDA, CBT	Oils classification in accordance with harvest, was high (94%); classification in terms of geographical origin was reduced to 85%. Inclusion of both the harvesting year and geographical origin resulted in a high classification (90%) for the EVOO samples grouped into the four cultivars.	81
Chemometric approach ^{1}H NMR fingerprint (bulk oil and unsaponifiable fraction)	^{1}H/500 MHz	99 commercial samples	Italy, Greece, Spain, Tunisia, Turkey, Cyprus, Syria	LDA, PLS-DA, SIMCA, CART	Application of PLS-DA to ^{1}H NMR resonances of the bulk olive oil and unsaponifiable fraction provided the best geographical prediction.	25

Method	Samples	Origin	Statistical methods	Comments	Ref.
1H NMR fingerprint; hydrogen and carbon isotopic ratios (IRMS methods)	963 PDO and VOO samples; three harvest periods (2004–2006)	Italy, Greece, Spain, France, Turkey, Cyprus, Syria	ANOVA, LDA, PLS-DA	PLS-DA afforded the best model for prediction of olive oil samples according to harvest year and geographical origin. Isotopic ratios improved the classification results.	104
1H NMR fingerprint	Samples from different Mediterranean areas; three harvest periods (2003, 2004) (896 samples)	Greece, Italy, Spain, Tunisia, Turkey	LDA, PLS-DA, ANN,	Use the whole spectrum. Four data sets were treated statistically. Classification varied (35–92%) depending on geographical origin and harvest year.	105
13C NMR fingerprint	12 Italian mono- and multi-cultivar oils	13 Italian areas	PCA	Use the whole spectrum. DEPT sequence improved sensitivity. Good classification according to cultivar within a particular area.	108
1H NMR fingerprint	19 Italian and 16 Greek mono-cultivars	Italy: Apulia (Dauno, Terra di Barri, Terra d'Otranto) Greece: islands of Kefalonia, Kerkira, Lefkada, Zakinthos	PCA, CA, MANOVA MC	Experiment with multiple saturation to observe minor signals. Correct prediction probabilities of 78% were achieved and region correct predictions between 53% and 100% depending of the spectral region examined.	111

[a]ANN = artificial neural networks; ANOVA = analysis of variance; CA = canonical analysis; CART = classification and regression trees; CBT = classification binary trees; DA = discriminant analysis; HCA = hierarchical cluster analysis; LDA = linear DA; LMR = linear multiple regression; MA = Monte Carlo validation approach; MANOVA = multivariate analysis of variance; PCA = principal component analysis; PCR = principal component regression; PLS = partial least squares; SCDA = stepwise canonical DA; SDA = stepwise DA; SIMCA = soft independent modeling and class analogy; TCA = tree cluster analysis.

milks derived from different cattle species or milks of different origin (see below). ^{1}H, ^{13}C, and ^{31}P NMR spectroscopy has been employed for the analysis of lipids in milk fat.[112–118] Partial assignments of the ^{1}H NMR spectra of the apolar part of butter (dissolved in $CDCl_3$) confirmed the presence of short and long chain saturated and unsaturated fatty acids, and in particular the isomer cis-9, trans-11 18:2 acid, also known as rumenic acid.[116] This acid constitutes the 80–90% of the total conjugated linoleic acids (CLA), which apparently are related to anti-carcinogenic effects of milk and dairy products. ^{1}H NMR analysis of the polar part of butter (dissolved in D_2O) showed the present of organic acids (formic, lactic, acetic acid, citric, free butyric acid, and sorbic acid) and lactose.[116] The levels of organic acids and lactose as determined by integration of the appropriate ^{1}H signals can be used as potential biomarkers for the quality and the production process of milk. Andreotti et al.[113] investigated the composition and the distribution of fatty acids in TAG from cows' and buffalos' milk fat by ^{13}C NMR spectroscopy. The ^{13}C NMR spectra of both fats extracted from milks with chloroform/methanol mixture were assigned with the aid of TAG standards. Quantitative analysis performed upon integration of the best-resolved signals in the spectrum allowed the determination of fatty acid composition for the two milk fats. Application of principal component analysis (PCA) on the acyl group composition as determined by ^{13}C NMR clearly distinguished the two milks. In a second study by the same group,[114] two additional milk fats from sheep and goat were added to those of caws and buffalos. The corresponding ^{13}C NMR spectra were assigned and quantified as before, and the composition and distribution of fatty acids were determined. The most important differences found in the fatty acid composition of milk fat amongst the four species involved the short chain acids butyric, caproic, caprylic and capric. Subsequent treatment of the fatty acid content with multivariate statistical analysis succeeded to differentiate the goat milk from sheep milk and both of these milks from cow and buffalo, the latter two milks being clustered in a single group.

Classification of cow and buffalo milk was attempted by Brescia et al.[112] using ^{1}H NMR spectroscopy. Assignment of ^{1}H signals of TAG molecules was performed by COSY experiments and 2D heteronuclear correlations through HMQC and HMBC experiments with known ^{13}C chemical shifts of previous study.[112] Unsupervised statistical analysis by PCA and HCA carried out on the same variables permitted a satisfactory classification of the two types of milk. Although ^{1}H NMR gives the same information as ^{13}C NMR regarding the differentiation of cow and buffalo milk, it is more promising as a screening method because of its speed and much higher sensitivity.

Amongst the minor compounds that exist in the milk fat in milk at large, only the phospholipids have studied systematically by ^{31}P NMR. Phospholipids in the milk fat of cows and ewes were determined by ^{31}P NMR spectroscopy.[117,118] The procedure followed was similar to that described in the literature with one exception; instead of the standard solvent chloroform/methanol/water-EDTA, the mixture dimethylformamide/triethylamine guanidinium hydrochloride solvent mixture was used. The new solvent mixture overcome the partition problems with the previous biphasic system and extended the range of ^{31}P chemical shifts, increasing thus the resolution of the experiment. Figure 7.16 shows the ^{31}P NMR spectrum of cow and buffalo milk fat along with the assignment of the various signals.[117] The good resolution of the ^{31}P spectra allowed for an easy quantification of the phospholipids for both types of milk fat.

An important parameter for the food industry is the solid fat content (SFC). This can be measured by the same methodology used for the oil content determination in oilseeds. Actually, the first commercial TD NMR application to foodstuff was the assessment of the SFC parameter,[119] which is recognised as an International Standard method.[120] Two versions of the same TD NMR experiment were used the direct and the indirect SFC methods. Both methods are based on the different relaxation rates of protons in the solid and liquid fat. The protons in the solid fat relax faster ($\sim$10 µs) back to their equilibrium state after their excitation by the 90° pulse than the protons in the liquid fat ($\sim$70 µs). The magnitude of the TD NMR signal (FID) as a function of time at a given temperature is shown schematically in Figure 2.35. The direct methodology separately measures the NMR signals from proton nuclei in the solid and liquid phases as shown in Figure 2.35. The solid fat content is calculated from the ratio $SFC = (F \times S_{S+L} - S_L)/(F \times S_{S+L})$, where F is a correction factor that allows the prediction of the FID amplitude S immediately after the pulse ($S_S = F \times S_{S+L}$), which is impossible to observe directly during the dead time of the receiver. The F-factor is determined by measuring samples of known SFC. The indirect method samples only one point on the FID, 70 µs after the end of the pulse, assuming that all magnetisation of solid components has relaxed. By repeating this measurement at a temperature where all fat is molten, one can determine the SFC. It is necessary to weigh the sample and a correction needs to be made since the NMR measurements are temperature sensitive. The direct method is the preferred one for fat analysis amongst the food technologists since it is more accurate and less complicated than the indirect method. However, both methods do not exploit all the information inherent in the whole line shape of the FID; the direct method samples only two points in the FID, whereas the indirect method just one point. It has been demonstrated in a number of studies[121–123] that a detailed analysis of the protons transverse relaxation in a TD NMR experiment provides valuable information about the crystalline and mesomorphic states of lipids during crystallisation and/or melting.

7.2.2 Margarines and Hydrogenated Fats

Margarine is a substitute of butter containing about 80% of fat blended with water, and containing vitamins and other ingredients. In old times, margarines were prepared from fats of animal origin (lard or beef tallow). Vegetables oils hardened by proper modification including partial hydrogenation or inter-esterification procedures manufacture today's margarines. Most margarines are formatted with soybean, palm, corn, rapeseed oil or even with partially

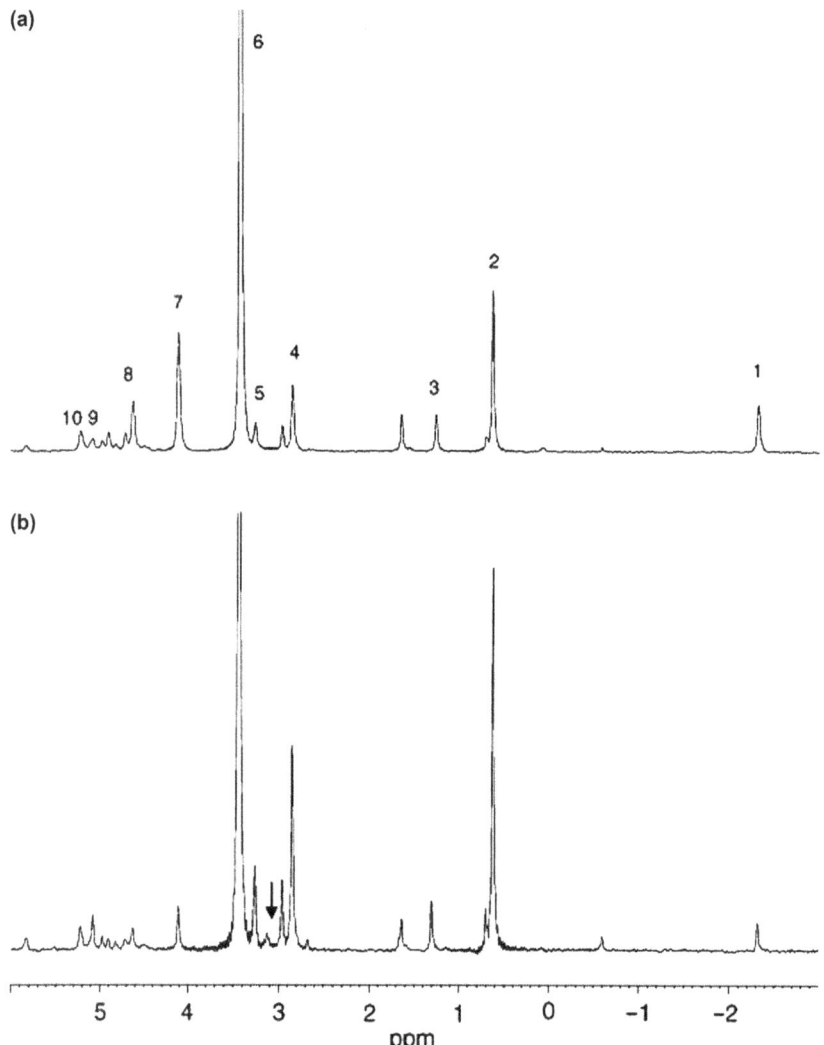

Figure 7.16 ^{31}P-NMR spectra of a ten-fold concentrated buffalo (a) and cow (b) milk
ultra filtrates at pH 9.4. The assignments were as follow: phosphocreatine
(used as an internal reference), −2.32ppm (peak 1); glycerophosphorylcholine, 0.62 ppm (peak 2); glycerophosphorylethanolamine,
1.26ppm (peak 3); *N*-acetylglucosamine-1-phosphate, 2.83 ppm (peak 4);
galactose-1-phosphate, 3.25 ppm (peak 5); inorganic phosphate,
3.44ppm (peak 6); phosphorylcholine, 4.11 ppm (peak 7); phosphoryl-
ethanolamine, 4.63ppm (peak 8); glycerol-1-phosphate, 5.09 ppm (peak
9); and glucose-6-phosphate, 5.21ppm (peak 10). The arrow in panel B
indicates lactose-1-phosphate and/or glucose-1-phosphate.
(Reprinted from ref. 117. Copyright (2006), with permission from
Elsevier.)

hydrogenated fish oils. The partial hydrogenation of oils and fats is accomplished by catalytic addition of hydrogen to the double bonds of the unsaturated acylglycerol chains. This procedure induces a partial reduction of the degree of unsaturation and a concomitant increase of the higher melting saturated chains. Also, hydrogenation process is always accompanied by substantial formation of the more stable *trans* double bonds upon isomerisation of the *cis* bonds. Since *trans* fatty acids act as coronary artery disease risk factor, alterative processes (*e.g.* chemical or enzymatic inter-esterification) have been applied for the preparation of margarines. A preliminary ^{13}C NMR study on hydrogenated oils was performed by Gunstone.[124] The spectra of hydrogenated oils contained a large number (50–100) of signals were partially assigned on the basis of different synthetic triacylglycerol molecules containing the (*trans*) elaidic acid in addition to palmitic, steraric and oleic acid with different glycerol positional distributions and information from the literature. Allylic signals distinguished between *cis* and *trans* fatty acids allowed the determination of the proportions of total *cis* to total *trans* isomers. Based on the interpretation of high-resolution ^{13}C NMR spectra, Gunstone[125] examined the presence (or absence) of milk fat or lauric oils (coconut oil and palm kernel oil), and hydrogenated fats in 15 samples of spreading fats, baking fats and vegetable creams. The tabulated ^{13}C NMR data[125] interpreted in a semi-quantitative manner provides an insight into the composition of the samples examined. Analysis of margarines by high resolution ^{1}H NMR has been performed in one case.[116] The ^{1}H NMR spectra of margarine samples of this study were similar to those of milk fats and permitted the quantification of organic and fatty acids in the polar and apolar part, respectively. As expected, no rumenic acid was detected in the margarine samples studied. The addition of preservatives in margarines is permitted, and therefore signals of benzoic acid and sorbic acid were observed in the spectrum.

A thorough quantitative analysis of hydrogenated and inter-esterified/ fractionated margarines was carried out by Sacchi *et al.*[126] using ^{13}C NMR spectroscopy. They managed to determine the composition of the acylglycerols and fatty acids, as well as the positional distribution of the acyl chains on glycerol. The signals of glyceryl carbons in the region 60–73 ppm indicated that margarine samples consisted mainly of mono- and di-acylglycerols. The signals of the methylene carbons in the 20–40 ppm region allowed for the determination of the fatty acids composition (saturated, *cis*-monoenic, *cis*-dienic, and *cis*-trienic acids). The large chemical shift differences between the *cis* and *trans* allylic and olefinic carbon resonances permitted the determination of *trans* fatty acids. No *trans* signals were found in the spectra of inter-esterified/fractionated margarines, while margarines containing hydrogenated fats showed the presence of *trans* isomers. It is apparent that inter-esterification does not isomerise the fatty acids double bonds and does not affect their degree of unsaturation. Finally, the carbonyl resonances provided information about the glycerol positional distribution of the acyl chains. Margarines produced by inter-esterification showed a high level of saturated fatty acids in the *sn*-2 position contrary to what is normally observed in vegetable oils. The carbonyl resonances of the hydrogenated margarine showed a complex profile with partially overlapped signals in *sn*-1 and *sn*-2 glycerol positions due to the presence of the newly formed *trans* components. ^{13}C carbonyl resonances were used, also, to identify the positional distribution of stearic acid introduced into canola oil during the process of producing *trans*-free margarine.[61] It was found that this fatty acid was incorporated mainly at the *sn*-1(3) positions, whereas no new *trans* fatty acids were formed.

7.2.3 Fish Oils

Like other oils and fats, fish oils may be divided into their lipid fractions such as tri-acylglycerol, di-acylglycerol, mono-acylglycerol, phospholipids, cholesterol, and free fatty acids. Fish oils are rich in polyunsaturated fatty acids (PUFAs), and particularly in omega-3 (*ω*-3) or *n*-3, such as docosahexaenoic acid (DHA) 22:6(*n*-3) and eicosapentaenoic acid (EPA) 22:5 (*n*-3), which are considered important constituents for the human health. In addition to DHA and EPA, other PUFAs exist in fish oils, such as 20:4 and other *n*-6 fatty acids (18:2, 22:4), *n*-9 fatty acids (18:1, 20:1, 22:1), and *n*-7 (16:1) fatty acids. A further important characteristic of fish lipids, which is very complex mixture of metabolites from an analytical point of view, is the distribution of fatty acids on the glycerol backbone in tri-acylglycerols and phospholipids. Extremely important is the finding that the quantity of total lipids in fish lipids may differ between various tissues and organs of the same species, and also between different species. The type and content of fatty acids present as free acid or as lipid is also influenced by the rearing history (hatchery-reared *versus* wild fish) of the species. Finally, perchloric acid extracts of fish tissues contain a number of bioactive compounds (*e.g.* amino acids, sugars, hypoxanthine) that help to protect cells against osmotic stresses and prevent oxidative damage.

Employment of NMR spectroscopy has provided information on all these issues. Relevant NMR studies from literature dealing with the fish oils characterisation along with useful information are summarised in Table 7.5.

These NMR studies were divided into three groups depending on the purpose of the analysis. The first group involves analyses of fish oils targeting to specific metabolites or class of metabolites, *e.g.* cholesterol, fatty acids, phospholipids or bioactive compounds. Studies of the second group identify a number of selected metabolites that are quantified by NMR spectroscopy and used for subsequent classification studies. This method has been described previously as targeted profiling method (see section 6.3). The third group does not require identification/quantification of individual metabolites, but it uses the whole spectrum (spectral fingerprinting) for subsequent statistical analysis. This methodology is similar to chemometric approach described in section 6.4.

Figure 7.17 shows a typical ^{1}H NMR spectrum of extracted lipid from Atlantic salmon.[127] Similar spectra were obtained from other marine species. Signal assignments have been performed by several authors[127–130] and are summarised in Table 7.6. Mannina *et al.*[130] has provided a more detailed assignment of the ^{1}H NMR spectrum of fish oil. The spectrum is consisted of 16

Table 7.5 Selected studies on the application of NMR spectroscopy in the analysis of fish oils.

Type of analysis	Metabolites or spectral fingerprinting	Nucleus/frequency	Application	Comments	Ref.
Target analysis	ω-3 polyunsaturated fatty acids	^{1}H/270, 400 MHz	Spectral assignment, quantification	Samples of fresh, cooked and canned tuna were analysed. Raw and cooked samples showed significantly higher levels of ω-3 polyunsaturated fatty acids.	128
	DHA and ω-3 fatty acids	^{1}H/500 MHz	Quantification	11 refined fish oils from bonito, salmon, and tuna were used. Analysis similar to that of vegetable oils.	134
	Hypoxanthine, amino acids, anserine, lactate, fatty acids	^{1}H/500 MHz	Identification, quality assessment	Use of the MR images of the whole farmed Atlantic salmon to select a specific volume element within the fish white muscle and run localized *in vivo* ^{1}H NMR spectra.	127
	Anserine, lactate	^{13}C/100.6 MHz	Quantification	Samples of minced of Atlantic salmon. Information about the muscle pH was derived from the pH-dependent chemical shifts.	140
	EPA, DHA, ω-3 fatty acids	^{1}H/400 MHz	Acyl chain composition	Fish oils extracted from trout heads, spines and viscera by two extraction techniques with solvents. This study examined the possibility of exploiting trout waste as a source of ω-3 rich oil.	133
	Fatty acids, DHA, EPA, sn-1 MAG, sn-2 and sn-3 DAG, trans FFA, free glycerol, cholesterol, vitamins A, E	^{1}H/400, 600 MHz, ^{13}C/150.9 MHz	Quantification, quality assessment	Encapsulated marine cod liver oil supplements were analysed. Discrimination of natural cod liver oils from those subjected to chemical modification with synthetic derivatives of EPA and DHA.	141

Table 7.5 (*Continued*)

Type of analysis	Metabolites or spectral fingerprinting	Nucleus/frequency	Application	Comments	Ref.
	Fatty acids	^{13}C/125.7 MHz	Quantification	Lipids extracted from the white muscle of Atlantic salmon	142
	ω-3 fatty acids	^{13}C/75.5 MHz	Spectral assignment	Complete assignment using 2D NMR techniques.	143
	Fatty acids	^{13}C/67.9, 100.6 MHz	Spectral assignment, quantification, acyl chains positional distribution	Lipids extracted from the white muscle of Atlantic tuna. Quantification was carried using the glyceridic resonances.	144
	Phospholipids	^{13}C/100.6 MHz	Spectral assignment, acyl chains positional distribution	Commercial samples of tuna were examined. Using of enzymatic hydrolysis of phospholipids facilitated identification of the acyl chains and their positional distribution.	145
	Phospholipids, cholesterol, PUFAs, TAG, MAG, DAG, FFA	^{1}H/600 MHz, ^{13}C/150.9 MHz	Quantification, acyl chains positional distribution	Samples of cod roe and milt.	146
	DHA, EPA	^{13}C/100.6 MHz	Spectral assignment	Commercial EPA and DHA standards were examined.	147
	Wax esters, TAG, phospholipids. FFA, PUFAs, cholesterol	^{13}C/75.5 MHz	Quantification	Commercial bottarga samples were examined.	148
	DHA, EPA, cholesterol, saturated and unsaturated acyl chains	^{1}H/400 MHz	Quantification, acyl chain composition	Evaluation of the proportions of concentrations of some acyl groups including DHA and EPA in cod liver oil. Comparison of the NMR with the IRFT data.	129

	Compounds	Nucleus/Frequency	Study	Description	Ref.
	DHA, different saturated and unsaturated acyl chains including ω-3 PUFA	^{1}H/300 MHz	Quantification, acyl chain composition, oxidative stability	Monitoring the oxidative stability of farmed salmon samples based on changes of various acyl chain proportions in lipids submitted to oxidative conditions. The influence of salting on the oxidation process was examined.	131
Targeted profiling	DHA, EPA, ω-3 PUFA	^{1}H/400, 600 MHz	Quantification, oxidation study	Monitoring the oxidation process of four types of fish oils oxidised under different conditions.	132
Chemometric approach	Chloroform and water soluble metabolites	^{1}H/600 MHz	Grouping with PCA	Discrimination of wild and farmed sea bass species.	130
	1D ^{1}H and 2D ^{1}H J-resolved NMR fingerprints	^{1}H/500 MHz	Comparative toxicity study	Study of metabolite changes in smolts of Chinook salmon induced by crude and dispersed oil.	137
	^{1}H NMR fingerprints	^{1}H/500 MHz	Comparative toxicity study with multivariate statistical analysis	Study of metabolite changes in of Chinook salmon pre-smolts induced by crude and dispersed oil.	138
	Bioactive compounds in perchloric acid extracts	^{1}H/500 MHz	Comparative study	Investigation of changes in bioactive compounds in extracts from white cod muscle under stresses, such as boiling, frying, freezing, freezing after thawing, and rehydration.	135
	^{1}H NMR fingerprints of perchloric acid extracts	^{1}H/400 MHz	Fish species discrimination according to aquaculture	Discrimination of gilthead sea bream according to three different growth system on the basis of robust bio-markers determined in perchloric acid extracts, which in addition provide evidences the aquaculture systems affect the metabolic processes during storage.	139
	Carbonyl region finger-prints of TAG molecules	^{13}C/150.9 MHz	Fish species discrimination	The ^{13}C NMR carbonyl region of TAG molecules and in particular differences in *sn*-2 positional distribution of the ω-3 fatty acid was used to distinguish oils from three different species (Atlantic salmon, mackerel, and herring).	151

Table 7.5 (Continued)

Type of analysis	Metabolites or spectral fingerprinting	Nucleus/frequency	Application	Comments	Ref.
	Carbonyl region finger-prints of phospholipids	^{13}C/150.9 MHz	Fish species discrimination	The ^{13}C NMR carbonyl region of phospholipids and in particular differences in *sn*-2 positional distribution of the ω-3 fatty acid was used to distinguish oils from three different gadoid species.	152
	^{13}C chemical shifts	^{13}C/125.7 MHz	Wild/farmed fish species discrimination and geographical origin classification	Cod liver oils authentication according to wild/farmed and geographical origin (Norway and Scotland) using GC and ^{13}C NMR data.	155
	^{13}C chemical shifts	^{13}C/125.7 MHz	Wild/farmed fish species discrimination	Identification of wild/farmed Atlantic salmon and the farm origin of farmed salmon using GC and ^{13}C NMR data.	156
	^{13}C NMR fingerprints	^{13}C/125.7 MHz	Wild/farmed fish species discrimination and geographical origin classification	Discrimination between wild/farmed Atlantic salmon; discrimination between different geographical locations (Norway, Scotland, Canada, Iceland, Ireland, Faeroes Islands, Tasmania); verification of the origin of market samples.	157

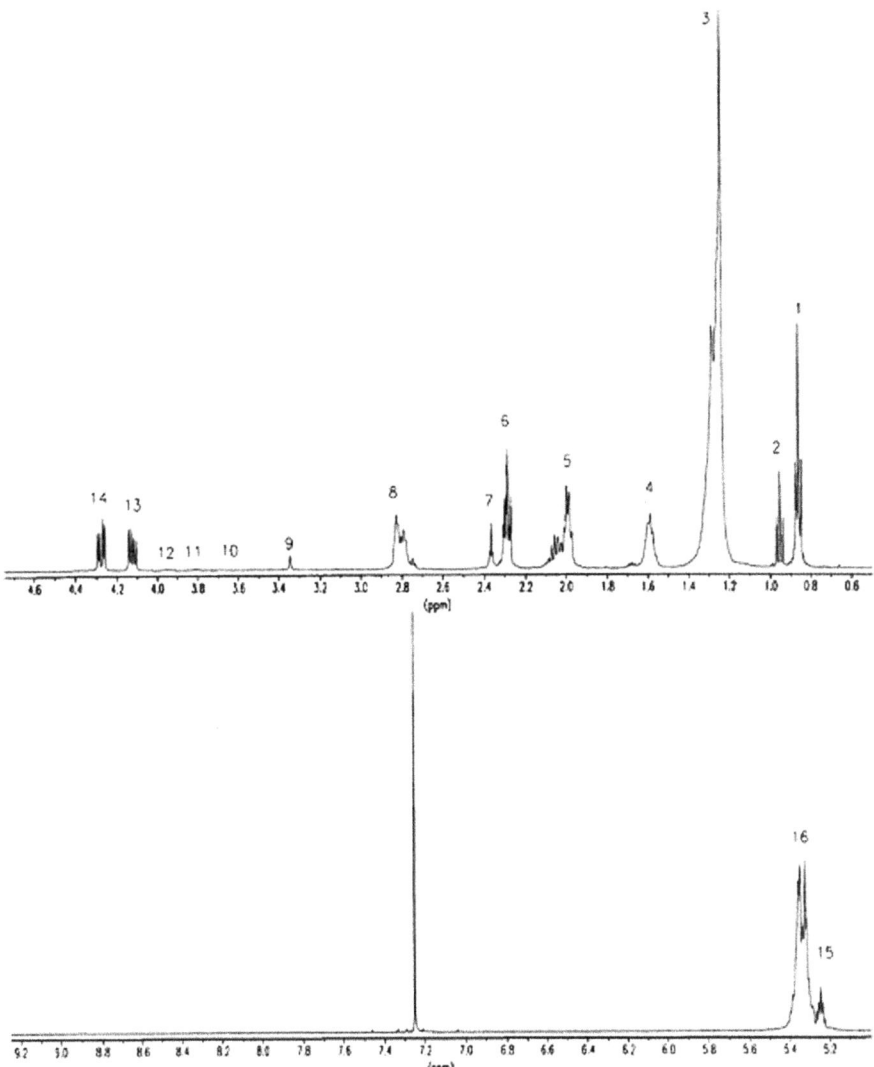

Figure 7.17   *In vitro* ^{1}H MR spectrum of lipid extracted from white muscle of Atlantic
salmon (*Salmo salar*). Assignments for the numbered resonances are
given in Table 7.6.
(Reprinted from ref. 127. Copyright (2005), with permission from Elsevier.)

signals attributed to saturated and unsaturated fatty acids and phospholipids of
fish lipid. Integration of these signals provides detailed information about the
fatty acid composition and in particular about the main ω-3 fatty acids, DHA
and EPA. The spectrum in Figure 7.17 gives information about the relative
proportions of mono-, di-, and triacylglycerols, and estimates of the mean
unsaturation index (UI) (integral ratio of olefinic protons over terminal methyl

Table 7.6 Peak assignments for ^{1}H MR spectrum of lipid extracted from white muscle of Atlantic salmon (*Salmo salar*).[127]

Signal	Compound	Proton	Chemical shifts/ppm
1	All fatty acids except *n*-3 fatty acids	–CH_3	0.85–0.89
2	*n*-3 fatty acids	–CH_3	0.95–0.98
3	All fatty acids except 20 : 5 and 22 : 6	–(CH_2)$_n$–	1.25–1.29
4	All fatty acids except 22 : 6	–CH_2–CH$_2$–COOH	1.60
5	Unsaturated fatty acids	–CH_2–CH=CH–	1.99–2.07
6	All fatty acids except 22 : 6	–CH_2–COOH	2.28
7	22 : 6	=CH–CH_2–CH_2– COOH	2.38
8	Polyunsaturated fatty acids	=CH–CH_2–CH=	2.81–2.84
9	Phosphatidylcholine	–N(CH_3)$_3$	3.35
10	Phospholipid	–	3.64
11	Phospholipid	–	3.80
12	Phosphatidylcholine, phosphatidylethanolamine	Glyceryl moiety	4.00
13	Glyceryl	*sn*-1(3) protons	4.11
14	Glyceryl	*sn*-1(3) protons	4.31
15	Glyceryl	*sn*-2 protons	5.26
16	Unsaturated fatty acids	–CH=CH–	5.34–5.36

The chemical shifts are referenced indirectly to TMS using the peak of CHCl$_3$ (δ 7.26 ppm). (Reprinted from ref. 127. Copyright (2005), with permission from Elsevier.)

protons of all acyl chains), which is a good indicator of the oxidative stability of oil (see section 7.1.3.1). As in vegetable oils, the degradation of the different acyl groups and the formation of oxidation products (hydroxyperoxides and aldehydes) can be observed simultaneously in the ^{1}H spectrum, monitoring thus the oxidation process occurring in fish oils.[131,132] An important industrial application of ^{1}H NMR spectroscopy has been proposed[132] concerning the determination of fatty acids profile and lipid classes of fish waste (heads, spines and viscera), exploiting thus fish waste as a source of ω-3 rich oil. Good agreement was observed between the ^{1}H NMR data and those obtained with gas chromatography.[128,132–134] ^{1}H NMR spectroscopy has been used for the examination of metabolites extracted with perchloric acid. These metabolites belong to different classes, *i.e.* sugars, amino acids, dipeptides, organic acids, *etc.*[127,130,135–138] Assignment of the relevant ^{1}H spectra was performed by using model compounds, 2D NMR experiments and PFG diffusion techniques. Mannina *et al.*[130] has provided the most detailed assignment of these compounds. The role of perchloric acid metabolites to physiological activity of fish has been discussed.[127] These compounds have been used as biomarkers to monitor biochemical changes occurring in seafood submitted to various processing conditions[135] (boiling, frying, freezing, freezing after thawing, rehydration). Also, they have been utilized to study the metabolic effects of dispersed crude oil spills on health and survival of salmon pre-smolt.[137,138] ^{1}H NMR metabolite profiling of lipid and/or perchloric acid extracts, together with suitable statistical analysis were used to discriminate between wild and farmed fish species,[135,136] or between fish farmed with different systems (see below).[139]

^{13}C NMR spectroscopy has been much more used than ^{1}H NMR in studies on fish oils, because of the better dispersion of the ^{13}C chemical shifts, the smaller line widths in proton-decoupled spectra and consequently improved resolution. However, as mentioned in section 5.2, one should be cautious and take the appropriate measures whenever ^{13}C NMR is used for quantitative analysis. A number of studies have used this technique to provide insight into the fatty acid profile, phospholipids and other lipid classes (TAG, DAG, MAG, FFA) in fish oils.[140-148] The ^{13}C NMR spectra of fish oils are much more complicated compared to those of vegetable oils. This is mainly due to the higher complexity of fatty acids and the presence of significant amounts of phospholipids. Figure 7.18 shows the ^{13}C NMR spectrum of lipid extracted from the white muscle of albacore tuna indicating the spectral regions where the various types of carbons resonate.[144]

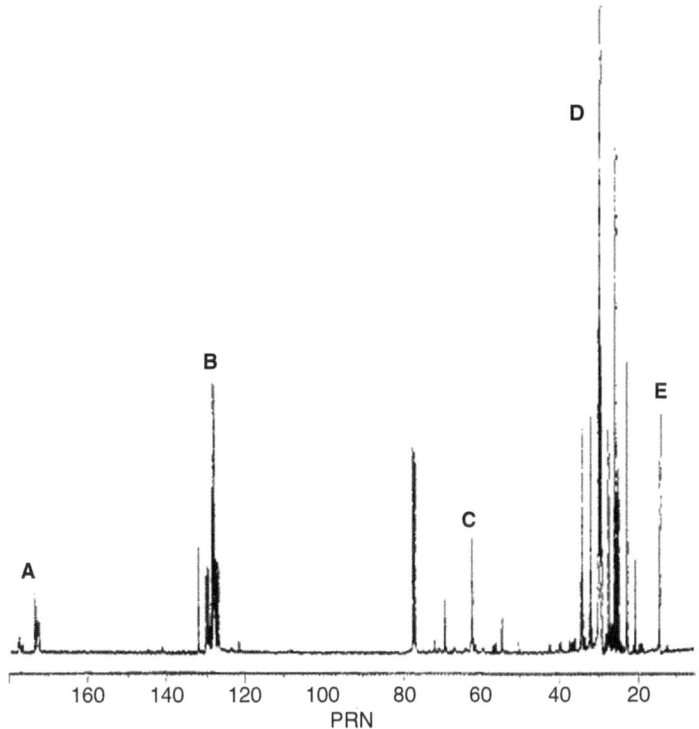

Figure 7.18   ^{13}C NMR spectrum of lipid extracted from the white muscle of alba-core tuna *(T. alalunga)*. Spectral regions are indicated as follows: A = carbonyls; B = olefinic carbons; C = glyceryl carbons; D = methylene carbons; E = methyl carbons.
(Reprinted from ref. 144. Copyright (1993), with permission from American Chemical Society.)

Assignment of the carbon signals, which often are close or higher than hundred, has been attempted in several instances using different approaches, such as standard model compounds, substituent shift parameters, T_1 measurements, chemical shift reagents, and comparison with compositional data from GC analyses.[142,144,145,149] Nevertheless, chemical shift assignment was most successful with the aid of 2D NMR spectroscopy.[143]

Another feature of fish oils analysis is related to positional distribution of the acyl chains on the glycerol backbone of TAG molecules.[144,150,151] Careful examination of the carbonyl and glyceridic regions of the ^{13}C NMR spectra confirmed that DHA showed preferential location in the *sn*-2 glycerol position, but EPA was found more randomly distributed between the *sn*-1 and *sn*-2 positions. The mono-unsaturated fatty acids were found preferentially esterified in the *sn*-1(3) position. Also, ^{13}C NMR spectroscopy provided information about the positional distribution of acyl chains in fish phospholipids.[145,146,152] From the interpretation of the carbonyl spectral region, it was concluded that ω-3 PUFAs was almost totally attached at the *sn*-2 position, of the phospholipids and in particular to the more abundant phosphatidyl choline and phosphatidyl ethanolamine, whereas, saturated and mono-unsaturated fatty acids were esterified at the *sn*-1 position. The spectral fingerprints in the carbonyl region of phospholipids has been used for the authentication of fish species (see below).[152]

7.2.3.1 *Quality Assessment and Authenticity*

Most of the analyses by high-resolution NMR presented previously in brief contribute largely to the quality assessment of the particular fish oils under normal conditions or under stress. However, some specific problems related to quality assessment and authentication can only be resolved, either by using different NMR specialties other than high-resolution NMR, or upon introducing statistical methods in the treatment of NMR metabolic profiles acquired from different sources (*e.g.* oils from different fish species). MRI has been suggested as an effective tool to assess the quality of intact tissues. MR images of the whole salmon were used to localise a specific volume within the intact white muscle of fish and run for this particular volume element *in vivo* ^{1}H NMR spectrum. This spectrum obtained with water-suppression revealed specific metabolites (*e.g.* anserine, lactate) that can provide information about metabolic alterations during growth under different conditions.[127] Discrimination between a freshly killed trout and one which has been frozen and then thawed was accomplished using MRI.[153] The use of various types of MRI experiments enabled the visualisation of the anatomic details of fish, which has favourable implications regarding fish quality. Specific T_1 and T_1^{sat}-weighted MRI experiments, and the observed changes in the measured magnetisation transfer (MT) parameters were more sensitive to the effects of freeze thawing, indicating a method that could be used to check the fish freshness. Figure 7.19 shows a series of MR images of the head of the rainbow trout obtained under different image contrast regimes.[153]

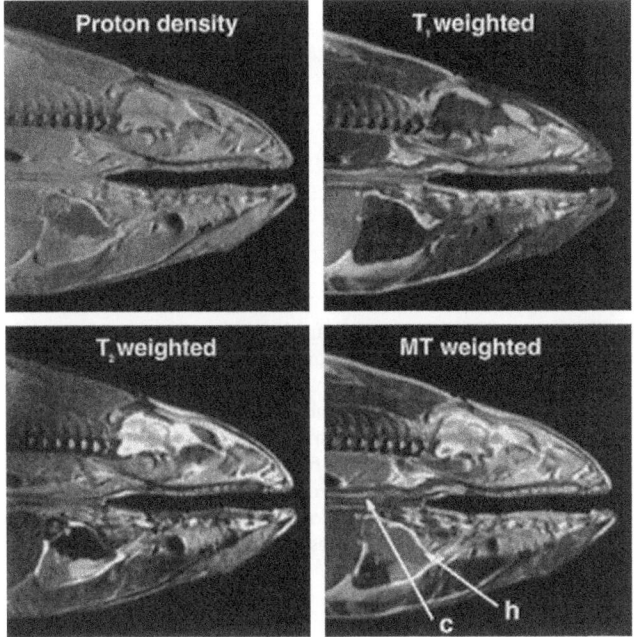

Figure 7.19 Sagittal images of the head of a rainbow trout under various imaging contrasts; proton density, T_1-weighted, T_2-weighted, MT-weighted Anatomical features: (c) oesophagus; (h) heart.
(Reprinted from ref. 153. Copyright (1999), with permission from Elsevier.)

The T_1-weighed image highlighted the lipids with respect to the water containing tissues owing to their shorter T_1 values. The T_2- and MT-contrasted images, showed the greatest distinction between different muscle tissues; one image shows at least three of the four heart chambers. The proton density map gave little tissue contrast because all the tissues contain approximately the same concentration of water. Another possibility to study fish tissues without oil extraction is the employment of the high-resolution magic angle spinning (HR-MAS) technique. Castejón *et al.*[154] developed a method using the HRMAS methodology to rapidly detect the presence of the main components of intact muscle of smoked salmon. The HRMAS spectra were fully assigned by using 2D NMR techniques. Apart from fatty acids, metabolites such as carbohydrates, nucleoside derivatives, osmolytes, amino acids, dipeptides and organic acids, in total 160 metabolites, were determined in intact muscle. The sufficient resolution of the HRMAS spectra allowed for the first time the semi-quantitative determination in intact muscle of the highly polyunsaturated DHA, whereas the ^{1}H-HRMAS NMR metabolite profiling was tested to identify changes of some bioactive compounds during smoked Atlantic salmon storage. The same methodology was employed to investigate the metabolic profiles from intact muscles and livers of Atlantic salmon.[135]

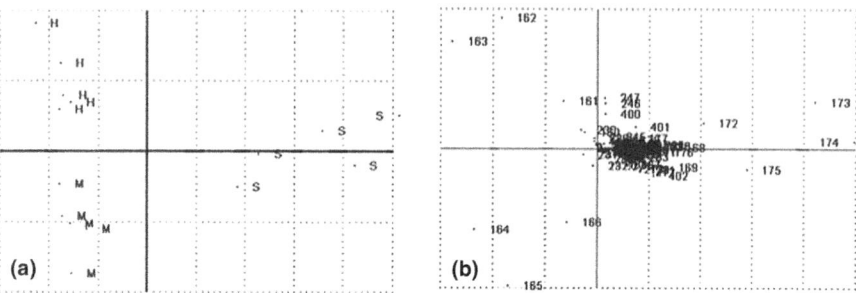

Figure 7.20   PC1 *versus* PC2 scores (a) and loadings plot (b) obtained from 400 data
points in the carbonyl-region of the ^{13}C-NMR spectra of lipids from
salmon (S), mackerel (M) and herring (H). The two-first principal
components explained 49 and 27% of the variance in the dataset. Variable number 160–175 corresponds to the region 172.9–172.8 ppm where
resonances from monounsaturated and saturated fatty acids (MUFAs
and SFAs) in the *sn*-2 position dominate.
(Reprinted from ref. 151. Copyright (2009), with kind permission from
Springer Science and Business Media.)

Several NMR methodologies for authentication of fish oils use NMR
metabolic profiles or NMR fingerprints in combination with multivariate statistical analyses. The general trend in these studies is the discrimination between
fish oils extracted from different fish species or differentiation of fish oils
according to wild/farmed species and geographical origin. The ^{13}C NMR fingerprints of carbonyl region of TAG molecules were used to distinguish oils
from three different species (Atlantic salmon, mackerel, and herring).[151] The
region specific analysis showed that there were significant differences in *sn*-2
positional distribution of the ω-3 fatty acid among the different species; this
finding was confirmed by subsequent ANOVA calculations. Application of
PCA on the carbonyl fingerprint succeeded to group the sample of the three fish
species as shown in Figure 7.20.[151]

Following the same methodology, but using the phospholipid fingerprint in
the carbonyl region of muscle lipids obtained by ^{13}C NMR spectroscopy,
authentication of different gadoid fish species (north-east arctic cod, Norwegian coastal cod, haddock, saithe, and pollack) was attempted.[152] The PCA
treatment of the data set (chemical shifts) was partially successful in separating
the five gadoic species. Subsequent application of LDA showed that 21 samples
out of 27 samples were correctly classified (78%), whereas the more powerful
statistical method of probabilistic neural network gave 100% correct
classification.

Mannina *et al.*[136] attempted to discriminate between wild and cultured sea
bass using the chloroform (liposoluble) and water-soluble metabolites determined in the ^{1}H NMR spectra. Satisfactory discrimination between wild and
farmed fish was obtained upon application of PCA to 29 NMR signal intensities selected by ANOVA. The plot of PCA loadings indicated that 13

variables were significantly different in wild and cultured sea bass including DHA, EPA, PUFA, and phospholipids. Standal *et al.*[155] employing GC and ^{13}C NMR techniques made possible the discrimination of cod liver oils extracted from wild and farmed fish originated from different locations in Norway and Scotland. PCA performed first on 123 ^{13}C chemical shifts and secondly on GC compositional data showed both a clear separation between wild and farmed fish and between the different farms and catch areas. Subsequent LDA application with cross-validation to the most discriminating chemical shifts selected by partial least squares resulted in wild/farmed classification rates 97 and 100% for GC and NMR data, respectively. However, better separation according to geographical origin was achieved by ^{13}C NMR (95%) rather than GC (63%). Martinez *et al.*[156] performed an analogous study in order to identify wild and farmed Atlantic salmon. ^{13}C NMR and GC data of 59 specimens from four different farms in Norway and 17 free-living fish were subjected to statistical treatment by four statistical methods. Among these, the supervised methods of probabilistic neural network (PNN) and the support vector machines (SVM) were the best methods to classify correctly 58 and 56 out of 59 samples according to their farm of origin using GC and NMR data, respectively. Statistical analysis of GC and NMR data obtained from the wild fish were less successful. In a subsequent publication,[157] the same group using ^{13}C NMR fingerprints of Atlantic salmon originated from seven countries (Norway, Scotland, Canada, Ireland, Faroe island, and Tasmania) in combination with the statistical methods PNN and SVM succeeded in discriminating wild and farmed fish with classification rates ranging from 98.5 to 100%, respectively. Predictions regarding the geographical origin of fish specimens were found to be somewhat more difficult, obtaining 82% correct classification with the PNN statistical method and 99% with the SVM method. Finally, the ^{1}H NMR metabolic profile of gilthead sea bream from three different aquaculture systems determined in combination with chemometrics was utilised in order to understand and interpret metabolic changes occurring as a consequence of the farming system.[139] Metabolites such as inosine, inosine 5'-monophosphate, histidine, glycine, alanine and glycogen proved to be reliable biomarkers for distinguishing amongst the farming systems and for providing information about the fish storage time.

References

1. *Vegetable Oils in Food Technology: Composition, Properties and Uses*, ed. F. D. Gunstone, Blackwell Publishing, Oxford, UK, 2002.
2. M. Bockisch, *Fats and Oils Handbook*, AOCS Press, Champaign, IL, USA, 1998, ch. 4.
3. M. D. Guillèn and A. Ruiz, *Trends Food Sci. Technol.*, 2001, **12**, 328.
4. M. D. Guillèn and A. Ruiz, *Eur. J. Lipid Sci. Technol.*, 2003, **105**, 688.
5. G. Knothe and J. A. Kenar, *Eur. J. Lipid Sci. Technol.*, 2004, **106**, 88.
6. R. Sacchi, M. Patumi, G. Fontanazza, P. Barone, P. Fiordiponti, L. Mannina, E. Rossi and A. Segre, *J. Am. Oil Chem. Soc.*, 1996, **73**, 747.

7. G. Vigli, A. Philippidis, A. Spyros and P. Dais, *J. Agric. Food Chem.*, 2003, **51**, 5715.

8. A. L. Segre and L. Mannina, *Recent Res. Dev. Oil Chem.*, 1997, **1**, 297.

9. Y. Miyake, K. Yokomizo and N. Matzuzaki, *J. Am. Oil Chem. Soc.*, 1986, **75**, 1091.

10. M. D. Guillèn and A. Ruiz, *J. Sci. Food Agric.*, 2003, **83**, 338.

11. P. Dais, A. Spyros, S. Christophoridou, E. Hatzakis, G. Fragaki, A. Agiomyrgianaki, E. Salivaras, G. Siragakis, M. Tasioula-Margari and M. Brenes, *J. Agric. Food Chem.*, 2007, **55**, 577.

12. F. D. Gunstone, in *Advances in Lipid Methodology*, ed. W. W. Cristie, The Oily Press, Dundee, UK, 1993, pp. 1.

13. L. Mannina, C. Luchinat, M. C. Emanuele and A. L. Segre, *Chem. Phys. Lipids*, 1999, **103**, 47.

14. A. D. Shaw, A. di Camillo and G. Vlahov, *Anal. Chim. Acta*, 1997, **348**, 357.

15. R. Zamora, V. Alba and F. J. Hidalgo, *J. Am. Oil Chem. Soc.*, 2001, **78**, 89.

16. L. Mannina, C. Luminat, M. Patumi, M. C. Emanuele and A. L. Segre, *Magn. Reson. Chem.*, 2000, **38**, 886.

17. S. Simonova, G. Ivanova and S. L. Spassov, *Chem. Phys. Lipids*, 2003, **126**, 167.

18. F. H. Mattson and R. A. Volpenhein, *J. Lipid Res.*, 1963, **4**, 392.

19. G. Vlahov, *Progr. NMR Spectrosc.*, 1999, **35**, 341.

20. P. Dais, in *Handbook of Olive Oil*, ed. Ramon Aparicio, ch. 11, in press.

21. K. F. Wollenberg, *J. Am. Oil Chem. Soc.*, 1990, **67**, 487.

22. S. Ng and W. L. Ng, *J. Am. Oil Chem. Soc.*, 1983, **60**, 1266.

23. L. Retief, J. M. McKenzie and K. R. Koch, *Magn. Reson. Chem.*, 2009, **47**, 271.

24. L. Mannina and A. Segre, *Grasas Aceites*, 2002, **53**, 22.

25. R. M. Alonso-Salces, K. Hiberger, M. V. Holland, J. M. Moreno-Rojas, C. Mariani, G. Bellan, F. Reniero and C. Guillou, *Food Chem.*, 2010, **118**, 956.

26. E. Hatzakis, A. Agiomyrgianaki, S. Kostidis and P. Dais, *J. Am. Oil Chem. Soc.*, 2011, **88**, 1695.

27. R. Sacchi, L. Paolillo, I. Giudicianni and F. Addeo, *Ital. J. Food Sci.*, 1991, **4**, 245.

28. P. Sacchi, F. Addeo, I. Giudicianni and L. Paolillo, *Riv. Ital. Sostanze Grasse*, 1990, **67**, 245.

29. G. Vlahov, *J. Am. Oil Chem. Soc.*, 1996, **73**, 1201.

30. F. D. Gunstone, *Chem. Phys. Lipids*, 1991, **58**, 219.

31. P. Dais and A. Spyros, *J. Agric. Food Chem.*, 2000, **48**, 802.

32. P. Fronimaki, A. Spyros, S. Christophoridou and P. Dais, *J. Agric. Food Chem.*, 2002, **50**, 2207.

33. F. M. Dayrit, O. E. M. Buenafe, E. T. Chainani and I. M. S. de Vera, *J. Agric. Food Chem.*, 2008, **56**, 5765.

34. E. Hatzakis, G. Dagounakis, A. Agiomyrgianaki and P. Dais, *Food Chem.*, 2010, **122**, 346.

35. R. W. Owen, W. Mier, A. Giacosa, W. E. Hull, B. Spiegelhalder and H. Bartsch, *Clin. Chem.*, 2000, **46**, 976.

36. R. W. Owen, R. Haubner, W. Mier, A. Giacosa, W. E. Hull, B. Spiegelhalder and H. Bartsch, *Food Chem. Toxicol.*, 2003, **41**, 703.

37. R. Limiroli, R. Consonni, G. Ottolina, V. Marsilio, G. Bianchi and L. Zetta, *J. Chem. Soc., Perkin Trans*, 1995, 1519.

38. G. Montedoro, M. Servili, M. Baldioli, R. Selvaggini, E. Miniati and A. Macchioni, *J. Agric. Food Chem.*, 1993, **41**, 2228.

39. P. Garibaldi, G. Jommi and L. Verrota, *Phytochemistry*, 1986, **25**, 865.

40. A. Bianco and N. Uccella, *Food Res. Int.*, 2000, **33**, 475.

41. S. Christophoridou, P. Dais, L.-H. Tseng and M. Spraul, *J. Agric. Food Chem.*, 2005, **53**, 4667.

42. S. Christophoridou and P. Dais, *Anal. Chim. Acta*, 2009, **633**, 283.

43. S. Christophoridou and P. Dais, *J Agric. Food Chem.*, 2006, **54**, 656.

44. S. Christophoridou, A. Spyros and P. Dais, *Phosphorus Sulfur Silicon Relat. Elem.*, 2001, **170**, 139.

45. J. M. Pearce and R. A. Komoroski, *Magn. Reson. Med.*, 1993, **29**, 724.

46. P. Meneses and T. Glonek, *J. Lipid Res.*, 1988, **29**, 679.

47. A. Puppato, D. B. DuPre, N. Stolowich and M. C. Yappert, *Chem. Phys. Lipids*, 2007, **150**, 176.

48. E. Hatzakis, A. Koidis, D. Boskou and P. Dais, *J. Agric. Food Chem.*, 2008, **56**, 6232.

49. G. Helmerich and P. Koehler, *J. Agric. Food Chem.*, 2003, **51**, 6645.

50. H. Saito, *Agric. Biol. Chem.*, 1987, **51**, 3432.

51. C. Skiers, P. Steliopoulos, T. Kuballa, U. Holzgrabe and B. Diehl, *J. Am. Oil Chem. Soc.*, 2012, **89**, 1383.

52. Y. Miyake, K. Yokomizo and N. Matzuzaki, *J. Am. Oil Chem. Soc.*, 1998, **75**, 15.

53. S. Ng and P. T. Gee, *Eur. J. Lipid Sci. Technol.*, 2001, **103**, 223.

54. M. D. Guillèn and E. Goicoechea, *J. Agric. Food Chem.*, 2007, **55**, 10729.

55. M. D. Guillèn and A. Ruiz, *Food Chem.*, 2006, **56**, 665.

56. M. D. Guillèn and A. Ruiz, *Eur. J. Lipid Sci. Technol.*, 2005, **107**, 36.

57. M. D. Guillèn and A. Ruiz, *Eur. J. Lipid Sci. Technol.*, 2004, **106**, 680.

58. A. W. D. Claxon, G. E. Hawkes, D. P. Richardson, D. P. Naughton, R. M. Haywood, C. L. Chander, M. Atherton, E. J. Lynch and M. C. Grootveld, *FEBS Lett.*, 1994, **355**, 81.

59. E. N. Frankel, *Lipid Oxidation*, The Oily Press, Dundee, UK, 1998.

60. AOCS, *The Lipid Library. Lipid Chemistry, Biology, Technology & Analysis*, ed. W. W. Cristie, http://lipidlibrary.aocs.org/index.html.

61. F. J. Hidalgo, G. Gómez, J. L. Navarro and R. Zamora, *J. Agric. Food Chem.*, 2002, **50**, 5825.

62. R. Zamora, G. Gómmez and F. J. Hidalgo, *J. Am. Oil Chem. Soc.*, 2002, **79**, 267.

63. R. N. Alonso-Salces, M.V. Holland and C. Guillou, *Food Control*, 2011, **22**, 2041.

64. E. Hatzakis and P. Dais, *J. Agric. Food Chem.*, 2008, **56**, 1866.

65. R. Sacchi, F. Adeo and L. Paollilo, *Magn. Reson. Chem.*, 1997, **35**, S133.
66. A. Spyros, A. Philippidis and P. Dais, *J. Agric. Food Chem.*, 2004, **52**, 157.
67. AOCS, *Official Methods and Recommended Practices of the American Oil Chemists' Society*, ed. D. Firestone, AOCS Press, Champaign, IL, USA, 1999, Recommended Practice Ak 4–95; AOCS, Oil content in oilseeds, *ibid*, Official method Am 2–93; AOCS, Oil content of rapeseed by nuclear magnetic resonance, *ibid*, Recommended Practice Ak 3–94.
68. P. N. Tiwari, P. N. Gambhir and T. S. Rajan, *J. Am. Oil Chem. Soc.*, 1974, **51**, 104.
69. P. N. Tiwari, *Bruker Minispec Applications Notes*, Bruker Analytische Messtechnik, Karlsruhe, Germany, 1999.
70. P. N. Gambhir, *Trends Food Sci. Technol.*, 1992, **3**, 191.
71. H. Todt, W. Burk, G. Guthausen, A. Guthausen, A. Kamlowski and D. Schmalbein, *Eur. J. Lipid Sci. Technol.*, 2001, **103**, 835.
72. J. G. Seland, G. H. Sorland, H. W. Anthonsen and J. Krane, *Appl. Magn. Reson.*, 2003, **24**, 41.
73. H. T. Pedersen, L. Munck and S. B. Engelsen, *J. Am. Oil Chem. Soc.*, 2000, **77**, 1069.
74. X. Shao and Y. Li, *Food Bioprocess Technol.*, 2012, **5**(5), 1817.
75. J. J. Kotyk, M. D. Pagel, K. L. Deppermann, R. F. Colletti, N. G. Hoffman, E. J. Yannakakis, P. K. Das and J. J. H. Ackerman, *J. Am. Oil Chem. Soc.*, 2005, **82**, 855.
76. S. M. D. Shaarani, A. Cárdenas-Blanco, M. H. Gao Amin, N. G. Soon and L. D. Hall, *Int. J. Agric. Biol.*, 2010, **12**, 101.
77. M. A. Brescia, T. Pugliese, E. Hardy and A. Sacco, *Food Chem.*, 2007, **105**, 400.
78. M. Gussoni, F. Greco, R. Consonni, H. Molinari, G. Zannoni, G. Bianchi and L. Zetta, *Magn. Reson. Imag.*, 1993, **2**, 259.
79. J. R. Morello, M. P. Romero and M. J. Motilva, *J. Am. Oil Chem. Soc.*, 2006, **83**, 683.
80. M. P. Romero, M. J. Tovar, T. Ramos and M. J. Moltilva, *J. Am. Oil Chem. Soc.*, 2003, **80**, 423.
81. A. Agiomyrgianaki, P. V. Petrakis and P. Dais, *Food Chem.*, 2012, **135**, 2561.
82. M. D'Imperio, M. Gobbino, A. Picanza, S. Constanzo, A. D. Corte and L. Mannina, *J. Agric. Food Chem.*, 2010, **58**, 11043.
83. T. Mavromoustakos, M. Zervou, E. Theodoropoulou, D. Panagiotopoulos, G. Bonas, M. Day and A. Helmis, *Magn. Reson. Chem.*, 1997, **35**, S3.
84. C. Fauhl, F. Reniero and C. Guillou, *Magn. Reson. Chem.*, 2000, **38**, 436.
85. G. Fragaki, A. Spyros, G. Siragakis, E. Salivaras and P. Dais, *J. Agric. Food Chem.*, 2005, **53**, 2810.
86. A. Agiomyrgianaki, P. V. Petrakis and P. Dais, *Talanta*, 2010, **80**, 2165.
87. R. Zamora, J. L. Navarro and F. J. Hidalgo, *J. Am. Oil Chem. Soc.*, 1994, **71**, 361.

88. L. Mannina, L. M. Patumi, P. Fiordiponti, M. C. Emanuele and A. L. Segre, *Ital. J. Food Sci.*, 1999, **11**, 139.
89. G. Vlahov, *Anal. Chim. Acta*, 2006, **577**, 281.
90. D. L. García-González, L. Mannina, M. D'Imperio, A. L. Segre and R. Aparicio, *Eur. Food Res. Technol.*, 2004, **219**, 545.
91. L. Mannina, M. D'Imperio, D. Capitani, S. Rezzi, C. Guillou, T. Mavromoustakos, M. D. M. Vichez, A. H. Fernandez, F. Thomas and R. Aparicio, *J. Agric. Food Chem.*, 2009, **57**, 11550.
92. G. Vlahov, *Magn. Reson. Chem.*, 1997, **35**, S8.
93. D. Šmejkalová and A. Piccolo, *Food Chem.*, 2010, **118**, 153.
94. L. Mannina, M. Patumi, N. Proietti, D. Bassi and A. L. Segre, *J. Agric. Food Chem.*, 2001, **49**, 2687.
95. L. Mannina, G. Fontanazza, M. Patumi, G. Ansanelli and A. L. Segre, *Grasa Aceites*, 2001, **6**, 380.
96. L. Mannina, M. Patumi, N. Proietti and A. L. Segre, *Ital. J. Food Sci.*, 2001, **13**, 53.
97. L. Mannina, G. Dugo, F. Salvo, L. Cicero, G. Ansanelli, C. Calcagni and A. L. Segre, *J. Agric. Food Chem.*, 2003, **51**, 120.
98. R. Sacchi, L. Mannina, P. Fiordiponti, P. Barone, L. Paollilo, M. Patumi and A. L. Segre, *J. Agric. Food Chem.*, 1998, **46**, 3947.
99. G. Vlahov, P. Del Re and N. Simone, *J. Agric. Food Chem.*, 2003, **51**, 5612.
100. L. Mannina, M. D'Imperio, R. Lava, E. Schievano and S. Mammi, *Riv. Ital. Sostanza Grasse*, 2005, **LXXX11**, 59.
101. M. D'Imperio, L. Mannina, D. Capitani, O. Bidet, E. Rossi, F. M. Buccarelli, G. B. Quaglia and A. L. Segre, *Food Chem.*, 2007, **105**, 1256.
102. P. V. Petrakis, A. Agiomyrgianaki, S. Christophoridou, A. Spyros and P. Dais, *J. Agric. Food Chem.*, 2008, **56**, 3200.
103. L. Mannina, F. Marini, M. Gobbino, A. P. Sobolev and D. Capitani, *Talanta*, 2010, **80**, 2141.
104. R. M. Alonso-Salces, J. M. Moreno-Rojas, M. V. Holland, F. Reniero, C. Guillou and K. Hèrberger, *J. Agric. Food. Chem.*, 2010, **58**, 5586.
105. S. Rezzi, D. E. Axelso, K. Hèberger, F. Reniero, C. Mariani and C. Guillou, *Anal. Chim. Acta*, 2005, **552**, 13.
106. G. Vlahov, A. D. Shaw and D. B. Kell, *J. Am. Oil Chem. Soc.*, 1999, **76**, 1223.
107. A. Sacco, M. A. Brescia, V. Liuzzi, F. Reniero, C. Guillou, S. Ghelli and P. van der Meer, *J. Am. Oil Chem. Soc*, 2000, **77**, 619.
108. G. Vlahov, C. Schiavone and N. Simone, *Magn. Reson. Chem.*, 2001, **39**, 689.
109. M. A. Brescia, G. Alviti, V. Liuzzi and A. Sacco, *J. Am. Oil Chem. Soc.*, 2003, **80**, 945.
110. E. Schievano, I. Arosio, R. Lava, V. Simionato, S. Mammi and R. Consonni, *Riv. Ital. Sostanza Grasse*, 2006, **LXXX111**, 14.
111. F. Longobardi, A. Ventrella, C. Napoli, E. Humpfer, B. Schótz, H. Schofer, M. G. Kontominas and A. Sacco, *Food Chem.*, 2012, **130**, 177.

112. M. A. Brescia, V. Mazzilli, A. Sgaramella, S. Ghelli, F. P. Fanizzi and A. Sacco, *J. Am. Oil Chem. Soc.*, 2004, **81**, 411.

113. G. Andreotti, E. Trivellone, R. Lamanna, A. Di Luccia and A. Motta, *J. Dairy Sci.*, 2000, **83**, 2432.

114. G. Andreotti, R. Lamanna, E. Trivellone and A. Motta, *J. Am. Oil Chem. Soc.*, 2002, **79**, 123.

115. P. Kalo, A. Kemppinen and I. Kilpelainen, *Lipids*, 1996, **31**, 331.

116. J. Schripsema, *J. Agric. Food Chem.*, 2008, **56**, 2547.

117. G. Andreotti, E. Trivellone and A. Motta, *J. Food Comp. Anal.*, 2006, **19**, 843.

118. S. Murgia, S. Mele and M. Monduzzi, *Lipids*, 2003, **38**, 585.

119. K. van Putte and J. van den Enden, *J. Am. Oil Chem. Soc.*, 1973, **51**, 318.

120. AOCS, Solid fat content (SFC) by low-resolution magnetic resonance, *Official Method Cd 16b*, 1993.

121. E. Trezza, A. M. Haiduc, G. J. W. Goudappel and J. P. M. van Duynhoven, *Magn. Reson. Chem.*, 2006, **44**, 1023.

122. B. Breitschuh and E. J. Windhab, *J. Am. Oil Chem. Soc.*, 1998, **75**, 897.

123. J. van Duynhoven, I. Dubourg, G.-J. Goudappel and E. Roijers, *J. Am. Oil Chem. Soc.*, 2002, **79**, 383.

124. F. D. Gunstone, *J. Am. Oil Chem. Soc.*, 1993, **70**, 965.

125. F. D. Gunstone, *J. Am. Oil Chem. Soc.*, 1993, **70**, 361.

126. R. Sacchi, F. Addeo, S. Spagna Musso, L. Paolillo and I. Giudicianni, *Ital. J. Food Sci.*, 1995, **1**, 27.

127. I. S. Gribbestad, M. Aursandc and I. Martinez, *Aquaculture*, 2005, **250**, 445.

128. R. Sacchi, I. Medina, S. P. Aubourg, F. Addeo and L. Paolillo, *J. Am. Oil Chem. Soc.*, 1993, **70**, 225.

129. M. D. Guillèn, I. Carton, E. Coicoechea and P. S. Uriarte, *J. Agric. Food Chem.*, 2008, **56**, 9072.

130. L. Mannina, A. P. Sobolev, D. Capitani, N. Iaffaldano, M. P. Rosato, P. Ragni, A. Reale, E. Sorrentino, I. D'Amico and R. Coppola, *Talanta*, 2008, **7**, 433.

131. M. D. Guillèn and A. Ruiz, *Food Chem.*, 2004, **86**, 297.

132. C. E. Tyl, L. Brecker and K.-H. Wagner, *Eur. J. Lipid Sci. Technol.*, 2008, **110**, 141.

133. L. Fiori, M. Solana, P. Tosi, M. Manfrini, C. Strim and G. Guella, *Food Chem.*, 2012, **134**, 1088.

134. T. Igarashi, M. Aursan, Y. Hirata, I. S. Gribbestad, S. Wada and M. Nonaka, *J. Am. Oil Chem. Soc.*, 2000, **77**, 737.

135. I. Martinez, T. Bathen, I. B. Standal, J. Halvorsen, M. Aursand, I. S. Gribbestad and D. E. Axelson, *J. Agric. Food Chem.*, 2005, **53**, 6889.

136. J. Bankefors, M. Kaszowska, C. Schlechtriem, J. Pickova, E. Brönnös, L. Edebo, A. Kiessling and C. Sandström, *Food Chem.*, 2011, **129**, 1397.

137. C. Y. Lin, B. S. Anderson, B. M. Phillips, A. C. Peng, S. Clark, J. Voorhees, H.-D. I. Wu, M. J. Martin, J. McCalld, C. R. Todd, F. Hsieh,

D. Crane, M. R. Viant, M. L. Sowby and R. S. Tjeerdema, *Aquat. Toxicol.*, 2009, **95**, 230.

138. A. R. Van Scoy, C. Y. Lin, B. S. Anderson, B. M. Phillips, M. J. Martin, J. McCalld, C. R. Todd, D. Crane, M. L. Sowby, M. R. Viant and R. S. Tjeerdema, *Ecotox. Environ. Safe.*, 2010, **73**, 710.

139. F. Savorani, G. Picone, A. Badiani, P. Fagioli, F. Capozzi and S. B. Engelsen, *Food Chem.*, 2010, **120**, 907.

140. M. Arsand, L. Jørgensen and H. Grasdalen, *Comp. Biochem. Physiol.*, 1995, **112B**, 315.

141. N. Siddiqui, J. Sim, C. J. L. Silwood, H. Toms, R. A. Iles and M. Grootveld, *J. Lipid Res.*, 2003, **44**, 2406.

142. M. Aursand and H. Grasdalen, *Chem. Phys. Lipids*, 1992, **62**, 239.

143. J.-M. Vatèle, B. Fenet and T. Eynard, *Chem. Phys. Lipids*, 1998, **94**, 239.

144. R. Sacchi, I. Medina, S. P. Aubourg, I. Giudicianni, L. Paolillo and F. Addeo, *J. Agric. Food Chem.*, 1993, **41**, 1247.

145. I. Medina and R. Sacchi, *Chem. Phys. Lipids*, 1994, **70**, 53.

146. E. Falch, T. R. Storseth and M. Aursand, *Chem. Phys. Lipids*, 2006, **144**, 4.

147. R. Sacchi, I. Medina, L. Paolillo and F. Addeo, *Chem. Phys. Lipids*, 1994, **69**, 65.

148. P. Scano, A. Rosa, F. C. Marincola, E. Locci, M. P. Melis, M. A. Dessi and A. Lai, *Chem. Phys. Lipids*, 2008, **151**, 69.

149. F. D. Gunstone, *Chem. Phys. Lipids*, 1990, **56**, 201; F. D. Gunstone, *Chem. Phys. Lipids*, 1990, **56**, 227; F. D. Gunstone, *Chem. Phys. Lipids*, 1991, **59**, 83; F. D. Gunstone, *Chem. Phys. Lipids*, 1990, **56**, 201.

150. F. D. Gunstone and S. Seth, *Chem. Phys. Lipids*, 1994, **72**, 119.

151. I. B. Standal, D. E. Axelson and M. Aursand, *J. Am. Oil Chem. Soc.*, 2009, **86**, 401.

152. I. B. Standal, D. E. Axelson and M. Aursand, *Food Chem.*, 2010, **121**, 608.

153. K. P. Nott, S. D. Evans and L. D. Hall, *Magn. Reson. Imag.*, 1999, **17**, 445.

154. D. Castejón, P. Villa, M. M. Calvo, G. Santa-María, M. Herraiz and A. Herrera, *Magn. Reson. Chem.*, 2010, **48**, 693.

155. I. B. Standal, A. Praël, L. McEvoy, D. E. Axelson and M. Aursand, *J. Am. Oil Chem. Soc.*, 2008, **85**, 105.

156. I. Martinez, I. B. Standal, D. E. Axelson, B. Finstad and M. Aursand, *Food Chem.*, 2009, **116**, 766.

157. M. Aursad, I. B. Standal, A. Praël, L. McEvoy, J. Irvine and D. E. Axelson, *J. Agric. Food Chem.*, 2009, **57**, 3444.

CHAPTER 8
Wine and Beverages

8.1 Alcoholic Beverages

8.1.1 Wine

From a chemical point of view, wine is essentially a concentrated solution of ethanol in water that also contains a wide range of organic compounds, such as organic acids, amino acids, sugars and aromatic molecules in varying concentrations. It is not surprising therefore that the first applications of NMR spectroscopy in wine analysis were attempts to measure the amount of ethanol in wine and other spirits by ^{1}H NMR, which provided good correlation with density[1] or chemical analysis.[2] Obtaining the ^{1}H NMR spectrum of wine is quite simple and requires only the addition of a small amount of D_2O (100 mL in 400–600 μL of wine) for locking. Figure 8.1 presents the ^{1}H NMR spectra of several different bottled wines from China, namely a dry red (a) and vertically enlarged in (b), a dry white (c), a medium dry white (d) and a blended wine (e).[3]

Although the water peak is successfully saturated, the spectra are dominated by the two huge ethanol peaks, since the samples were studied directly, and without any pre-concentration or pH adjustment step. Major peaks originating from methanol, succinate, acetate, α- and β-glucose, sucrose and hippurate (in Figure 8.1(e)) are labelled on the spectra. Hippuric acid and sucrose are additives and are only present in the spectrum of the blended wine. A total of 39 different compounds,[3] including organic acids, amino acids and alcohols were identified in the ^{1}H spectra of wine with the help of 2D NMR spectroscopy. These compounds form the basis set for wine metabolomics applications, as discussed in detail later in this chapter. Figure 8.2 presents in greater detail the assignment of the major amino acids identified in the ^{1}H NMR spectrum of wine.[4]

The determination of chemical compounds present in low concentration in wine is facilitated by the use of multiple signal suppression techniques, such as WET (water suppression enhanced through T_1 effects), that help to take

RSC Food Analysis Monographs No. 10
NMR Spectroscopy in Food Analysis
By Apostolos Spyros and Photis Dais
© Apostolos Spyros and Photis Dais 2013
Published by the Royal Society of Chemistry, www.rsc.org

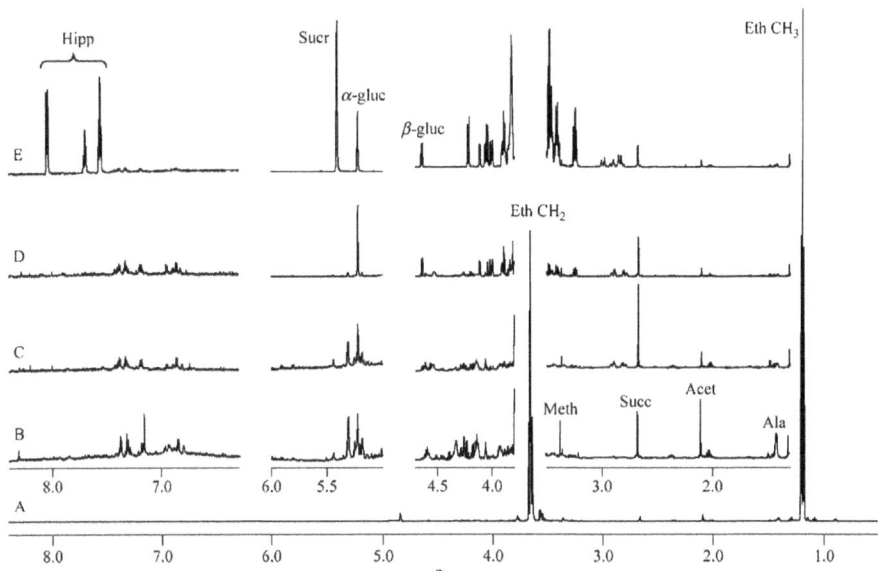

Figure 8.1 ^{1}H NMR spectrum of dry red wine (a), and partially enlarged ^{1}H NMR spectra of dry red wine (b), dry white wine (c), medium dry white wine (d) and blended wine (e). Assignment of major resonances is labelled: Eth = ethanol; Acet = acetic acid; Ala = alanine; Succ = succinic acid; Meth = methanol; α/β-gluc = α/β-glucose; Sucr = sucrose; Hipp = hippurate.
(Reprinted with from ref. 3. Copyright (2007), with permission from John Wiley and Sons.)

advantage of the full dynamic range of the spectrometer receiver and allow the weak proton signals of these compounds to be better digitised.[5] Figure 8.3 presents the ^{1}H NMR spectra of wine without any saturation (a) and using the WET sequence to saturate the water, ethanol and glycerol signals (b).

Both spectra were acquired and processed under the same conditions and with the same vertical scale amplification, demonstrating the increased analytical capabilities of multiple pre-saturation. A further methodological improvement is the use of freeze drying in order to reduce the concentrations of water, ethanol and other volatile compounds in wine and pre-concentrate the sample. Pre-concentration increases the S/N ratio of minor component signals, facilitates the use of quantitative ^{1}H NMR (through an internal standard such as DSS or TSP) and has the added advantage that only the residual water signal might need to be presaturated.[6] The external adjustment of sample pH to low values (2 or 3) is another common practice in wine NMR studies, used to avoid chemical shift deviations caused by small changes in the sample's original pH. Care should be taken though when comparing NMR data obtained at different pH, since the effect can be large, especially for ^{13}C NMR shifts.

Pre-concentration steps, such as vacuum distillation, freeze-drying or drying under an inert gas flow, such as nitrogen or argon provide a suitable means to

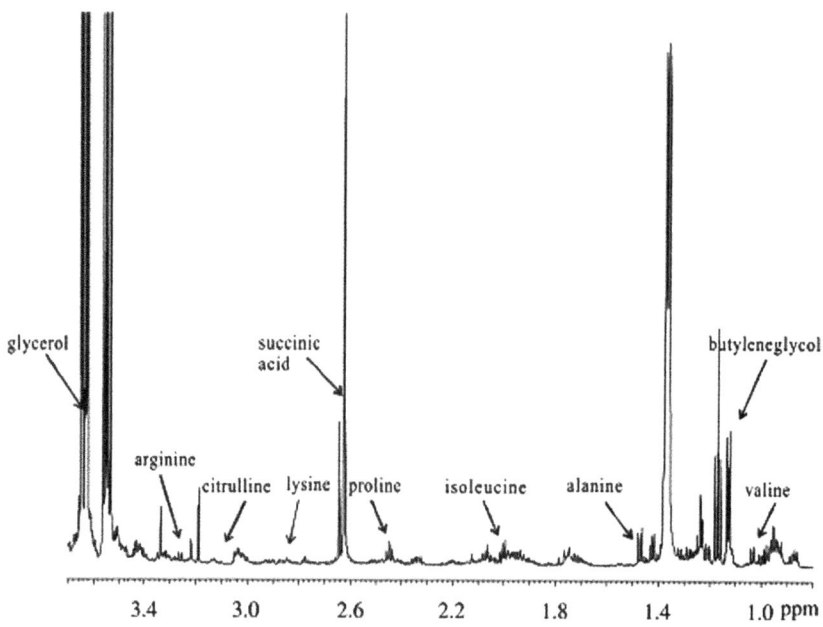

Figure 8.2   Low frequency region of the ^{1}H NMR spectrum of a wine sample (Welsch Riesling, Coastal wine-growing region, 1999) indicating the signals of selected amino acids.
(Reprinted from ref. 4. Copyright (2002), with permission from Elsevier.)

obtain information on minor wine components that are not directly observable in straightforward ^{1}H NMR spectra, such as polyphenols. Phenolic compounds (catechins, flavonols, anthocyanins) are important bioactive wine components and are responsible for some important organoleptic and sensory character-istics of wine, for example anthocyanins are flavonoids responsible for the rich colour of red wine. Vacuum distillation at room temperature has been used to pre-concentrate wine samples for anthocyanin NMR analysis,[7] while another approach has been to study chromatographically separated anthocyanin frac-tions.[8] The phenolic content of wines can be studied successfully by employing a different approach for sample pre-concentration through the use a suitable resin that selectively binds phenolic compounds, which are then washed off by ethanol.[9] Figure 8.4 depicts the aromatic region of the ^{1}H NMR spectra of two Greek wines, indicating the spectral position of several phenolic compounds that were identified.

Solid phase extraction has been recently reported as a superior extraction procedure for the analysis of phenolics in grape juice.[10] It should be mentioned that pre-concentration by vacuum distillation or freeze drying is not able to completely remove all ethanol present in wine, and it has been proposed that this may be a source of poor spectral reproducibility.[11]

Due to the low sensitivity of the ^{13}C nucleus, access to the rich compositional information contained in the ^{13}C NMR spectrum of wine needs either the use of

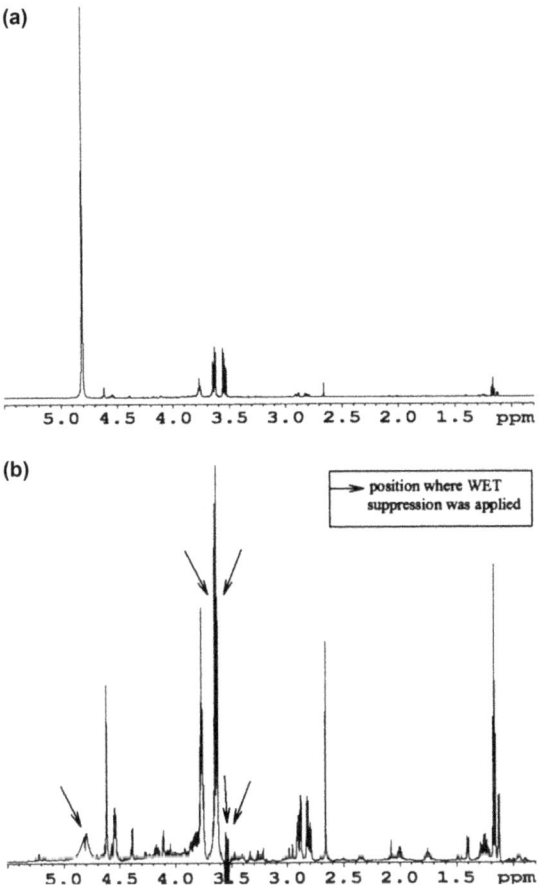

Figure 8.3 Parts of ^{1}H 1D spectra of wine (Sauvignon, Coastal wine-growing region, 1994) without suppression (a) and with WET suppression of water at 4.80 ppm, of ethanol at 3.64 ppm, of glycerol at 3.62 ppm, and of unassigned signals at 3.55 and 3.53 ppm (b). Both spectra were acquired and processed under the same conditions and with the same vertical scale amplification. (Reprinted from ref. 5. Copyright (2001), with permission from American Chemical Society.)

pre-concentrated samples[12] or extremely long acquisition times.[6] 2D heteronuclear NMR spectroscopy represents a more powerful approach for the full assignment of the ^{1}H and ^{13}C chemical shifts of organic compounds, and has been used extensively for the full assignment of all major amino acids present in wine.[5,6] Figure 8.5 presents the heteronuclear ^{1}H–^{13}C HSQC and HMBC 2D NMR spectra of a sample of freeze-dried wine, indicating the assignment of several compounds. The cross peaks in the HSQC spectrum originate from C–H pairs that are directly bonded, whereas in the HMBC spectrum the cross peaks are due to long range J coupling between protons and distant (2 or 3 bonds) carbons.

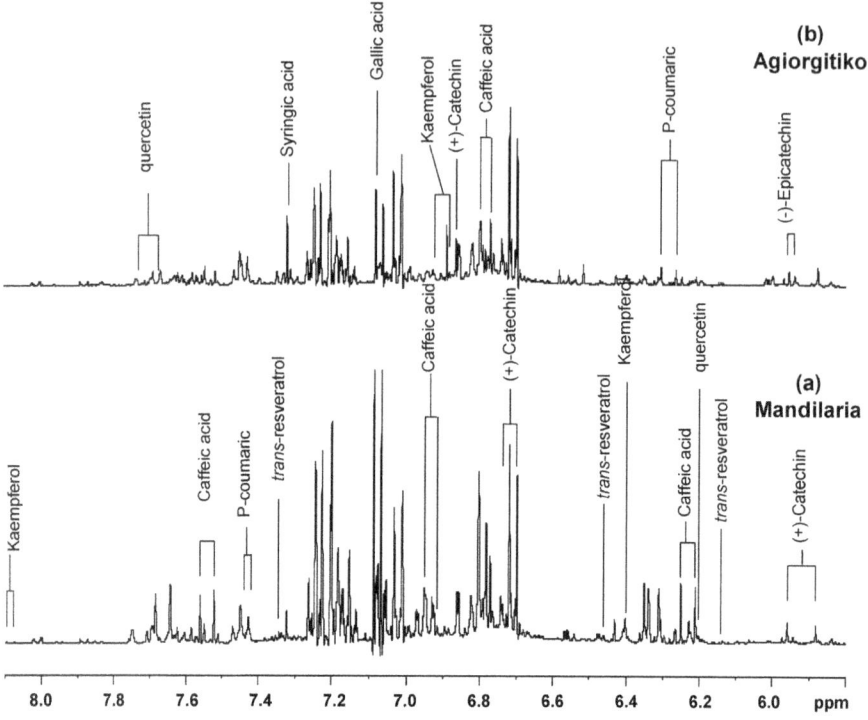

Figure 8.4   Representative 400 MHz ^{1}H NMR spectra (δ 5.80–8.10 ppm) of wine
extracts of (a) *Mandilaria* and (b) *Agiorgitiko* wines along with the assignment of resonances resulting following spiking of 10 standard polyphenols.
(Reprinted from ref. 9. Copyright (2009), with permission from American
Chemical Society.)

The HMBC spectrum of Figure 8.5 shows coupling to carbonyl groups and
is very useful in wine analysis, since wine contains a lot of compounds bearing a
carboxyl group, such as organic acids and amino acids. A similar heteronuclear
2D NMR analysis was performed for wines at pH = 2 in order to assign all the
amino acids present in wine.

With the help of an internal standard, NMR spectra can in principle provide
directly the concentration of organic compounds in a sample of wine. For
example, ^{13}C NMR spectra were used successfully to measure quantitatively
the amino acid concentration of several European wines, with the help of 1,3-
propanodiol (PD) as the internal standard.[12] However, food researchers usually
resort to ^{1}H NMR for the quantitative determination of the chemical constituents of foods, because the proton nucleus has higher sensitivity and shorter
T_1 relaxation times. With the use of an internal standard (TSP) it is possible to
quantify by ^{1}H NMR ten different amino acids, organic acids and alcohols in
wine without any sample pre-treatment at a magnet field strength of 400 MHz[13]
and surely this number will increase by going to higher field strengths.

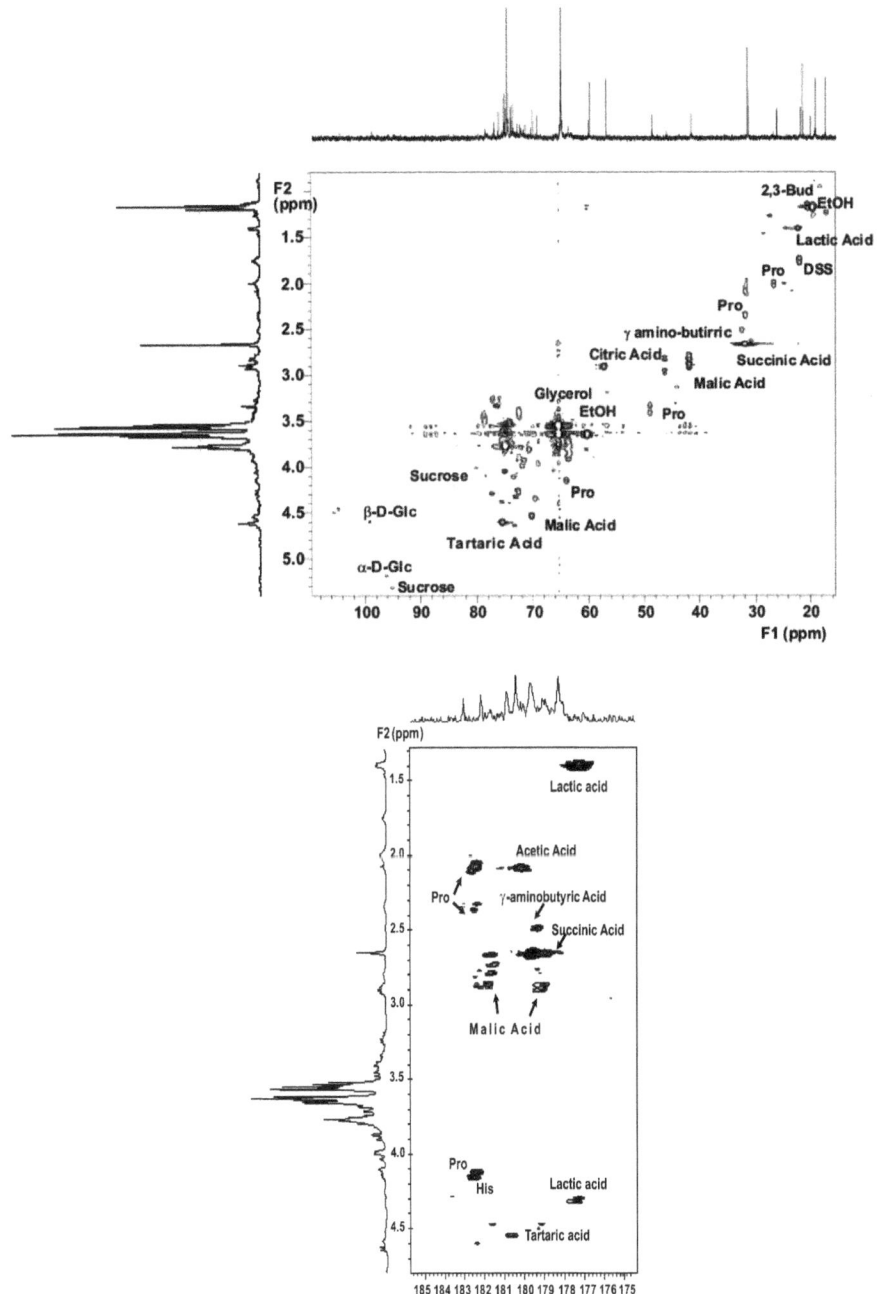

Figure 8.5 Part of a 2D ^1H–^{13}C gHSQC spectrum (top), and expansion of the carbonyl region of a 2D ^1H – ^{13}C gHMBC spectrum (bottom) of wine. Abbreviations used are as follows: Glc = glucose; 2,3-Bud = 2,3-butane-diol; standard three-letter abbreviations for amino acids.
(Reprinted from ref. 6. Copyright (2008), with permission from American Chemical Society.)

In this way NMR spectroscopy has allowed the identification of a wide range of minor compounds in wine, including amino acids, alcohols and organic acids.

Since proton NMR spectroscopy provides easy access to wine metabolome, there is a significant body of work that uses NMR as a metabolomic tool in wine analysis, examining factors such as grape variety, geographical origin, fermentation practices, and pedoclimatic conditions.[14,15] In fact, it is possible to use NMR metabolomics to obtain information about all stages of the wine manufacturing procedure, from the vineyard to the bottle.[14] Separating the effect of different factors on wine characteristics is certainly not an easy task, however this is not a wine-specific problem, but rather a common source of complication in all metabolomics studies. In most cases, it is recommended to isolate the different effects by studying, for example, monovarietal wines or wines produced in the same geographical area, at least for initial exploratory studies. The high added value of wine is, among other factors, related to the geographical origin of production, thus it is not surprising that origin authentication based on the wine metabolome has been studied extensively. It has been possible to separate wines produced from the same grape (Cabernet Sauvignon) that have been harvested in the continental areas of Australia, France, and California,[16] Italian (Apulian) and Slovenian wines,[17] and wines originating from different regions within Italy, Basilicata-Campania[6] and Calabria-Campania.[13] Chinese,[3] Greek,[9] German,[18] Brasilian,[19] and Romanian[20] wines have been the subject of metabolomics studies, along with efforts to establish connections between the metabolome and vintage,[21] sensory properties,[18, 22–24] grape variety[25] and pedoclimatic conditions.[26,27] As a representative example, Figure 8.6 presents the PCA analysis of wines of different grape variety that led to successful cultivar discrimination.[25]

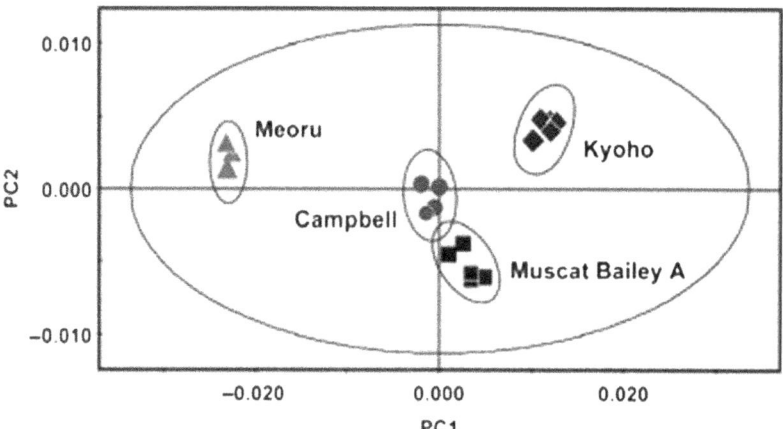

Figure 8.6   PCA scores plot derived from the ¹H NMR spectra of wines vinified from different grape cultivars. All wines were fermented with *Saccharomyces bayanus* PC, for alcoholic fermentation.
(Reprinted from ref. 25. Copyright (2009), with permission from Elsevier.)

Recently, the understanding of the role of tannins in red wine flavour and taste has benefited, among other techniques, from NMR spectroscopy.[28]

Apart from studies of the final product, NMR spectroscopy can be applied to all stages of winemaking, including studies of grapes, must fermentation and aging. The effect of vineyard conditions on the chemical composition of grapes (pulp, skin and seeds) and the resulting wines by ^{1}H NMR has been reported.[29] The effect of several other factors on the metabolic profile of grapes, such as pedoclimatic conditions,[30,31] geographical origin,[32] and culti- var,[32–34] can be examined using ^{1}H NMR metabolomics. ^{1}H NMR is an efficient tool to monitor wine fermentation,[35] the evolution of malic and lactic acids during fermentation of grape must,[36] and examine the role of yeasts[37] and bacteria.[38] ^{13}C NMR was used very recently to study the metabolic transformation of amino acids to higher alcohols responsible for wine aroma, by incorporating ^{13}C-labelled amino acids into the grape must before fermentation.[39]

A basic (at least) understanding of the effect of several factors on wine quality is a prerequisite in order to tackle more complicated analytical problems in wine authentication, such as the identification of blending, either at the grape or the wine level. An initial attempt using ^{1}H NMR profiling and chemometrics (LDA and artificial neural networks) has been recently realised,[40] and will undoubtly be further explored in wine blend analysis.

Some more advanced NMR experiments have been performed in order to elucidate wine composition and assist analysis and authentication. The hyphenated LC-NMR-MS analysis of a wine phenolic extract provided information on the major phenolic compounds contained that was not available by NMR spectroscopy alone.[41] Diffusion ordered spectroscopy (DOSY) was applied recently to Port wine and provided useful information on compositional differences between wines of different ages by differentiating compounds based on diffusivity.[42] The relative amount of an unidentified possible disaccharide, and large aromatic species was found to decrease sig- nificantly in the oldest wine, as expected from the known formation and precipitation of anthocyanin-based polymers during red wine aging. Solid state CP-MAS ^{13}C NMR spectroscopy can be useful for the analysis of the insoluble, lacquer-like pigmented deposits that adhere to the inner glass surface of bottled red wines,[43] although this line of research has not received much attention so far.

Wine is a valuable and expensive commodity, thus in several cases it is not possible to obtain a sample for NMR analysis without destroying it. The possibility of using NMR spectroscopy for the analysis of intact wine bottles has thus a great potential for the wine market, and is also a suitable target for on-line control of industrial fermentation processes. Such experiments require the use of solenoid-type imaging magnets to allow room for the bottle and home made systems for sample positioning and NMR excitation and acquisi- tion. However, ^{1}H and even ^{13}C NMR spectra of full bottles can be obtained this way, and allow applications such as the detection of wine spoilage by oxidation leading to elevated concentrations of acetic acid[44,45] or the

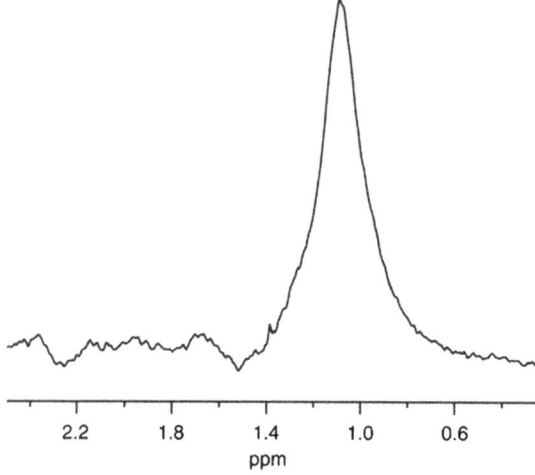

Figure 8.7   Full-bottle low-field suppressed ^{1}H NMR spectrum for the 1970 *Chateau Latour*. The lack of any methyl group triplet resolution reflects the limited time spent on room temperature shimming.
(Reprinted from ref. 45. Copyright (2006), with permission from Elsevier.)

measurement of the CO_2 content of champagne and sparkling wine[46] within intact bottles. Building an optimised NMR system that will allow the detection of metabolites of lower concentration is expected to broaden the range of possible applications and could enable the study of vintage effects and authenticity of collectible wines. Figure 8.7 presents the aliphatic region of the ^{1}H NMR spectrum of a 1970 *Chateau Latour* intact bottle. The lack of any methyl group peak at 2.1 ppm from acetic acid indicates no spoilage is present.

8.1.2   Vinegar

Vinegar is a product obtained through the acetic fermentation of alcoholic liquids and can be prepared from a variety of raw materials, such as grapes, fruits, honey, *etc*. Although generally considered as a cheap product, several specialty vinegars such as Italian balsamic (BV) and traditional balsamic (TBV) have a high market value, and there is a growing interest in replacing sensorial analysis with instrumental analytical methods, such as NMR, for authenticity verification. The ^{1}H NMR spectrum of vinegar contains information on the levels of low molecular weight compounds present (organic acids, sugars, amino acids and alcohols) and can be obtained quite easily in either D_2O[47,48] (see Figure 8.8 for the spectrum of a balsamic vinegar) or DMSO-d_6[49] using water suppression.

^{1}H NMR in either solvent cannot reliably quantify amino acids without a pre-concentration step because of their low concentration in vinegar. 5-Hydroxymethyl-2-furfural (HMF) and acetoxymethylfurfural (AcMF) are sugar degradation products characteristic of the cooked must used in TBV

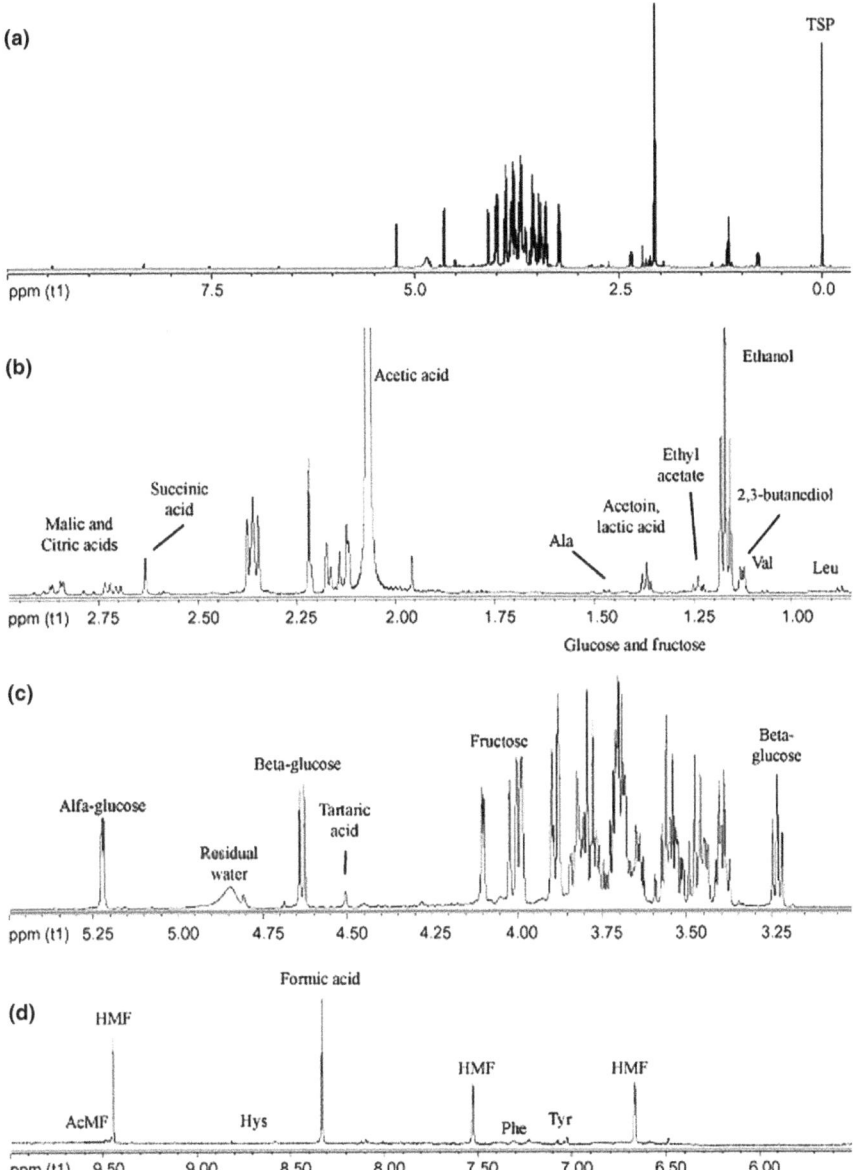

Figure 8.8   (a) ^{1}H NMR spectra recorded at 600 MHz of balsamic vinegar; vertical expansions of (b) the aliphatic/alcoholic region, (c) the sugar region and (d) the aromatic/aldehyde regions.
(Reprinted from ref. 48. Copyright (2007), with permission from Elsevier.)

preparation, and are absent in simple vinegars. The experimental (relaxation delay, sample concentration, water suppression effect) and analytical (accuracy, precision, repeatability, reproducibility) aspects of the ^{1}H NMR methodology for the quantitative analysis of organics in vinegar have been examined in detail, and very good correlation was obtained with traditional methods such as GC-MS and titration.[47]

The study of aging of balsamic vinegars is very important, since traditional balsamic vinegar is aged for a quite large period of time, between 12–25 years, while balsamic vinegar is only moderately aged up to 3 years, and the production procedure for the two types of balsamic vinegar is quite different. ^{1}H NMR analysis showed that these two types of vinegar can be differentiated by the effect of aging on the concentration of several metabolites.[47,49,50] Glucose, fructose and 5-hydroxymethyl-2-furfural increase in concentration during aging, while the levels of acetic acid, lactic acid, acetoin and ethanol decrease. Multivariate statistical models (PCA, PLS-DA) based on metabolite concentrations were very successful in classifying known TBV and BV samples and correctly predicting the balsamic vinegar type of unknown samples.[50] In other studies, ^{1}H NMR was used to suggest glucose and fructose acetates as aging markers for balsamic vinegars,[51] and identify a previously unknown dihydroflavonol, (+)-dihydrorobinetin, as an aging marker characteristic of vinegars aged in acacia (*Robinia pseudoacacia*) wood barrels.[52] ^{1}H NMR in combination with chemometrics has also been used successfully to classify vinegars obtained from different raw materials.[47,48] The study of the compositional characteristics of Brazilian commercial vinegars made it is possible to discriminate between honey, orange, pineapple, and rice with respect to wine vinegars and suggested some chemical markers that could be used for the NMR quality control of vinegars.[48]

The ^{13}C NMR spectrum of vinegar can be used to obtain information on the distribution of the main sugar components (glucose and fructose) that is not available through ^{1}H NMR because of the complicated peak pattern in the sugar region of the ^{1}H NMR spectrum. Consonni *et al.*[53] showed that during the must cooking of TBV of Modena under limited amounts of water the natural distribution pattern of different pyranosidic and furanosidic forms of glucose and fructose (αGP and βGP, αFP, βFP, αFF and βFF) is altered. This alteration of tautomeric sugar forms is retained in non-protic solvents such as DMSO-d_6, but not in water and can be 'read' in the anomeric region of the spectrum. By utilising a set of reference values for fructose forms distribution, ^{13}C NMR is suggested as an objective instrumental method for the authentication of traditional balsamic vinegar of Modena, and has been filed for patenting.[53]

[1]H relaxometry may also prove to be an important tool in the authentication of balsamic and traditional balsamic vinegars of Modena. Shorter [1]H T_1 spin–lattice relaxation times at a single magnetic field strength for acetic acid and β-glucose protons were measured for a series of balsamic vinegars of increasing aging time,[54] and were associated with increasing solid soluble concentration typically observed for older vinegar samples. Inclusion of T_1 data in statistical models based on the quantitation of several aging markers of vinegar improved significantly the differentiation of balsamic vinegars. Continuing along this line of research, field cycling NMR relaxometry was used to measure the ^{1}H nuclear magnetic resonance dispersion (NMRD) profiles of

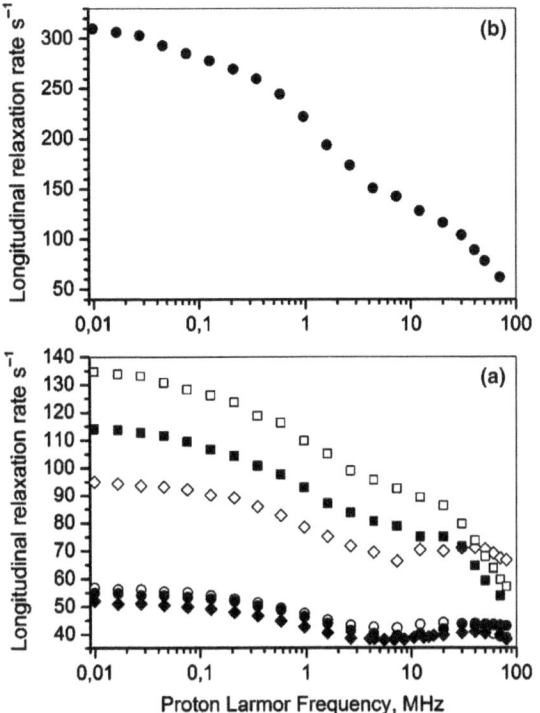

Figure 8.9 NMR dispersion profiles of samples of traditional balsamic vinegar of Modena. (a) Samples 9 (○) and 10 (●) with an aging process of more than 12 years, samples 11 (◇) and 12 (◆) with an aging process of 20 years, sample 13 (□) with an aging process of more than 25 years, and sample 14 (■) with an aging process of 34 years. (b) Sample 15 (●) with an aging process of more than 50 years.
(Reprinted from ref. 55. Copyright (2009), with permission from American Chemical Society.)

water protons over an extended range of Larmor frequencies (0.01 to 80 MHz).[55] Figure 8.9 depicts the NMRD profile of several genuine TBVM samples as a function of aging time. It was shown that counterfeit TBV specimens could be identified on the basis of the comparison of their T_1 and T_2 (transverse relaxation time) values with respect to the corresponding values of genuine samples, while a relationship that relates the observed T_1 to the age of the genuine vinegars was reported.

8.1.3 Beer

Beer is probably the oldest fermented beverage known to man. It is obtained from malted grains (mostly barley), hops and yeast and its main ingredients are water, ethanol and carbohydrates. Figure 8.10 presents a typical ^{1}H NMR spectrum of a lager beer obtained without any sample pre-treatment other than

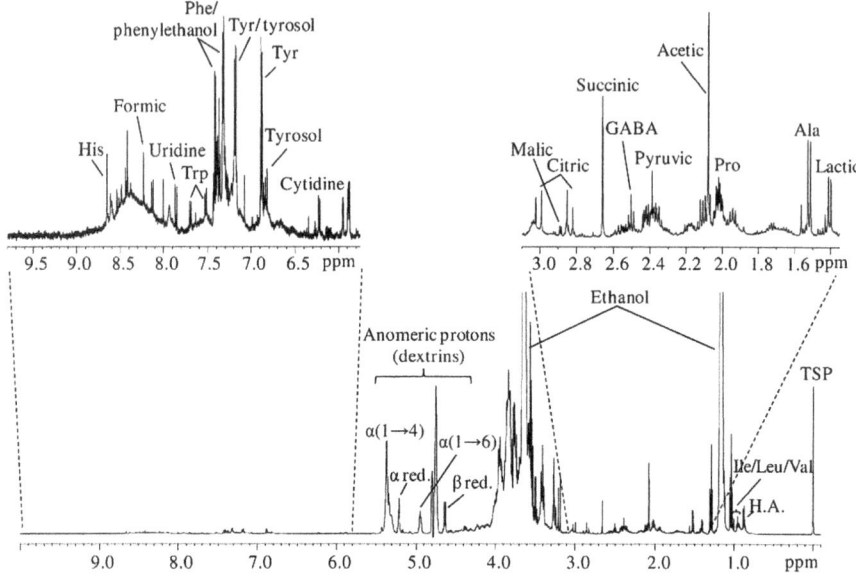

Figure 8.10 Typical 500 MHz ^{1}H NMR spectrum of whole lager beer sample, with insets shown for the high-field (right) and low-field (left) regions of the spectrum. Standard three-letter abbreviations are used for amino acids. (Reprinted from ref. 56. Copyright (2011), with permission from John Wiley and Sons.)

CO_2 degassing, where the most important low molecular weight compounds are indicated.[56]

The assignment of the ^{1}H NMR spectrum of beer required extensive use of 2D homo- and heteronuclear NMR experiments,[57,58] since it contains several broad features, especially in the aromatic and sugar region, originating from high molecular weight compounds present in beer. The assignment of the aromatic region of the spectrum was accomplished by the use of hyphenated LC-NMR-MS spectroscopy that led to the identification of several compounds, including the pairs of aromatic amino acids and corresponding alcohols phenylalanine/phenylethanol, tryptophan/tryptophol, and tyrosine/tyrosol.[41] The concentration of these compounds depends on fermentation and wort composition, while aromatic alcohols are additionally important flavour components of beer.[56] LC-NMR-MS was also necessary to unravel the structure of oligomeric dextrins present in beer, because of the heavy overlap in the sugar region of the ^{1}H NMR spectrum. Figure 8.11 depicts the LC-NMR chromatograph of an ale beer (A) and the NMR and MS spectra obtained for the specified elution times corresponding to maltose (retention time (RT) 16.6 min), maltotriose, maltotetraose, and higher oligomers (up to 9 glucose units) at decreasing RT.[59]

It is important to note that the two different beer samples (ale, lager) investigated by LC-NMR-MS were found to have significantly different oligosaccharide composition, reflecting the different production conditions

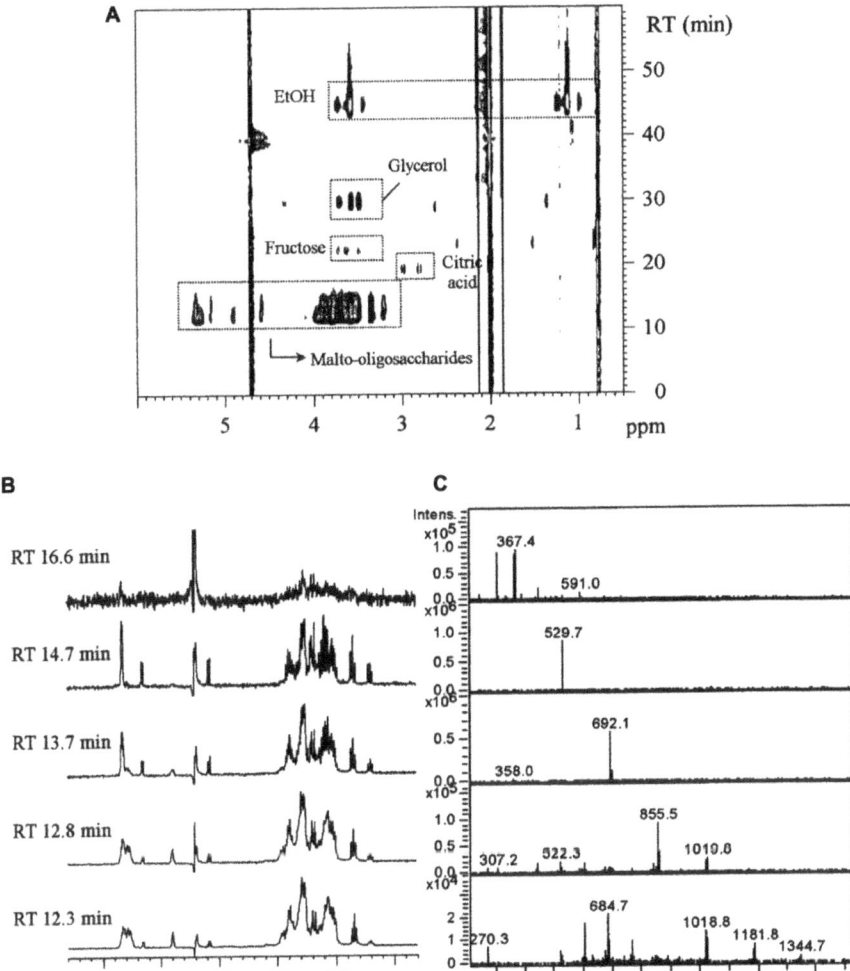

Figure 8.11 (a) On-flow LC-NMR record obtained for a sample of ale beer. The labels identify the main separated fractions, (b) rows extracted from the on-flow record, (c) MS spectra acquired concurrently with the NMR data, using positive-ionisation.
(Reprinted from ref. 59. Copyright (2003), with permission from American Chemical Society.)

employed. A preliminary attempt at using diffusion-ordered spectroscopy (DOSY) to aid the analysis of the complex ¹H NMR spectrum of beer and beer concentrates has also been made.[60]

The ¹H NMR quantification of amino acids and organic acids in beer has received attention, since these compounds are important beer metabolites and their compositional profile depends both on wort composition and yeast metabolism.[61,62] The results of both direct spectral integration and PLS regression of ^{1}H NMR data were compared to reference methods such as HPLC, capillary electrophoresis and enzymatic assays, showing good to excellent quantification accuracy for the majority of metabolites.

Metabolic profiling by ^{1}H NMR represents a rapid and promising method for the characterisation of beer in terms of type, brewing site, and production date.[63] PCA of 2D *J*-resolved spectra of six different brands of pilsner beer allowed their discrimination based on compositional differences in nucleic acid derivatives (adenine, uridine and xanthine), amino acids (tyrosine and proline), organic acids (succinic and lactic acid), alcohols (tyrosol and isopropanol), cholines and carbohydrates.[58] The effect of brewing site and production date on the compositional profile of lager beer samples of the same brand (4% v/v alcohol), originating from three different countries and production dates was studied by the PCA analysis of their ^{1}H NMR spectra.[64] All three production sites were successfully distinguished by their contents in lactic and pyruvic acids, aromatic components and linear/branched carbohydrate content, while production date variability was satisfactory in two of the three sites. Along the same lines, an attempt to explore the applicability of ^{1}H NMR as a tool for the quality control and authenticity assessment of beer in an official food control environment was made.[65] A high throughput flow-injection NMR system was used for sample handling, allowing fast spectral acquisition and leading to the construction of a comprehensive (80 samples) database of beer spectra that

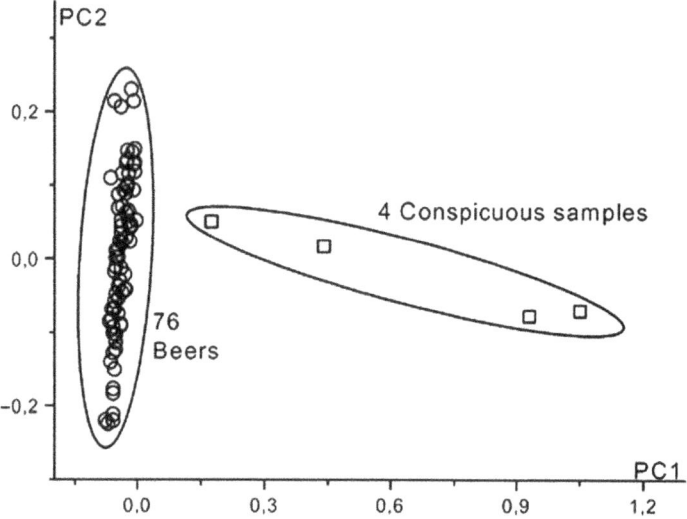

Figure 8.12   Scatter plot of the PCA scores obtained from bucketing (9–0.4 ppm) of the ^{1}H NMR spectra of eighty beer samples.
(Reprinted from ref. 65 with permission from Springer Science and Business Media.)

were used for PCA classification. Figure 8.12 shows how this approach can be used for the fast identification of conspicuous samples.[65]

Lactic acid was identified as responsible for the variance in the PCA scores of the four samples, and conventional chemical and microbiological analyses verified that the deviating samples suffered beer spoilage. Furthermore, it was possible to distinguish between beers made with barley malt and wheat malt, or produced in different brewing sites, while using partial least squares (PLS) NMR spectra were successfully correlated with results from reference methods for calculating the original gravity, ethanol and lactic acid concentration of beer samples.

^{1}H NMR spectroscopy can be used for assessing the compositional changes occurring in beer during storage, which may result in the loss of some desirable product properties, and thus are of prime importance to the brewing industry.[66] PLS-DA and spectral integration showed that a clear aging trend is observed during forced aging (45 °C up to 18 d) of beer. A steady increase in 5-hydroxymethylfurfural was the most significant age marker for beer, while γ-aminoburytic acid was suggested as a particularly important substrate for Maillard reactions during storage by 2D correlation analysis of the NMR data.

The NMR analysis of beer has also benefited from studying beer extracts,[56] a practice necessary to get structural and compositional information on, for example, hop bitter acids that although present in low concentration in beer, are responsible for important beer sensory properties such as aroma and bitterness. Hop bitter α- and β-acids and iso-α-acids, the latter obtained from thermal isomerisation of α-acids during wort boiling, have been the subject of several investigations, with NMR playing an important role in the study of the degradation of iso-α-acids during storage.[67–69] The structure of the degradation products of β-iso-acids has also been studied.[70]

8.1.4 Spirits

Spirits are alcoholic beverages obtained by distillation of alcohol produced by fermented grains, fruits or vegetables, and include popular products such as whisky, vodka, rum, gin, *etc.* The term 'spirit' generally also includes liqueurs, although the latter differ markedly from spirits in that they contain added sugar and flavourings. Spirits contain very high amounts of ethanol and thus the problem of concurrent suppression of unwanted signals of water and ethanol becomes more pronounced when compared to wine, for example. This factor has undoubtedly played an important role in the limited amount of investigations that utilise ^{1}H NMR spectroscopy for the analysis of spirits, compared to wine and beer, until recently.

A ^{1}H NMR study (using water pre-saturation) to establish the authenticity of *Zivania*, a traditional Cypriot alcoholic beverage, and compare it to other spirits of similar alcoholic content obtained from different countries provided a good degree of prediction and classification between two types of *Zivania* and other spirits,[71] although improved classification was obtained by combining NMR data with chromatographic data.[72]

Recently, a novel NMR experiment involving the application of a shaped pulse sequence during the relaxation delay that suppresses very efficiently the eight ^{1}H NMR frequencies of water and ethanol (the OH singlet of both water and ethanol, as well as the CH_2 quartet and CH_3 triplet of ethanol) was proposed for the analysis of spirits.[73] Validation of the analytical NMR methodology was performed successfully for the quantification in spirits of methanol, diethyl phthalate, acetaldehyde, 2-phenyl ethanol, benzaldehyde and ethyl acetate. Figure 8.13 compares the ^{1}H NMR spectra of a high ethanol spirit

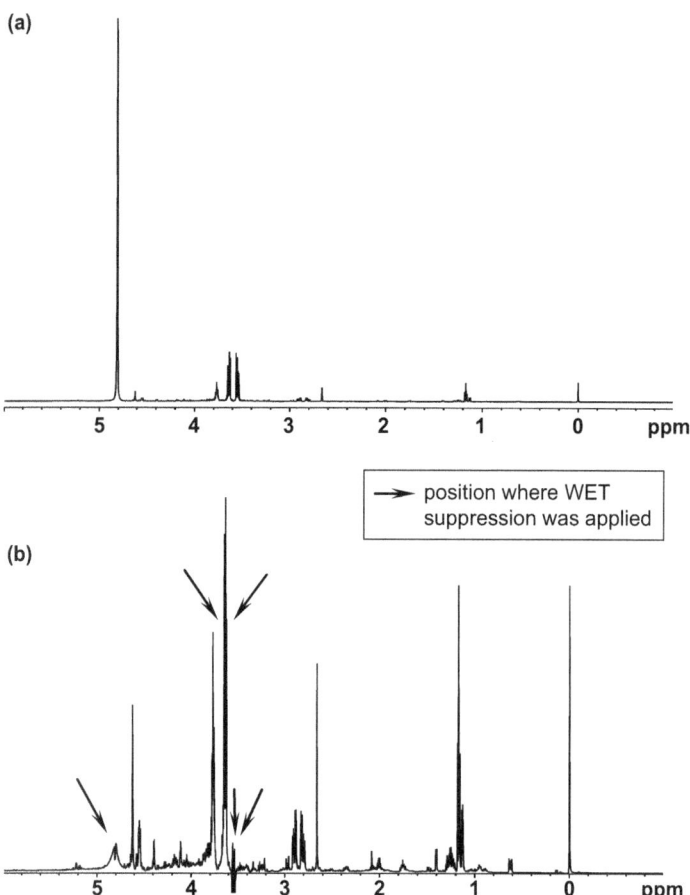

Figure 8.13 ^{1}H NMR spectrum of a homemade spirit (69.8% vol.) recorded with a water pre-saturation pulse sequence (a) and with eight-fold suppression of ethanol and water signals (b). The inserts show details in the 2.5–1.5 ppm region.
(Reprinted from ref. 73. Copyright (2011), with permission from John Wiley and Sons.)

(*Samogon*, homemade spirit from Russia with 69.8% vol. alcohol) obtained using simple water pre-saturation (a); and the new eight-fold shaped pulse suppression scheme (b); demonstrating the significant increase in signal intensity of several minor spirit compounds that could not be observed at all using water pre-saturation.

Using the above methodology for multiple peak suppression, it was possible to quantify several minor compounds in diverse spirit samples, including diethyl phthalate (DEP) and polyhexamethylene guanidine (PHMG) in surrogate alcohols (*i.e.* illegal alcohol not originally intended for human consumption) from Russia,[74] and total thujone (a bicyclic monoterpene ketone) in absinthe.[75] ^{1}H NMR, using eight-fold suppression of water and ethanol in combination with PCA, was used for the non-targeted analysis of a large number of spirits (304 samples) obtained from 14 different countries. Seven conspicuous samples were identified by PCA, and turned out to contain hazardous compounds, while several minor spirit components were quantified in all samples by PLS regression using GC analysis as the reference.[76]

Another possibility suggested for the analysis of minor compounds in foods is the use of F2-selective 2D NMR spectroscopy, a technique that was recently demonstrated to increase the number of compounds detected in the ^{1}H NMR spectrum of Japanese sake.[77]

^{1}H NMR-obtained metabolite data were successfully used to differentiate between three typical flavour types (light, strong, and sauce flavour) of *Daqu*, a fermentation starter and substrate complex that is used to initiate fermentations for the production of Chinese liquor (alcoholic spirit),[78] by principal components analysis. The metabolic profile changes during the various stages of fermentation of a specific starter, *Fen Daqu*, used for the production of a typical Chinese light-flavour liquor, Fen liquor, were also studied using PCA analysis of ^{1}H NMR data.[79]

NMR spectroscopy has been used extensively for the characterisation of non-volatile taste-active compounds that migrate to spirits during aging of bourbon whisky in oak wooden barrels,[80] lignins,[81] and phenolic compounds in brandy.[82]

The nature of hydrogen-bonding between water and ethanol and the dynamics of proton exchange in various spirits (Japanese shoshu,[83] sake,[84] whisky,[85] vodka[86]) has been studied by ^{1}H NMR, ^{17}O NMR[87] and other spectroscopic techniques, since they are affected by the presence in spirits of minor compounds possessing labile protons.

8.2 Non-Alcoholic Beverages

8.2.1 Coffee

Coffee in its various preparations is one of the most popular non-alcoholic beverages consumed worldwide. Coffee can be *green* (before roasting), *roasted*, or *soluble*, the latter term referring to the dried water-soluble form provided to consumers for beverage preparation, and of the *Coffea arabica (arabica)* or

(a)

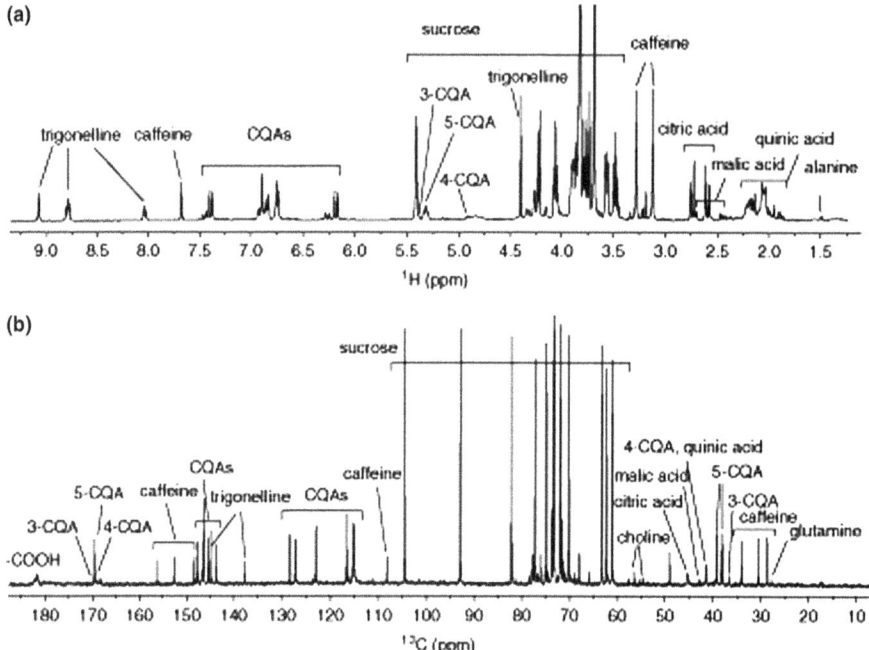

(b)

Figure 8.14   1D (a) ^{1}H and (b) ^{13}C NMR spectrum of green coffee bean extracts in
D$_2$O.
(Reprinted from ref. 73. Copyright (2011), with permission from John
Wiley and Sons.)

Coffea canephora (robusta) variety. All types of coffee are easily amenable to
high-resolution NMR analysis directly as the beverage itself, or by the
extraction of beans by boiling water. Figure 8.14 depicts the ^{1}H and ^{13}C NMR
spectra of a D$_2$O extract of green coffee beans of the *arabica* variety, indicating
signals from its main organic components, namely organic acids and amino
acids, chlorogenic acids (3-, 4- and 5-caffeoylquinic acid isomers), quinic acid,
sucrose, caffeine and trigonelline.[88]

Early ^{1}H NMR work involved the direct but qualitative analysis of three
different varieties of espresso coffee, and the study of the effect of roasting on
the concentration of several coffee metabolites, assigned by using 2D NMR
homonuclear ^{1}H–^{1}H COSY and TOCSY spectra.[89] Heteronuclear ^{1}H–^{13}C 2D
NMR spectroscopy was later applied for the characterisation of coffee metabolites, and a quantitative ^{1}H NMR methodology was proposed for the
determination of caffeine in *arabica* and *robusta* coffees.[90]

Recently, a quantitative ^{1}H NMR analytical methodology for the simultaneous determination of caffeine, formic acid, trigonelline and 5-(hydroxy-
methyl)furfural in soluble coffee has also been developed, validated by
comparison with standard HPLC and enzymatic methods and applied successfully to characterise several commercial coffee brands.[91] Quantification was

facilitated by the fact that all four compounds provide suitable singlet peaks in the low-field region of the ^{1}H NMR spectrum, however the high-field region of the spectrum of coffee is too crowded and makes the quantification of more metabolites difficult. Wei *et al.* solved this problem by resorting to the ^{13}C NMR spectrum of coffee and reported a comprehensive analysis of both green and roasted coffee beans of the *arabica* variety, using extensive 2D NMR experiments for the assignment of the coffee spectra.[88,92] Quantitative 1D ^{13}C NMR spectra were obtained by the inverse-gated decoupling method that eliminates the nuclear Overhauser enhancement of carbons due to protons, using copper sulfate as a relaxation agent to reduce the long T_1 spin – lattice relaxation values of carbon nuclei and hexamethyldisiloxane sealed in a capillary as a concentration standard. It is possible to quantify a total of twenty nine different metabolites in coffee beans using quantitative ^{13}C NMR spectroscopy, as depicted in Figure 8.15 that presents the differences in metabolite concentration between green and roasted *arabica* beans.[92]

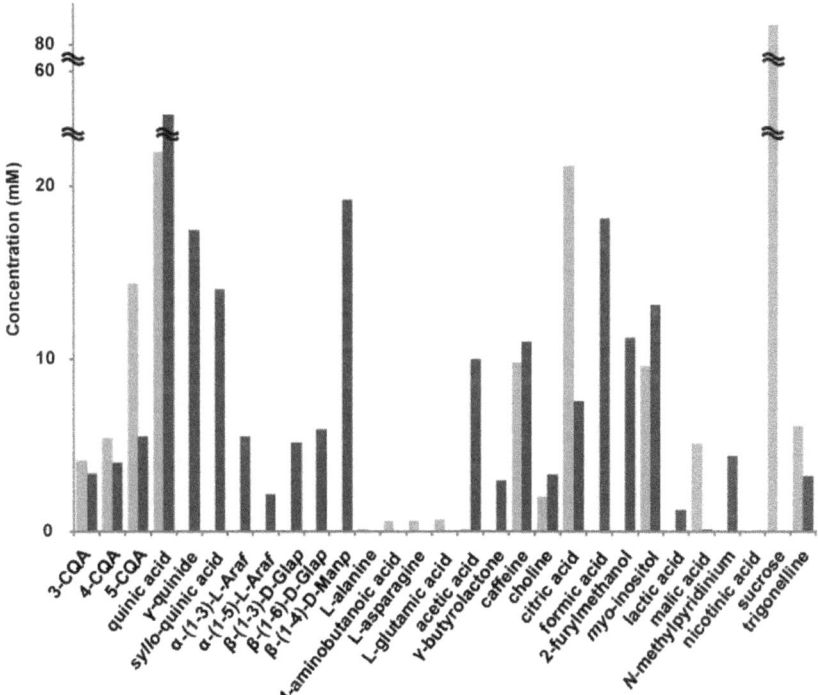

Figure 8.15 Concentrations of green (grey bars) and roasted (black bars) coffee bean extract components. α-(1–3)-L-Ara*f*, α-(1–5)-L-Ara*f*, β-(1–3)-D-Gal*p*, β-(1–6)-D-Gal*p*, and β-(1–4)-D-Man*p* indicate α-(1–3)-L-arabinofuranose, α-(1–5)-L-arabinofuranose, β-(1–3)-D-galactopyranose, β-(1–6)-D-galactopyranose, and β-(1–4)-D-mannopyranose, respectively.
(Reprinted from ref. 92. Copyright (2011), with permission from American Chemical Society.)

5-Chlorogenic acid was the most abundant isomer in both green and roasted coffee, although chlorogenic acids decreased after roasting, decomposing to quinic acid which is seen to increase. Polysaccharides are absent in green coffee, and it is suggested that they are formed during roasting by the decomposition of cellulose and glucoproteins. Sucrose and amino acids are completely elimi- nated during roasting possibly due to their participation in Maillard reactions. A qualitative study of the effect of roasting has also been presented by 1D ^{1}H HR-MAS NMR spectroscopy of ground green coffee beans.[93]

Because of the important effect of roasting on the chemical composition and the organoleptic and nutritional properties of coffee beverages, roasting of green *arabica* coffee beans at three different levels (light, medium and dark) was studied by using the above-described NMR methodology in greater detail, in order to follow the quantitative chemical evolution of thirty different coffee components as a function of roasting time, up to 9 min.[94] It was reported that during roasting, sucrose and chlorogenic acids were degraded while quinic acids, *N*-methylpyridinium, and water-soluble polysaccharides were formed. Therefore, these compounds could be used as chemical markers of roasting degree, while caffeine and myo-inositol were relatively thermally stable. Unsupervised multivariate analysis of the NMR-obtained data by PCA was able to classify the coffee samples according to the different degrees of roasting.

^{1}H NMR metabonomics has also been used to study the authentication and quality control of coffee from the point of view of blending of different vari- eties,[95] production site[96] and geographical origin[97] discrimination. Several multivariate analysis models were used for the differentiation of blends of ground green coffee of *arabica* and *robusta* species based on ^{1}H NMR data.[95] The discrimination observed was satisfactory, and was attributed to the higher content of chlorogenic acids and caffeine in *robusta* coffee, however roasted coffee samples could not be differentiated, probably because of the severe chemical composition alterations due to roasting, which as already discussed above, mask varietal differences.

The successful classification of 98 instant spray-dried coffee samples origi- nating from three different manufacturers was accomplished by a combination of PCA and LDA models built from the respective ^{1}H NMR spectra of D_2O coffee extracts. Blind testing of the PCA model with 36 samples of instant coffee not used for model construction resulted in 100% correct sample identification from the three manufacturers, while 5-hydroxyfurfural was identified as a primary differentiation marker. This compound is a product of the coffee roasting procedure and is not contained in green coffee beans.[89,94]

An interesting application of NMR-based metabonomic in coffee analysis involves the metabolic fingerprinting of 40 roasted *arabica* samples from the three main coffee production areas, America, Africa and Asia.[97] Figure 8.16 presents the clear separation of samples according to their geographical origin based on an OPLS-DA model performed on ^{1}H NMR data, with fatty acids, chlorogenic acids, lactate, acetate and trigonelline being the metabolites mainly responsible for differentiating the American, African and Asian samples.

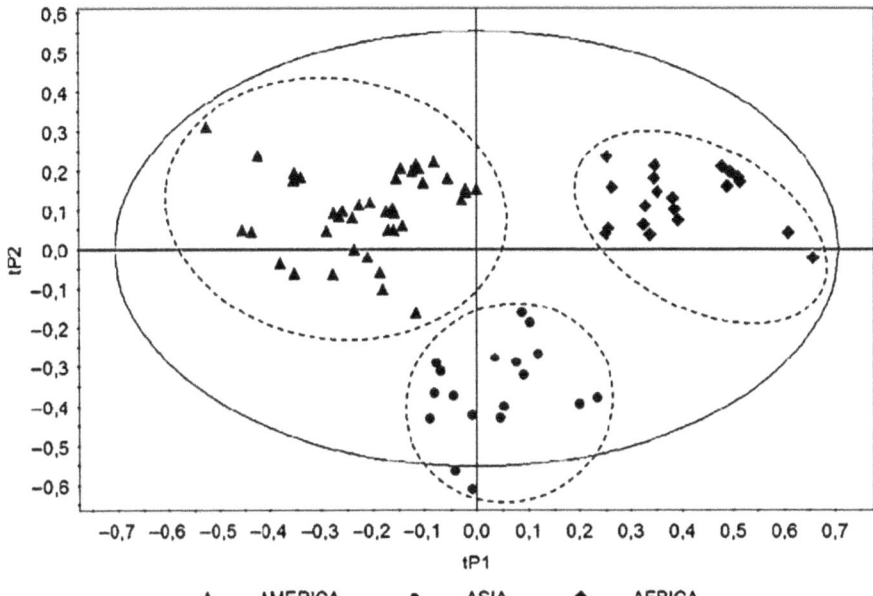

Figure 8.16 Score of OPLS-DA analysis of ^{1}H NMR spectra of a total of 40 samples of roasted coffee: filled triangles, dots and diamonds represent American, Asian and African roasted coffee samples, respectively. Overall goodness of fit $R^2Y = 81.5\%$ and overall cross validation coefficient $Q^2 = 69.7\%$. (Reprinted from ref. 97, Copyright (2012), with permission from Elsevier.)

NMR spectroscopy has played an integral role, along with other techniques, in the structural characterisation of minor coffee components that contribute to its textual, sensory and antioxidant properties. Although full coverage is out of the scope of the present book, reference to some recent work involving carbohydrates,[98–100] bitter compounds,[101–103] diterpenes,[104] and melanoidins[105] is provided for the interested reader.

Water mobility and distribution in coffee beans plays an important role in coffee processing. TD-NMR has been used to study the diffusion and types of water (free and bound) in green coffee beans and changes observed during roasting. ^{1}H NMR T_2 relaxometry pointed to the presence of two types of mobile water in arabica roasted coffee with 50% water content at 90 °C, attributed to water in cell wall polymers and in water filling cell's lumen.[106] The appearance of two separate proton pools of different mobilities at different coffee bean hydration levels may be correlated to the antiplasticiser (low hydration) and to the plasticiser (high hydration) effect of water on the coffee bean matrix.[107,108]

8.2.2 Cocoa

Hot cocoa is a popular beverage prepared from cocoa powder and water or milk. Cocoa powder is obtained after fermentation, drying and roasting of

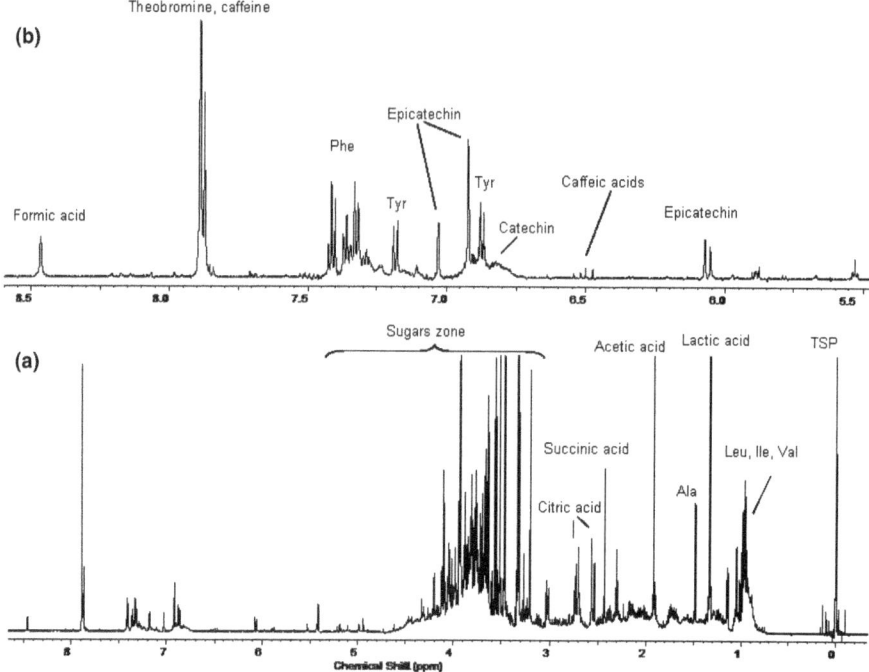

Figure 8.17 ¹H NMR spectrum (600 MHz, solvent D₂O/CD₃OD 8 : 2 v/v) of a
Trinitario cocoa bean hydroalcoholic extract: (a) full spectrum; (b)
expansion of the 5.5–8.5 ppm region.
(Reprinted from ref. 109. Copyright (2010), with permission from
American Chemical Society.)

cocoa beans (*Theobroma cacao* L.), as described for coffee beans above,
while cocoa butter and cocoa liquor (obtained from cocoa beans processing)
are important food components in the chocolate industry. The ¹H NMR
spectrum of a cocoa bean extract of the *Trinitario* variety is presented in
Figure 8.17, where also the major signals obtained from amino acids, organic
acids, polyalcohols, sugars, methylxanthins, catechins and phenols present are
labelled.[109]

A total of twenty-one different substances can be quantified in the ¹H NMR
spectrum of such extracts with the help of TSP as an internal standard. Three
different varieties of cocoa beans (*Forastero, Criollo,* and *Trinitario*) from different countries (Ecuador, Ghana, Grenada, and Trinidad) were analysed, and
significant differences were noted as a function of variety and geographical
origin, the latter indicating the importance of the fermentation and drying
processes on the chemical composition of cocoa beans. ¹H NMR has also been
used for the quantification of a few metabolites in raw and processed (fermentation, drying, roasting) cupuassu (*Theobroma grandiflorum* Spreng), a
species similar to cocoa in composition and application.[110]

8.2.3 Tea

Tea is a popular beverage obtained from the leaves of the plant *Camelia sinensis*, and can be divided into three basic types, green (unfermented), oolong (semi-fermented), and black (fermented). Catechins are the major phenolics in green tea, the structure of epicatechin (EC), epigallocatechin (EGC), epicatechin-3-gallate (ECG), and epigallocatechin-3-gallate (EGCG) are depicted in Figure 8.18.

Figure 8.19 depicts the ^{1}H NMR spectra of green tea infusions from leaves harvested at three different growing areas in South Korea, indicating catechins and other major green tea metabolites observed, such as carbohydrates, caffeine, quinic acid and amino acids, of which theanine is the most abundant.[111]

A recent NMR study of the metabolite profile of green, partially (oolong) and fully fermented (black) teas showed that the concentration of catechins is almost diminished in black tea infusions.[112] Theaflavins and thearubigins are major tea pigments produced by the oxidation of catechins during tea fermentation. ^{1}H, ^{13}C 1D and 2D NMR spectroscopy has been used for the structure elucidation of theaflavins,[113,114] thearubigins,[115] and to demonstrate the oligomeric structure of thearubigins with molecular weights of up to 2000 Da.[116]

As a plant extract, tea contains hundreds of chemical compounds in low concentrations. A recent study utilising the analytical power of hyphenated LC-MS-SPE-NMR spectroscopy successfully annotated 177 phenolic compounds in tea, mostly glycosylated and acetylated derivatives of flavan-3-ols

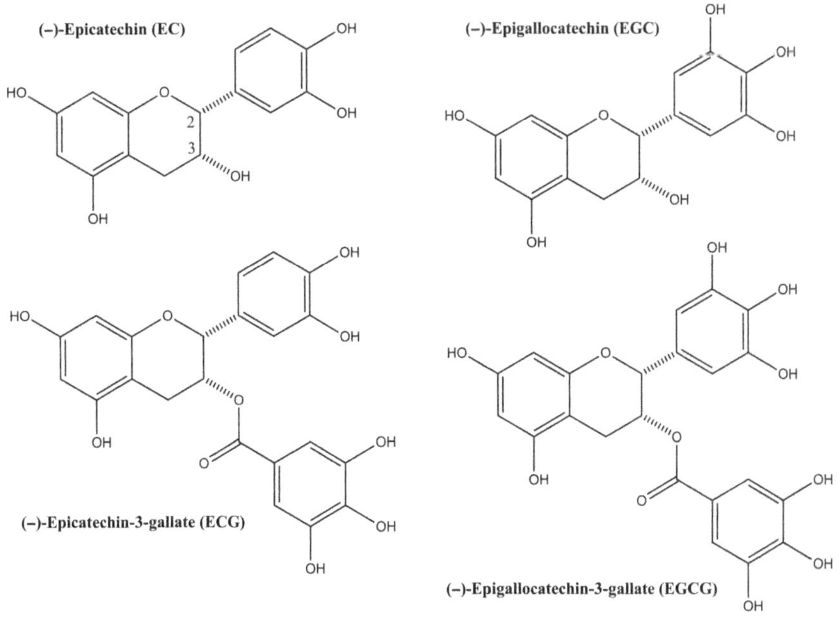

Figure 8.18 Chemical structure of major catechins in tea.

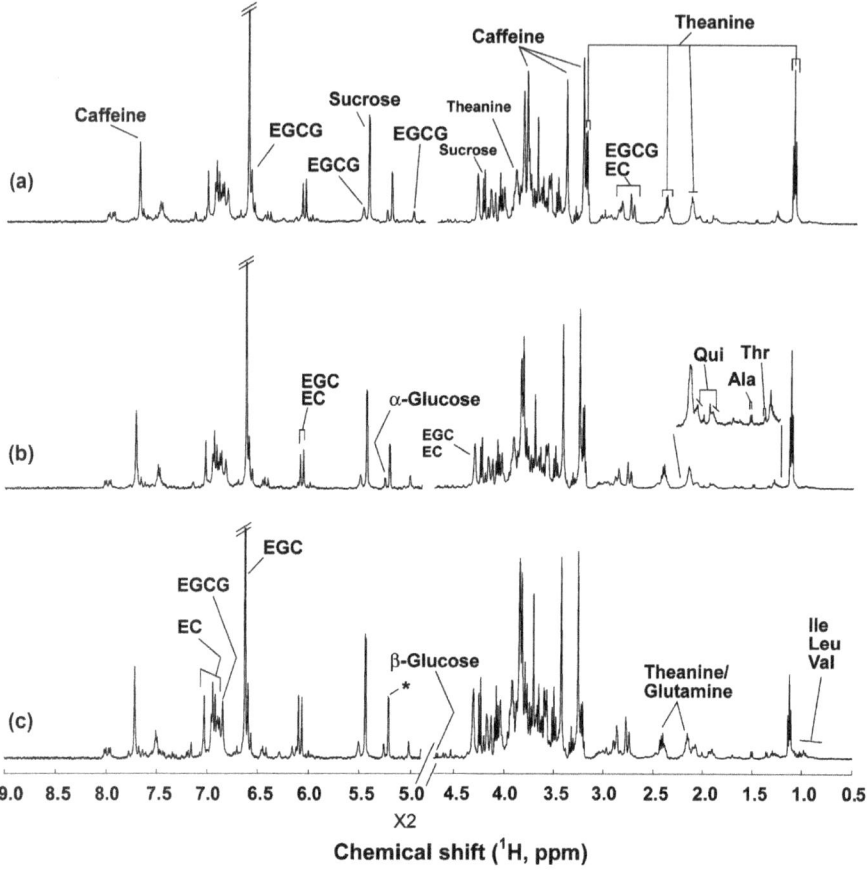

Figure 8.19 ¹H NMR spectra of green tea infusions from leaves harvested at three different growing areas in South Korea, Hannam (a), Dosun (b), and Seogwang (c). Ile = isoleucine; Leu = leucine; Val = valine; Ala = alanine; Thr = threonine; Qui = quinic acid; EC = (−)-epicatechin; EGC = (−)-epigallocatechin; EGCG = (−)-epigallocatechin-3-gallate.
(Reprinted from ref. 111. Copyright (2010), with permission from American Chemical Society.)

and flavonols.[117] Most traditional 1D and 2D NMR spectroscopy of tea extracts has of course been used extensively for structure elucidation purposes in tea. Anthocyanins,[118] polysaccharides,[119–123] catechins,[124–127] and various antioxidant compounds[128,129] in tea and herbal teas from all over the world have been studied. Herbal teas are infusions prepared by various traditional plants, and usually are associated with beneficial health effects. Some representative examples of herbal teas analysed by NMR spectroscopy include rosemary,[130] honeybush,[131,132] chamomile,[133] burrito[134] (a folk Paraguayan tea), macela[135] (traditional Brazilian tea), sage,[136] horehound[137] (Lithuanian herbal tea), rooibos tea[138] and several traditional Chinese herbal teas.[128,139,140]

Aluminium and some other metals can accumulate on tea leaves, and this prompted researchers to examine whether high tea consumption may lead to extended human exposure, a hypothesis that was not verified. ^{27}Al NMR spectroscopy was proposed as a quantitative method for the determination of aluminium in tea leaves and infusions,[141] and was used for aluminium speciation in tea infusions.[142,143] It was reported that most of the aluminium was bound to catechins, while some was bound to phenolic and organic acids. The interaction of catechins with other molecules, such as proteins,[144] caffeine[144] and lipid membranes has also been studied by ^{1}H NMR spectroscopy.

^{1}H NMR-based metabolomics has been used extensively for the analysis of tea, tea infusions and herbal teas. Table 8.1 summarises the available literature on tea NMR metabolomics studies that have been used to elucidate variations inflicted on the metabolic profile of tea leaves by a great number factors, such as tea type (green, oolong, black), harvest time, plucking position, growing altitude, climatic effects, fermentation, geographical origin,[145] and establish relations between tea metabolome and quality.[146–148]

As an example, Figure 8.20 depicts the excellent OPLS-DA discrimination between different black teas cultivated in Sri Lanka at different altitudes.[114]

Solid-state NMR has been used to aid the analysis of bioactive tea components and their interactions with biomacromolecules towards an understanding of their physiological role. Figure 8.21 depicts the solid-state ^{13}C CPMAS NMR spectrum of black tea, and the assignment of some characteristic flavanol signals.

Although the spectrum contains peaks from all components present in the solid tea leaves, including theaflavins, thearubigins, carbohydrates, terpenoids and hemicellulose, it was possible to differentiate green and black tea by their NMR spectra.[149] [13]C, ^{31}P and ^{2}H solid-state NMR studies provided direct experimental evidence that the EGCG molecule interacts with phospholipid bilayers,[150,151] multi-lamellar vesicles and bicelles.[152]

Table 8.1 NMR-based metabolomics studies on tea and herbal teas.

Factor examined	Tea type	Leaf extraction conditions	Multivariate analysis model	Ref.
Quality	Green	70% MeOD, 70 °C	PCA, HCA	146
Type, origin	All	Water, 75 °C	PCA, SIMCA	145
Astrigency	Black	Water, infusion	PLS	147
Quality	Green	Water, 60 °C	PCA, PLS	148
Type	All	60% MeOH, r.t.	PCA	117
Geographic, climatic	Green	Water, 60 °C	PCA, OPLS-DA	111
Growing altitude	Black	Water, boiling	PLS-DA	114
Fermentation	All	Water, 60 °C	PCA	112
Plucking position	Green	80% MeOD, r.t.	PCA, OPLS-DA	111
Quality	Sage tea	Water	PCA	136
Harvest time	Rosemary tea	Water, boiling	PCA, OPLS-DA	130
Geographic	Chamomile tea	Water	PCA	133

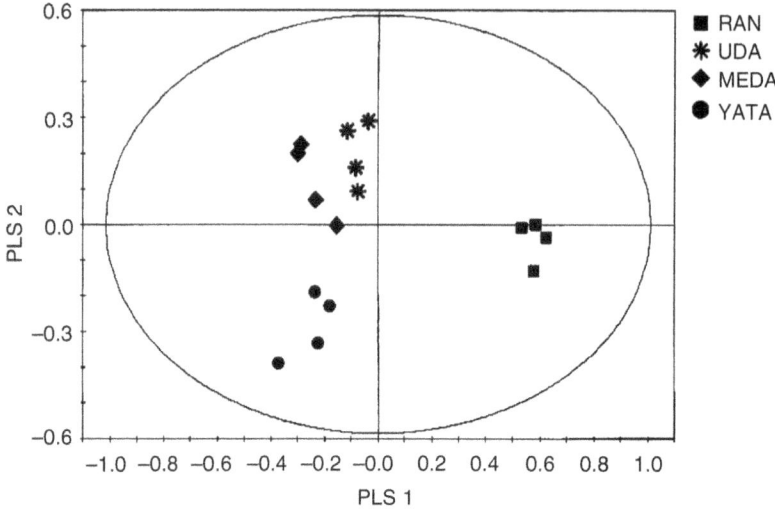

Figure 8.20   PLS-DA score plot derived from the ¹H NMR spectra of black teas cultivated at different altitudes in Sri Lanka: RAN (>1800 m), UDA (1200 m), MEDA (600–1200 m), and YATA (600 m).
Reprinted with permission from ref. 114. Copyright (2011) American Chemical Society.

8.2.4   Juices

Fruit juices are worldwide popular beverages prepared by direct mechanical extraction of a large variety of fruits, and because of their high nutrient content are associated with several health benefits. Obtaining liquid-state NMR spectra of most juices is straightforward with filtration or centrifugation being a necessary step to remove solid particles (pulp), especially in orange and pine-apple juices. An adjustment of the pH is usually employed *via* a buffer solution, especially in metabolomics studies where chemical shift stability demands are more rigorous, or in works aimed at the quantitative analysis of organic acids in juices. Figure 8.22 presents the ¹H NMR spectrum of apple juice, which is dominated by sucrose, glucose and fructose in the sugar region between 3 and 5.4 ppm.

Sugars are the main components of almost all fruit juices with other important organic compound categories being organic acids, amino acids and phenolics. For more details on the assignment of the ¹H NMR spectra of various types of juices, mostly using 2D NMR, the reader is directed to the references provided in Table 8.2.

¹H NMR spectroscopy can be used for the quantitative determination of organic acids and phenolics in juices. Analytical protocols for the quantifica-tion of epicatechin[153] and chlorogenic acid[154] in cider apple juice, formic acid in apple juice,[155] malic and citric acids[156] in apple, apricot, pear, kiwi, orange, strawberry and pineapple juices have been reported and validated. Phlorin

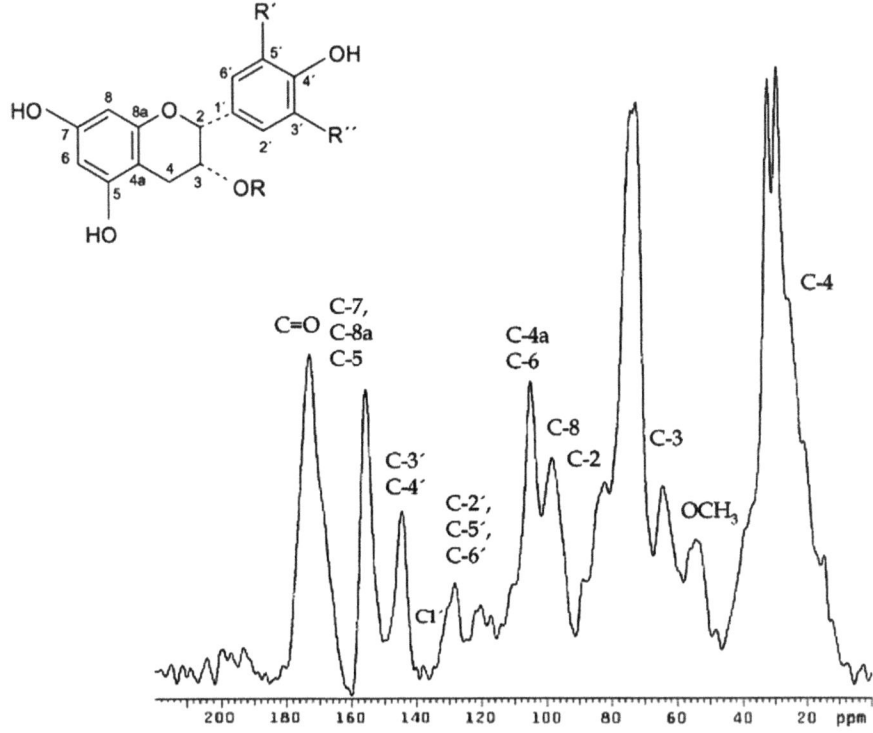

Figure 8.21 75 MHz CP-MAS ^{13}C NMR spectrum of black tea obtained with a con-
tact time of 1.5 ms, a repetition time of 4 s and a spinning rate of 4.25 kHz.
(Reprinted from ref. 145, Copyright (2003), with permission from
Elsevier.)

(3,5-dihydroxyphenyl β-D-glucopyranoside), a marker for peel over-extraction
in orange juice can be quantified by ^{1}H NMR spectroscopy.[157,158]

1D and 2D NMR spectroscopy has been used for the structural analysis of a
wide variety of organic compounds present in low concentrations in juices.
Usually, extracts of juices or chromatographically separated fractions are used
for such analyses. Representative examples of such type of studies include the
identification of pyranoanthocyanidins in black carrot[159] and cherry juice,[160]
polymeric procyanidins in apple[161] and pear juice,[162] and phenolics in pine-
apple,[163] lemon,[164] and noni juice.[165,166] Recently, a procedure that enhances
the efficiency of NMR-based profiling of phenolics in grape and pomegranate
juice has been proposed using solid-phase extraction.[10]

A different way towards spectral simplification for the analysis of minor
components of fruit juices without resorting to prior extraction is the use of
more advanced NMR methodologies, such as hyphenated LC-MS-NMR and
DOSY NMR. Figure 8.23 presents the continuous-flow NMR chromatogram
of mango juice and follows the elution of major sugars and organic acids.[167]

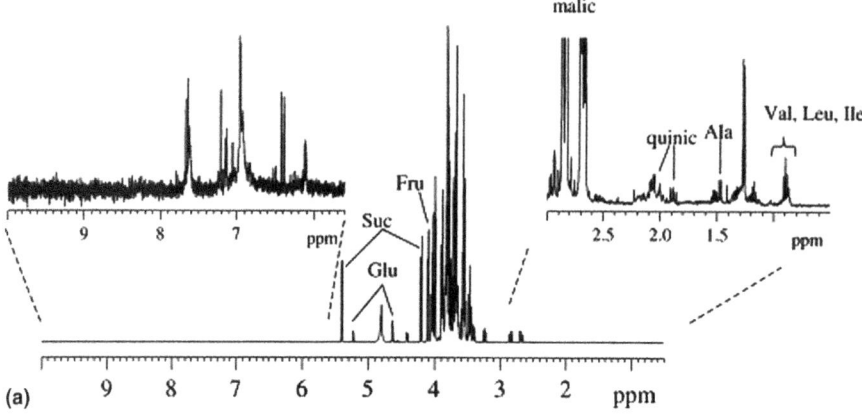

Figure 8.22 ¹H NMR spectrum of apple juice at a field of 500 MHz. Vertical expansions are shown for aliphatic and aromatic regions and some assignments are indicated.
(Reprinted from ref. 60. Copyright (2004), with permission from Elsevier.)

Table 8.2 NMR-based metabolomics studies on fruit juices.

Factor examined	Fruit	Conditions	Multivariate analysis model	Ref.
Variety	Apple	Centrifugation	PCA, CVA	174
Adulteration, pulp wash	Apple	Centrifugation, pH = 3.74	PCA, LDA, PLS	177
Variety	Cider apple	Centrifugation, pH = 2.74	HCA, DA	175
Elevation, rootstock, soil depth	Mandarin	Centrifugation, pH = 6.8	PCA, OPLS-DA	178
Adulteration	Orange	Centrifugation	PCA, DA	173
Differentiation	Grapefruit, orange	Centrifugation, pH = 4.0	EWZS, ICA	179
Sensory quality	Sour cherry	—	PCA	176
Variety	Mango	Centrifugation, pH = 4.2	PCA	170

LC-MS-NMR has also been used successfully for the characterisation of cinnamic acids and their derivatives in grape juice,[41] and carotenoids in tomato and mandarin juice.[168] Diffusion-ordered spectroscopy has been used for the analysis of mango,[167] apple,[60] and tomato juice.[169] A novel route towards the analysis of minor compounds in complex mixtures such as fruit juices involves the used of band-selective NMR spectroscopy[77] utilising the selective excitation of only part of the NMR spectrum. This way, large signals from major components (such as sugars in juices) are completely eliminated from the 1D and

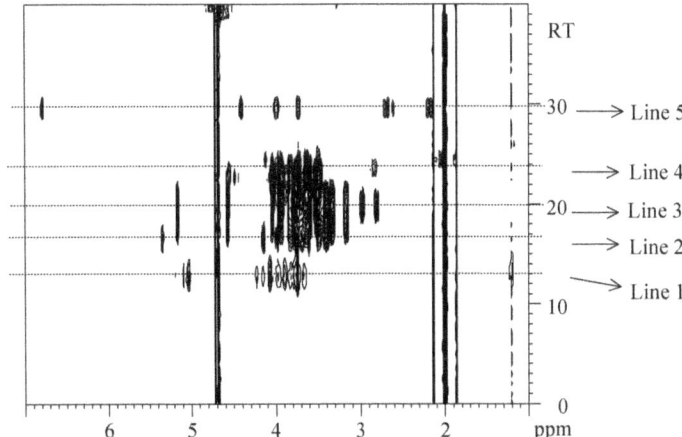

Figure 8.23 Continuous-flow ¹H NMR chromatogram resulting from the elution of mango juice on an ION-300 column (flow rate $0.3\,mL\,min^{-1}$). Compounds identified in lines 1 to 5 are as follows: (1) pectic fraction (no MS ionisation); (2) sucrose + glucose; (3) glucose + fructose + citric acid; (4) fructose + malic acid; and (5) shikimic acid.
(Reprinted from ref. 167 by with permission from Taylor & Francis Ltd.)

2D NMR spectra of the sample under study, allowing the analysis of minor components. Application of band-selective NMR has allowed the superior characterisation of the phenolic fraction of mango juice and the classification of various mango cultivars.[170]

The solid content (pulp) of juices, although it represents a problem during their liquid-state NMR analysis, is nevertheless an integral part of juices as beverages and quite important from a nutritional point of view. Juice pulp can be studied directly by solid-state HR-MAS NMR spectroscopy, and studies of both mango and tomato pulps have been published. Apart from the appearance of a few broad lipid signals, no significant difference was reported between the ¹H NMR spectra of tomato juice and pulp.[169] The HR-MAS analysis of mango pulps was performed in parallel with the liquid-state NMR analysis of mango juice as a function of fruit ripening.[171] Figure 8.24 presents the aliphatic region of the ¹H HR-MAS NMR spectra of mango pulp as a function of increased ripening. Ripening results in a significant decrease in the concentration of lactic acid (signal at δ 2.8 ppm) and an increase in alanine (δ 1.45 ppm), while other changes are also evident in the distribution of sugars and aromatic compounds in both pulp and juice samples.[171]

NMR spectroscopy has been used extensively for the quality evaluation and authentication of fruit juices. Examples include the detection of spoilage and microbial contamination of natural mango juice,[172] and additions of sucrose, beet medium invert sugar and sodium benzoate in orange juice.[173] The application of multivariate analysis models on the NMR-obtained metabolite profile of fruit juices has the ability to contribute significantly towards the

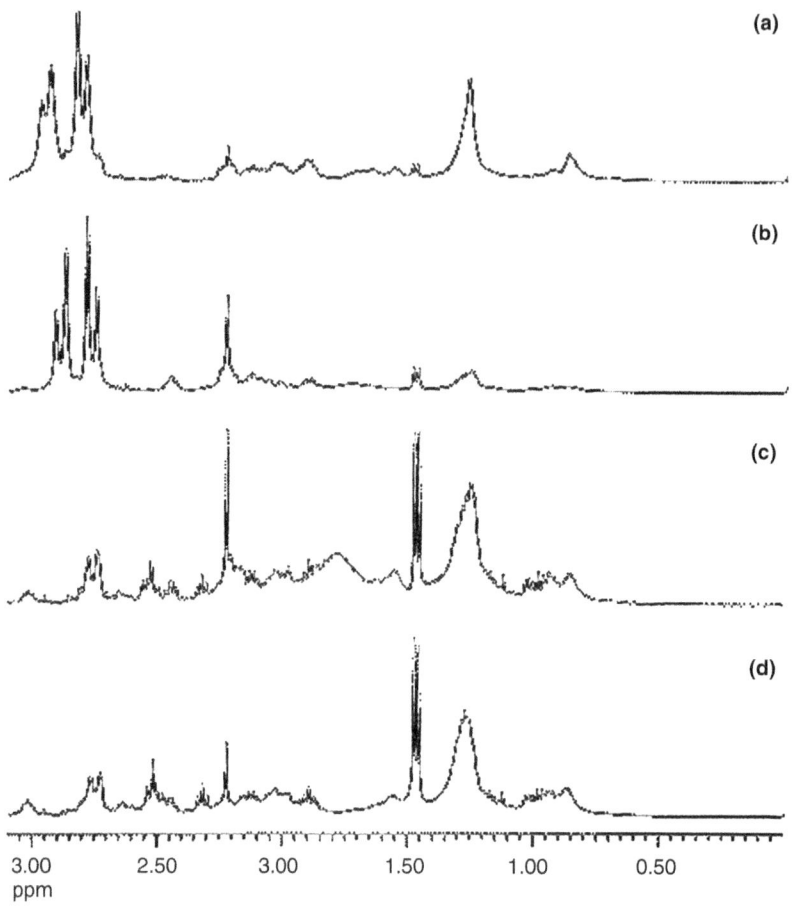

Figure 8.24   Aliphatic region of the ¹H HR-MAS NMR spectra of mango pulps at
different stages of ripening: (a) day 1, (b) day 5, (c) day 15 and (d) day 19.
Spinning rate used was 1200 Hz for all spectra.
(Reprinted from ref. 171. Copyright (2000), with permission from
American Chemical Society.)

authentication, varietal discrimination,[174,175] sensory analysis[176] and other
factors important to fruit juice analysis.[177–179] Metabolomics applications
dealing with fruit juices are summarised in Table 8.2.

The efficiency of NMR-based metabolic profiling as a tool in fruit juice
quality and authentication is highlighted by the introduction of SGF Profi-
ling™, an NMR-based screening method for the quality control of fruit juices in
a fully automated environment.[158] With the help of a large database of
authenticated juice samples, targeted analysis of selected markers is used for the
identification of several types of fraud, while untargeted analysis helps in
geographic origin, fruit content and blending identification.[180]

References

1. U. Anders, F. Tittgemeier and G. Hailer, *Z. Lebensm. Unters. Forsch.*, 1976, **162**, 21.
2. M. Guillou and C. Tellier, *Anal. Chem.*, 1988, **60**, 2182.
3. Y.-Y. Du, G.-Y. Bai, X. Zhang and M.-L. Liu, *Chin. J. Chem.*, 2007, **25**, 930.
4. I. J. Kosir and J. Kidric, *Anal. Chim. Acta*, 2002, **458**, 77.
5. I. J. Kosir and J. Kidric, *J. Agric. Food Chem.*, 2001, **49**, 50.
6. L. Viggiani and M. A. C. Morelli, *J. Agric. Food Chem.*, 2008, **56**, 8273.
7. E. Ferrari, G. Foca, M. Vignali, L. Tassi and A. Ulrici, *Anal. Chim. Acta*, 2011, **701**, 139.
8. I. J. Kosir, B. Lapornik, S. Andrenek, A. G. Wondra, U. k. Vrhovek and J. Kidric, *Anal. Chim. Acta*, 2004, **513**, 277.
9. M. Anastasiadi, A. Zira, P. Magiatis, S. A. Haroutounian, A. L. Skaltsounis and E. Mikros, *J. Agric. Food Chem.*, 2009, **57**, 11067.
10. A. K. Savage, J. P. M. Van Duynhoven, G. Tucker and C. A. Daykin, *Magn. Reson. Chem.*, 2011, **49**, S27.
11. F. M. Amaral and M. S. B. Caro, *Food Chem.*, 2005, **93**, 507.
12. M. V. Holland, A. Bernreuther and F. Reniero, in *Magnetic Resonance in Food Science*, ed. P. S. Belton, I. Delgadillo, A. M. Gil and G. A. Webb, The Royal Society of Chemistry, Cambridge, UK, 1995, pp. 136.
13. M. Caruso, F. Galgano, M. A. Castiglione Morelli, L. Viggiani, L. Lencioni, B. Giussani and F. Favati, *J. Agric. Food Chem.*, 2012, **60**, 7.
14. Y.-S. Hong, *Magn. Reson. Chem.*, 2011, **49**, S13.
15. S. Javier, *Trends Anal. Chem.*, 2010, **29**, 234.
16. H. S. Son, M. K. Ki, F. Van Den Berg, G. S. Hwang, W. M. Park, C. H. Lee and Y. S. Hong, *J. Agric. Food Chem.*, 2008, **56**, 8007.
17. M. A. Brescia, I. J. Kosir, V. Caldarola, J. Kidric and A. Sacco, *J. Agric. Food Chem.*, 2003, **51**, 21.
18. K. Ali, F. Maltese, R. Toepfer, Y. H. Choi and R. Verpoorte, *J. Biomol. NMR*, 2011, **49**, 255.
19. H. G. da Silva Neto, J. B. P. da Silva, G. E. Pereira and F. Hallwass, *Magn. Reson. Chem.*, 2009, **47**, S127.
20. M. C. Buzas, N. Chira, C. Deleanu and S. Rosca, *Rev. Chim. (Bucharest, Rom.)*, 2003, **54**, 831.
21. J. E. Lee, G. S. Hwang, F. Van Den Berg, C. H. Lee and Y. S. Hong, *Anal. Chim. Acta*, 2009, **648**, 71.
22. K. Skogerson, R. Runnebaum, G. Wohlgemuth, J. de Ropp, H. Heymann and O. Fiehn, *J. Agric. Food Chem.*, 2009, **57**, 6899.
23. S. Rochfort, V. Ezernieks, S. E. P. Bastian and M. O. Downey, *Food Chem.*, **121**, 1296.
24. O. Cala, N. Pinaud, C. Simon, E. Fouquet, M. Laguerre, E. J. Dufourc and I. Pianet, *FASEB J.*, 2010, **24**, 4281.
25. H. S. Son, G. S. Hwang, H. J. Ahn, W. M. Park, C. H. Lee and Y. S. Hong, *Food Res. Int*, 2009, **42**, 1483.

26. P. Mazzei, N. Francesca, G. Moschetti and A. Piccolo, *Anal. Chim. Acta*, 2010, **673**, 167.

27. M. C. Todasca, L. Fotescu, N. A. Chira, C. Deleanu and S. Rosca, *Rev. Chim. (Bucharest, Rom.)*, 2011, **62**, 131.

28. C. Absalon, S. Fabre, I. Tarascou, E. Fouquet and I. Pianet, *Anal. Bioanal. Chem.*, 2011, **401**, 1485.

29. H.-S. Son, G.-S. Hwang, K. M. Kim, H.-J. Ahn, W.-M. Park, F. Van Den Berg, Y.-S. Hong and C.-H. Lee, *J. Agric. Food Chem.*, 2009, **57**, 1481.

30. G. E. Pereira, J.-P. Gaudillere, C. van Leeuwen, G. Hilbert, M. l. Maucourt, C. Deborde, A. Moing and D. Rolin, *Anal. Chim. Acta*, 2006, **563**, 346.

31. E. Lopez-Rituerto, F. Savorani, A. Avenoza, J. H. Busto, J. M. Peregrina and S. B. Engelsen, *J. Agric. Food Chem.*, 2012, **60**, 3452.

32. G. E. Pereira, J.-P. Gaudillere, C. van Leeuwen, G. Hilbert, O. Lavialle, M. Maucourt, C. Deborde, A. Moing and D. Rolin, *J. Agric. Food Chem.*, 2005, **53**, 6382.

33. L. Forveffle, J. Vercauteren and D. N. Rutledge, *Food Chem.*, 1996, **57**, 441.

34. K. Ali, F. Maltese, A. M. Fortes, M. S. Pais, R. Verpoorte and Y. H. Choi, *Anal. Chim. Acta*, 2011, **703**, 179.

35. S. Clark, N. W. Barnett, M. Adams, I. B. Cook, G. A. Dyson and G. Johnston, *Anal. Chim. Acta*, 2006, **563**, 338.

36. A. Avenoza, J. H. Busto, N. Canal and J. M. Peregrina, *J. Agric. Food Chem.*, 2006, **54**, 4715.

37. A. Hanganu, M. C. Todasca, N. A. Chira and S. Rosca, *Rev. Chim. (Bucharest, Rom.)*, 2011, **62**, 689.

38. J.-E. Lee, Y.-S. Hong and C.-H. Lee, *J. Agric. Food Chem.*, 2009, **57**, 4810.

39. E. Lopez-Rituerto, A. Avenoza, J. H. Busto and J. M. Peregrina, *J. Agric. Food Chem.*, 2010, **58**, 4923.

40. G. Imparato, E. Di Paolo, A. Braca and R. Lamanna, *J. Agric. Food Chem.*, 2011, **59**, 4429.

41. A. M. Gil, I. F. Duarte, M. Godejohann, U. Braumann, M. Maraschin and M. Spraul, *Anal. Chim. Acta*, 2003, **488**, 35.

42. M. Nilsson, I. F. Duarte, C. u. Almeida, I. Delgadillo, B. J. Goodfellow, A. M. Gil and G. A. Morris, *J. Agric. Food Chem.*, 2004, **52**, 3736.

43. E. J. Waters, Z. Peng, K. F. Pocock, G. P. Jones, P. Clarke and P. J. Williams, *J. Agric. Food Chem.*, 1994, **42**, 1761.

44. A. J. Weekley, P. Bruins, M. Sisto and M. P. Augustine, *J. Magn. Reson.*, 2003, **161**, 91.

45. D. N. Sobieski, G. Mulvihill, J. S. Broz and M. P. Augustine, *Solid State Nucl. Magn. Reson.*, 2006, **29**, 191.

46. G. Autret, G. Liger-Belair, J. M. Nuzillard, M. Parmentier, A. D. De Montreynaud, P. Jeandet, B. T. Doan and J. C. Beloeil, *Anal. Chim. Acta*, 2005, **535**, 73.

47. A. Caligiani, D. Acquotti, G. Palla and V. Bocchi, *Anal. Chim. Acta*, 2007, **585**, 110.
48. E. F. Boffo, L. A. Tavares, M. M. C. Ferreira and A. G. Ferreira, *LWT–Food Sci. Technol.*, 2009, **42**, 1455.
49. R. Consonni and A. Gatti, *J. Agric. Food Chem.*, 2004, **52**, 3446.
50. R. Consonni, L. R. Cagliani, F. Benevelli, M. Spraul, E. Humpfer and M. Stocchero, *Anal. Chim. Acta*, 2008, **611**, 31.
51. M. Cirlini, A. Caligiani and G. Palla, *Food Chem.*, 2009, **112**, 51.
52. A. B. Cerezo, J. L. Espartero, P. Winterhalter, M. C. Garcia-Parrilla and A. M. Troncoso, *J. Agric. Food Chem.*, 2009, **57**, 9551.
53. R. Consonni, L. R. Cagliani, S. Rinaldini and A. Incerti, *Talanta*, 2008, **75**, 765.
54. R. Consonni and L. R. Cagliani, *Talanta*, 2007, **73**, 332.
55. S. Baroni, R. Consonni, G. Ferrante and S. Aime, *J. Agric. Food Chem.*, 2009, **57**, 3028.
56. J. E. Rodrigues and A. M. Gil, *Magn. Reson. Chem.*, 2011, **49**, S37.
57. I. Duarte, A. n. Barros, P. S. Belton, R. Righelato, M. Spraul, E. Humpfer and A. M. Gil, *J. Agric. Food Chem.*, 2002, **50**, 2475.
58. A. Khatib, E. G. Wilson, H. K. Kim, A. W. M. Lefeber, C. Erkelens, Y. H. Choi and R. Verpoorte, *Anal. Chim. Acta*, 2006, **559**, 264.
59. I. F. Duarte, M. Godejohann, U. Braumann, M. Spraul and A. M. Gil, *J. Agric. Food Chem.*, 2003, **51**, 4847.
60. A. M. Gil, I. Duarte, E. Cabrita, B. J. Goodfellow, M. Spraul and R. Kerssebaum, *Anal. Chim. Acta*, 2004, **506**, 215.
61. L. I. Nord, P. Vaag and J. Duus, *Anal. Chem.*, 2004, **76**, 4790.
62. J. E. A. Rodrigues, G. L. Erny, A. S. Barros, V. I. Esteves, T. Brandao, A. A. Ferreira, E. Cabrita and A. M. Gil, *Anal. Chim. Acta*, 2010, **674**, 166.
63. I. F. Duarte, A. Barros, C. Almeida, M. Spraul and A. M. Gil, *J. Agric. Food Chem.*, 2004, **52**, 1031.
64. C. Almeida, I. F. Duarte, A. Barros, J. Rodrigues, M. Spraul and A. M. Gil, *J. Agric. Food Chem.*, 2006, **54**, 700.
65. D. W. Lachenmeier, W. Frank, E. Humpfer, H. Schäfer, S. Keller, M. Mörtter and M. Spraul, *Eur. Food Res. Technol.*, 2005, **220**, 215.
66. J. A. Rodrigues, A. S. Barros, B. Carvalho, T. Brando and A. M. Gil, *Anal. Chim. Acta*, 2011, **702**, 178.
67. D. Intelmann and T. Hofmann, *J. Agric. Food Chem.*, 2010, **58**, 5059.
68. D. Intelmann, G. Kummerlowe, G. Haseleu, N. Desmer, K. Schulze, R. Frohlich, O. Frank, B. Luy and T. Hofmann, *Chem.–Eur. J.*, 2009, **15**, 13047.
69. L. I. Nord, S. B. Sørensen and J. Ø. Duus, *Magn. Reson. Chem.*, 2003, **41**, 660.
70. G. Haseleu, D. Intelmann and T. Hofmann, *J. Agric. Food Chem.*, 2009, **57**, 7480.
71. P. Petrakis, I. Touris, M. Liouni, M. Zervou, I. Kyrikou, R. Kokkinofta, C. R. Theocharis and T. M. Mavromoustakos, *J. Agric. Food Chem.*, 2005, **53**, 5293.

72. R. I. Kokkinofta and C. R. Theocharis, *J. Agric. Food Chem.*, 2005, **53**, 5067.
73. Y. B. Monakhova, H. Schafer, E. Humpfer, M. Spraul, T. Kuballa and D. W. Lachenmeier, *Magn. Reson. Chem.*, 2011, **49**, 734.
74. Y. B. Monakhova, T. Kuballa, J. Leitz and D. W. Lachenmeier, *Int. J. Anal. Chem.*, 2011, **2011**, 704795.
75. Y. B. Monakhova, T. Kuballa and D. W. Lachenmeier, *Int. J. Spectrosc.*, 2011, **2011**, 171684.
76. Y. B. Monakhova, T. Kuballa and D. W. Lachenmeier, *Appl. Magn. Reson.*, 2012, **42**, 343.
77. M. Koda, K. Furihata, F. Wei, T. Miyakawa and M. Tanokura, *Magn. Reson. Chem.*, 2011, **49**, 710.
78. X. H. Wu, X. W. Zheng, B. Z. Han, J. Vervoort and M. J. R. Nout, *J. Agric. Food Chem.*, 2009, **57**, 11354.
79. L. Van-Diep, X. W. Zheng, K. Ma, J. Y. Chen, B. Z. Han and M. J. R. Nout, *J. Inst. Brew.*, 2011, **117**, 516.
80. A. Glabasnia and T. Hofmann, *J. Agric. Food Chem.*, 2006, **54**, 3380.
81. N. Vivas, M. F. Nonier, I. Pianet, N. Vivas de Gaulejac and F. Fouquet, *C. R. Chim.*, 2006, **9**, 1221.
82. C. Gomez-Cordoves, B. Bartolome and M. L. Jimeno, *J. Agric. Food Chem.*, 1997, **45**, 873.
83. A. Nose, T. Hamasaki, M. Hojo, R. Kato, K. Uehara and T. Ueda, *J. Agric. Food Chem.*, 2005, **53**, 7074.
84. A. Nose, M. Myojin, M. Hojo, T. Ueda and T. Okuda, *J. Biosci. Bioeng.*, 2005, **99**, 493.
85. A. Nose, M. Hojo, M. Suzuki and T. Ueda, *J. Agric. Food Chem.*, 2004, **52**, 5359.
86. N. Hu, D. Wu, K. Cross, S. Burikov, T. Dolenko, S. Patsaeva and D. W. Schaefer, *J. Agric. Food Chem.*, 2010, **58**, 7394.
87. S. Okouchi, Y. Ishihara, S. Ikeda and H. Uedaira, *Food Chem.*, 1999, **65**, 239.
88. F. Wei, K. Furihata, F. Hu, T. Miyakawa and M. Tanokura, *Magn. Reson. Chem.*, 2010, **48**, 857.
89. M. Bosco, R. Toffanin, D. De Palo, L. Zatti and A. Segre, *J. Sci. Food Agric.*, 1999, **79**, 869.
90. L. A. Tavares and A. G. Ferreira, *Quim. Nova*, 2006, **29**, 911.
91. G. del Campo, I. Berregi, R. Caracena and J. Zuriarrain, *Talanta*, 2010, **81**, 367.
92. F. Wei, K. Furihata, F. Hu, T. Miyakawa and M. Tanokura, *J. Agric. Food Chem.*, 2011, **59**, 9065.
93. A. Ciampa, G. Renzi, A. Taglienti, P. Sequi and M. Valentini, *J. Food Qual.*, 2010, **33**, 199.
94. F. Wei, K. Furihata, M. Koda, F. Hu, T. Miyakawa and M. Tanokura, *J. Agric. Food Chem.*, 2012, **60**, 1005.
95. L. A. Tavares, A. G. Ferreira, M. M. C. Ferreira, A. Correa and L. H. Mattoso, in *Magnetic Resonance in Food Science*, The Royal Society of Chemistry, Cambridge, UK, 2005, pp. 80.

96. A. J. Charlton, W. H. H. Farrington and P. Brereton, *J. Agric. Food Chem.*, 2002, **50**, 3098.
97. R. Consonni, L. R. Cagliani and C. Cogliati, *Talanta*, 2012, **88**, 420.
98. F. M. Nunes, A. Reis, A. M. S. Silva, M. R. M. Domingues and M. A. Coimbra, *Phytochemistry*, 2008, **69**, 1573.
99. A. Oosterveld, G. J. Coenen, N. C. B. Vermeiden, A. G. J. Voragen and H. A. Schols, *Carbohydr. Polym.*, 2004, **58**, 427.
100. P. Capek, M. Matulov, L. Navarini and F. S. Liverani, *J. Food Nutr. Res.*, 2009, **48**, 80.
101. S. Kreppenhofer, O. Frank and T. Hofmann, *Food Chem.*, 2010, **126**, 441.
102. O. Frank, S. Blumberg, C. Kunert, G. Zehentbauer and T. Hofmann, *J. Agric. Food Chem.*, 2007, **55**, 1945.
103. R. H. Stadler, N. Varga, C. Milo, B. Schilter, F. Arce Vera and D. H. Welti, *J. Agric. Food Chem.*, 2002, **50**, 1200.
104. H. Scharnhop and P. Winterhalter, *J. Food Comp. Anal.*, 2009, **22**, 233.
105. D. Gniechwitz, N. Reichardt, J. Ralph, M. Blaut, H. Steinhart and M. Bunzel, *J. Sci. Food Agric.*, 2008, **88**, 2153.
106. M. L. Mateus, D. Champion, R. Liardon and A. Voilley, *J. Food Eng.*, 2007, **81**, 572.
107. P. Rocculi, G. Sacchetti, L. Venturi, M. Cremonini, M. Dalla Rosa and P. Pittia, *J. Agric. Food Chem.*, 2011, **59**, 8265.
108. G. Venditti, E. Schievano, L. Navarini and S. Mammi, *Food Biophys.*, 2010, **6**, 321.
109. A. Caligiani, D. Acquotti, M. Cirlini and G. Palla, *J. Agric. Food Chem.*, 2010, **58**, 12105.
110. I. M. Figueiredo, N. R. Pereira, P. Efraim, N. H. P. Garcia, N. R. Rodrigues, A. Marsaioli and A. J. Marsaioli, *J. Agric. Food Chem.*, 2006, **54**, 4102.
111. J. E. Lee, B. J. Lee, J. O. Chung, J. A. Hwang, S. J. Lee, C. H. Lee and Y. S. Hong, *J. Agric. Food Chem.*, 2010, **58**, 10582.
112. J. E. Lee, B. J. Lee, J. O. Chung, H. J. Shin, S. J. Lee, C. H. Lee and Y. S. Hong, *Food Res. Int.*, 2011, **44**, 597.
113. A. L. Davis, Y. Cai and A. P. Davies, *Magn. Reson. Chem.*, 1995, **33**, 549.
114. A. Ohno, K. Oka, C. Sakuma, H. Okuda and K. Fukuhara, *J. Agric. Food Chem.*, 2011, **59**, 5181.
115. A. L. Davis, J. R. Lewis, Y. Cai, C. Powell, A. P. Davis, J. P. G. Wilkins, P. Pudney and M. N. Clifford, *Phytochemistry*, 1997, **46**, 1397.
116. N. Kuhnert, *Arch. Biochem. Biophys.*, 2010, **501**, 37.
117. J. J. J. van der Hooft, M. Akermi, F. Y. Ünlü, V. Mihaleva, V. G. Roldan, R. J. Bino, R. C. H. de Vos and J. Vervoort, *J. Agric. Food Chem.*, 2012, DOI: 10.1021/jf300297y.
118. T. Saito, D. Honma, M. Tagashira, T. Kanda, A. Nesumi and M. Maeda-Yamamoto, *J. Agric. Food Chem.*, **59**, 4779.
119. Y. Wang, X. Wei and Z. Jin, *Food Res. Int.*, 2009, **42**, 739.
120. J. Xiao, J. Huo, H. Jiang and F. Yang, *Int. J. Biol. Macromol.*, 2011, **49**, 1143.
121. L. Yang, S. Fu, X. Zhu, L. M. Zhang, Y. Yang, X. Yang and H. Liu, *Biomacromolecules*, 2010, **11**, 3395.

122. T. R. Cipriani, C. G. Mellinger, L. M. De Souza, C. H. Baggio, C. S. Freitas, M. C. A. Marques, P. A. J. Gorin, G. L. Sassaki and M. Iacomini, *J. Nat. Prod.*, 2006, **69**, 1018.

123. P. Zhou, M. Xie, S. Nie and X. Wang, *Sci. China, Ser. C: Life Sci.*, 2004, **47**, 416.

124. M. H. Abd El-Razek, *Asian J. Chem.*, 2007, **19**, 4867.

125. R. Amarowicz and F. Shahidi, *Nahrung–Food*, 2003, **47**, 21.

126. P. Chen and Q. Z. Du, *Chin. J. Chem.*, 2003, **21**, 979.

127. X. Cao, Y. Tian, T. Zhang and Y. Ito, *J. Liq. Chromatogr. Relat. Technol.*, 2001, **24**, 1723.

128. G. M. She, D. Wang, S. F. Zeng, C. R. Yang and Y. J. Zhang, *J. Food Sci.*, 2008, **73**, C476.

129. J. W. Jhoo, *Korean J. Food Sci. Technol.*, 2008, **40**, 707.

130. C. Xiao, H. Dai, H. Liu, Y. Wang and H. Tang, *J. Agric. Food Chem.*, 2008, **56**, 10142.

131. A. Kokotkiewicz, M. Luczkiewicz, P. Sowinski, D. Glod, K. Gorynski and A. Bucinski, *Food Chem.*, 2102, **133**, 1373.

132. B. I. Kamara, E. V. Brandt, D. Ferreira and E. Joubert, *J. Agric. Food Chem.*, 2003, **51**, 3874.

133. Y. Wang, H. Tang, J. K. Nicholson, P. J. Hylands, J. Sampson, I. Whitcombe, C. G. Stewart, S. Caiger, I. Oru and E. Holmes, *Planta Med.*, 2004, **70**, 250.

134. A. L. Piccinelli, F. De Simone, S. Passi and L. Rastrelli, *J. Agric. Food Chem.*, 2004, **52**, 5863.

135. G. A. B. Vieira, M. A. S. Lima, A. M. E. Bezerra and E. R. Silveira, *J. Braz. Chem. Soc.*, 2006, **17**, 43.

136. S. G. Walch, D. W. Lachenmeier, T. Kuballa, W. Stohlinger and Y. B. Monakhova, *Anal. Chem. Insights*, 2012, **7**, 1.

137. A. Pukalskas, P. R. Venskutonis, S. Salido, P. D. Ward and T. A. Van Beek, *Food Chem.*, 2012, **130**, 695.

138. N. Krafczyk and M. A. Glomb, *J. Agric. Food Chem.*, 2008, **56**, 3368.

139. J. Gao, B. Liu, Z. Ning, R. Zhao, A. Zhang and Q. Wu, *J. Food Biochem.*, 2009, **33**, 808.

140. P. Chen, C. Li, S. Liang, G. Song, Y. Sun, Y. Shi, S. Xu, J. Zhang, S. Sheng, Y. Yang and M. Li, *J. Chromatogr., B: Anal. Technol. Biomed. Life Sci.*, 2006, **843**, 183.

141. K. R. Koch, *Analyst*, 1990, **115**, 823.

142. T. Nagata, M. Hayatsu and N. Kosuge, *Phytochemistry*, 1992, **31**, 1215.

143. D. Qi, J. Tong, Y. Sun, S. Chen and S. Luo, *Developments in Food Science,* Elsevier B.V., Amsterdam, 1995, Vol. 37, pp. 827.

144. C. Pascal, F. Pat, V. Cheynier and M. A. Delsuc, *Biopolymers*, 2009, **91**, 745.

145. M. Fujiwara, I. Ando and K. Arifuku, *Anal. Sci.*, 2006, **22**, 1307.

146. G. Le Gall, I. J. Colquhoun and M. Defernez, *J. Agric. Food Chem.*, 2004, **52**, 692.

147. M. P. Y. Lillo, P. N. Sanderson, E. L. Wantling and P. D. A. Pudney, *Dev. Food Sci.*, 2006, **43**, 537.

148. L. Tarachiwin, U. Koichi, A. Kobayashi and E. Fukusaki, *J. Agric. Food Chem.*, 2007, **55**, 9330.

149. A. Martinez-Richa and P. Joseph-Nathan, *Solid State Nucl. Magn. Reson.*, 2003, **23**, 119.

150. S. Kumazawa, K. Kajiya, A. Naito, H. Sait, S. Tuzi, M. Tanio, M. Suzuki, F. Nanjo, E. Suzuki and T. Nakayama, *Biosci., Biotechnol., Biochem.*, 2004, **68**, 1743.

151. Y. Uekusa, M. Kamihira-Ishijima, O. Sugimoto, T. Ishii, S. Kumazawa, K. Nakamura, K. I. Tanji, A. Naito and T. Nakayama, *Biochim. Biophys. Acta, Biomembr.*, 2011, **1808**, 1654.

152. K. Kajiya, S. Kumazawa, A. Naito and T. Nakayama, *Magn. Reson. Chem.*, 2008, **46**, 174.

153. I. Berregi, J. I. Santos, G. Del Campo and J. I. Miranda, *Talanta*, 2003, **61**, 139.

154. I. Berregi, J. I. Santos, G. Del Campo, J. I. Miranda and J. M. Aizpurua, *Anal. Chim. Acta*, 2003, **486**, 269.

155. I. Berregi, G. del Campo, R. Caracena and J. I. Miranda, *Talanta*, 2007, **72**, 1049.

156. G. Del Campo, I. Berregi, R. Caracena and J. I. Santos, *Anal. Chim. Acta*, 2006, **556**, 462.

157. L. M. M. Louche, E. M. Gaydou and J. C. Lesage, *J. Agric. Food Chem.*, 1998, **46**, 4193.

158. M. Spraul, B. Schutz, P. Rinke, S. Koswig, E. Humpfer, H. Schafer, M. Mortter, F. Fang, U. C. Marx and A. Minoja, *Nutrients*, 2009, **1**, 148.

159. M. Schwarz, V. Wray and P. Winterhalter, *J. Agric. Food Chem.*, 2004, **52**, 5095.

160. M. Rentzsch, P. Quast, S. Hillebrand, J. Mehnert and P. Winterhalter, *Innov. Food Sci. Emerg.*, 2007, **8**, 333.

161. L. Yeap Foo and Y. Lu, *Food Chem.*, 1999, **64**, 511.

162. N. E. Es-Safi, S. Guyot and P. H. Ducrot, *J. Agric. Food Chem.*, 2006, **54**, 6969.

163. L. Wen, R. E. Wrolstad and V. L. Hsu, *J. Agric. Food Chem.*, 1999, **47**, 850.

164. Y. Miyake, M. Mochizuki, M. Okada, M. Hiramitsu, Y. Morimitsu and T. Osawa, *Biosci., Biotechnol., Biochem.*, 2007, **71**, 1911.

165. S. C. Yang, J. D. Su, M. Y. Wang and T. C. Tsai, *Taiwanese J. Agric. Chem. Food Sci.*, 2009, **47**, 300.

166. V. Samoylenko, J. Zhao, D. C. Dunbar, I. A. Khan, J. W. Rushing and I. Muhammad, *J. Agric. Food Chem.*, 2006, **54**, 6398.

167. I. F. Duarte, B. J. Goodfellow, A. M. Gil and I. Delgadillo, *Spectrosc. Lett.*, 2005, **38**, 319.

168. C. Tode, T. Maoka and M. Sugiura, *J. Sep. Sci.*, 2009, **32**, 3659.

169. A. P. Sobolev, A. Segre and R. Lamanna, *Magn. Reson. Chem.*, 2003, **41**, 237.

170. M. Koda, K. Furihata, F. Wei, T. Miyakawa and M. Tanokura, *J. Agric. Food Chem.*, 2012, **60**, 1158.

171. A. M. Gil, I. F. Duarte, I. Delgadillo, I. J. Colquhoun, F. Casuscelli, E. Humpfer and M. Spraul, *J. Agric. Food Chem.*, 2000, **48**, 1524.

172. I. F. Duarte, I. Delgadillo and A. M. Gil, *Food Chem.*, 2006, **96**, 313.

173. J. T. W. E. Vogels, L. Terwel, A. C. Tas, F. Van Den Berg, F. Dukel and J. Van Der Greef, *J. Agric. Food Chem.*, 1996, **44**, 175.

174. P. S. Belton, I. J. Colquhoun, E. K. Kemsley, I. Delgadillo, P. Roma, M. J. Dennis, M. Sharman, E. Holmes, J. K. Nicholson and M. Spraul, *Food Chem.*, 1998, **61**, 207.

175. G. Del Campo, J. I. Santos, N. Iturriza, I. Berregi and A. Munduate, *J. Agric. Food Chem.*, 2006, **54**, 3095.

176. M. R. Clausen, B. H. Pedersen, H. C. Bertram and U. Kidmose, *J. Agric. Food Chem.*, 2011, **59**, 12124.

177. G. Le Gall, M. Puaud and L. J. Colquhoun, *J. Agric. Food Chem.*, 2001, **49**, 580.

178. X. Zhang, A. P. Breksa, D. O. Mishchuk and C. M. Slupsky, *J. Agric. Food Chem.*, 2011, **59**, 2672.

179. M. Cuny, E. Vigneau, G. Le Gall, I. Colquhoun, M. Lees and D. N. Rutledge, *Anal. Bioanal. Chem.*, 2008, **390**, 419.

180. M. Spraul, B. Schutz, E. Humpfer, M. Mortter, H. Schafer, S. Koswig and P. Rinke, *Magn. Reson. Chem.*, 2009, **47**, S130.

CHAPTER 9

Fruits and Vegetables

9.1 Fruits

NMR spectroscopy has found increased application in the characterisation of fruits as foods. Fruit growth, ripening (both pre- and post-harvest), storage, freezing and other processing factors on fruit texture and quality have been investigated by high resolution liquid- and solid-state NMR experiments, MRI and relaxometry.

9.1.1 Composition and Metabolite Profiling

The most common NMR methodology used for the study of the chemical composition of fruits is liquid-state NMR spectroscopy of their juices or aqueous extracts, as discussed in detail for several fruits in chapter 8. A plethora of low molecular weight compounds present in fruits can be identified and quantified in such studies, allowing the metabolite profiling of fruits and the study of the factors that affect it, such as cultivar, maturity, *etc*. Metabolic profiling studies on fruits, including kiwi,[1] melon,[2] apple,[3] citrus,[4] papaya[5] and tomato,[6–8] using high-resolution liquid-state NMR are summarised in Table 9.1, and evidently this methodology is enjoying increased popularity in recent years for fruit analysis.

The semi-solid texture of fruits, imposed by high water content, makes HR-MAS NMR spectroscopy a very promising technique for intact fruit analysis. Using this approach, the preparation of aqueous extracts or juices is unnecessary, since fruit tissues can be studied directly and without any sample preparation step. Initial studies demonstrated the ability of solid-state ^{13}C and ^1H HR-MAS spectroscopy to provide fruit spectra with narrow lines, suitable for the study of intact tissues from several types of fruit.[9–11] HR-MAS NMR was subsequently used to monitor compositional changes in intact mango pulp during ripening, in conjunction with liquid-state NMR experiments on mango

RSC Food Analysis Monographs No. 10
NMR Spectroscopy in Food Analysis
By Apostolos Spyros and Photis Dais
© Apostolos Spyros and Photis Dais 2013
Published by the Royal Society of Chemistry, www.rsc.org

Table 9.1 NMR-based chemometrics studies of fruits.

Factor examined	Fruit	NMR methodology	Multivariate analysis model	Ref.
Ripening	Kiwi	HR	PCA	1
Spatial localisation	Melon	HR	PCA	2
Cultivar	Apple	HR-MAS, HR	PCA, PLS-DA	3
Spatial localisation, maturity	Citrus	HR	PCA, PLS	4
Flesh gelling	Papaya	HR	PCA	5
Cultivar	Tomato	HR	PCA	6
Cultivar, fertilisation, ripening	Tomato	HR	PCA, PLS	7
Spatial localisation, maturity	Tomato	HR-MAS	PCA	14
Cultivar, maturity	Tomato	HR-MAS	PCA, ASA	15
Maturity	Tomato	MRI	PLS-DA	18
Growth origin	Cherry tomato	MRI	ANOVA	19
Mechanical damage	Tomato	MRI	PLS	20
Peeling outcome	Tomato	MRI	PLS-DA, SIMCA	21
Sensory attributes	Canned tomato	HR	HCA, PCA	8

juice, and the potential of HR-MAS for the study of the overall biochemistry of fruits non-invasively was recognised.[12] HR-MAS NMR spectra of fruit tissue contain signals from solid components of pulp, such as lipids, biopolymers and other macromolecules. For example, HR-MAS NMR was used for the characterisation of cutin, an insoluble biopolyester present in tomato cuticles.[13] Furthermore, the compositional variability in different tissue parts of a fruit can be revealed with HR-MAS NMR. Figure 9.1 presents the ^{1}H HR-MAS spectra of peel (a), flesh (b), purée (c) and the ^{1}H NMR spectrum of juice (d) obtained from mature red tomatoes.

Lipid peaks do not appear in the juice spectrum (d), while differences in the concentrations of amino acids and other metabolites are evident between peel and flesh spectra in the high-field region of the spectra.[14] Similar metabolite differences were observed in the spectra of apple juice and pulp obtained by HR-MAS NMR spectroscopy,[3] and were attributed to oxidative and fermentation processes during juice preparation, indicating the advantage of HR-MAS in providing an unperturbed metabolite image of intact fruit tissue. HR-MAS has been used successfully for metabolomic studies of apple[3] and tomato[14,15] (see Table 9.1).

Solid-state ^{13}C CP-MAS NMR spectroscopy can provide valuable information regarding high molecular weight compounds present in fruits. Cellulose, pectins, lignins, cutin-like polymers and condensed tannins were identified in aronia (chokeberry), bilberry, blackcurrant and apple by ^{13}C CP-MAS NMR spectroscopy,[16] and a combination of CP-MAS and HR-MAS NMR were used for the characterisation of starch obtained from melon and watermelon seeds.[17]

Finally, it should be mentioned that a number of studies, summarised in Table 9.1, have appeared recently that employ multivariate analysis of MR

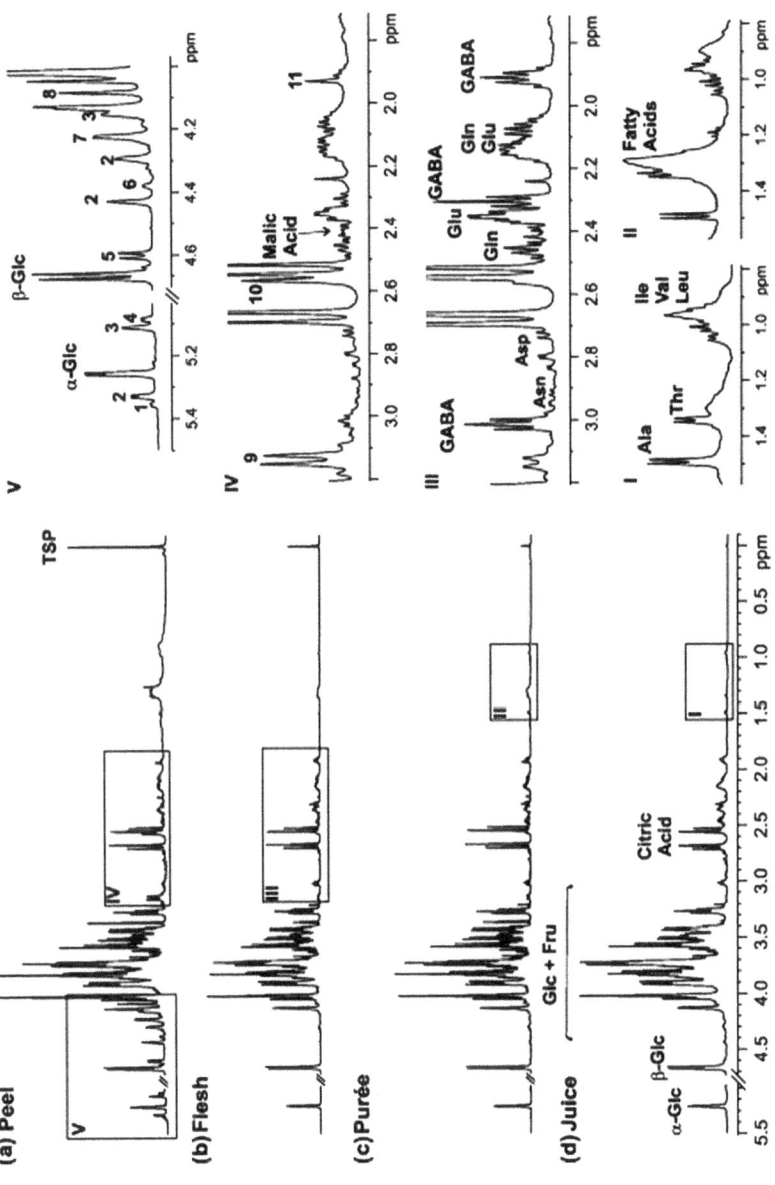

Figure 9.1 ¹H HR-MAS spectra of peel (a), flesh (b), purée (c) and ¹H NMR spectrum of juice (d) from mature red tomatoes. Resonances for selected metabolites are labelled. (Reprinted from ref. 14. Copyright (2010), with permission from Elsevier.)

images for classification purposes. Examples include the use of MRI for classifying tomato maturity,[18] growth origin,[19] and studies that uses multivariate image analysis to assess the mechanical damage[20] and predict the peeling outcome of tomatoes during processing.[21]

9.1.2 Growth and Ripening

We have described in section 9.1.1 some applications of NMR-based metabolomics for the characterisation of ripening in fruits. Another magnetic resonance methodology that provides important information on the physiological and morphological changes of fruits during growth and ripening (both of greater interest to food scientists) is MRI and low-field relaxation studies. The great advantage of MRI is of course its non-invasive character, allowing the examination of changes in the internal characteristics of whole fruits nondestructively.

9.1.2.1 Growth and Maturity

Fruit growth and ripening has received attention and has been studied by MRI and relaxometry by several workers. Figure 9.2 presents a spin-echo "morphological" MR image of a tomato, indicating the tissue nomenclature used.

Early work demonstrated that it was possible to discriminate the onset of ripening spectroscopically before it was obvious externally, based on an increase of MR image intensity of the locular tissue compared to the pericarp wall, attributed to increased locular water content.[22] In a study of cherry tomatoes of different ripening stages it was reported that the content of high mobility water in locular cavities, including seed envelopes, increased as ontogeny advanced, while low mobility water (short T_1) in the pericarp increased in quantity and mobility during maturation.[23] An MRI study of PGI

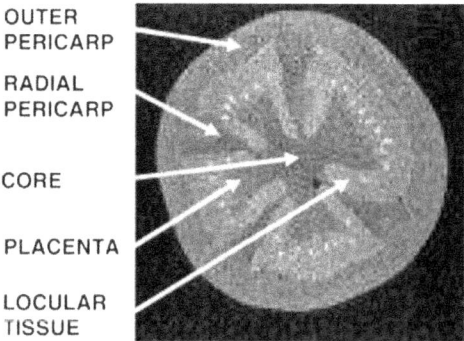

Figure 9.2   SE 'morphological' MR image of tomato, indicating the tissue nomenclature.
(Reprinted from ref. 58. Copyright (2009), with permission from Elsevier.)

Pachino cherry tomatoes (*L. esculentum*, cv. Shiren), harvested in different periods of the year at the same ripening stage showed both morphological and T_1/T_2 differences according to harvest period, with T_1 presenting higher variations.[24] Ripening-related changes in the cell wall's polysaccharide composition and mobility were also identified by solid-state NMR spectroscopy of mature-green and red ripe tomatoes.[25]

Ontogenetic changes of water status and accumulated soluble compounds in growing cherry fruits were examined by a combination of ^{1}H NMR imaging, ^{1}H NMR localised spectral imaging and ^{13}C NMR spectroscopy.[26] It was found that during growth, sugars accumulated in the pericarp with increasing water content, but disappeared with further ripening of the fruit, suggesting that over-maturation of cherries is not desirable because of a loss of sweetness and firmness.

In grape berries, it was possible to follow the ripening process by the MRI estimation of the degrees Brix (°Bx) values of berries for a period of three months using T_1 and T_2 relaxation, which were in good agreement with the refractometry results.[27] Shorter T_1 values were reported for ripe grape berries, associated with the higher sugar content in the ripe stage.

Kiwi fruit ripening has been followed by high-resolution NMR methods and by means of portable unilateral NMR monitoring directly the water status of kiwi fruits during growth by measuring the T_2 spin–spin relaxation time of water. An increase of the T_2 relaxation time was observed during the monitoring period.[1]

MRI and relaxometry for fruits containing a significant amount of oil, apart from water, have also been used to study growth and ripening. The structural distribution of oil and water in olive tissue, and the changes taking place due to ripening were studied by MRI and chemical shift imaging (CSI).[28,29] The large difference in the chemical shifts of water (close to δ 5.0 ppm) and oil (δ 1.2 ppm, fatty acids $-CH_2$) allows the use of selective pulses and the MRI study of the distribution of water and oil in the olive fruit independently. It was reported that during ripening, the oil content increases and is more homogeneously distributed in the pulp, while the water content, both in the seed and in the pulp, is reduced.[28] Another interesting observation was that different olive fruits from the same tree were not at exactly the same ripening stage judging by their MR images, indicating that a large numbers of samples need to be examined if results are to be generalised. MRI studies during growth have also been reported for the sapota fruit (*Manilkara achras* L.)[30] and the oil palm fruit.[31] In the sapota fruit it was found that water declined in seeds and locules but increased in fleshy tissues with ripening. The water content increase in the flesh was accompanied by an increase in sugars concentration.[30] In the oil palm fruit, the multi-exponential fitting of the T_2 relaxation data of the whole fruit resulted in three distinct components assigned to oil, water protons, and less mobile water protons of the shell.[31] Other interesting fruit ripening NMR studies that have been reported include 'Kensington Pride' mango fruit,[32] banana[33,34] and strawberry.[35]

9.1.2.2   Post-harvest Ripening

Post-harvest ripening of fruits is very important, since it critically affects fruit quality and is closely related to the long-term storage of fruits under a variety of conditions, which is a commonly followed practice to increase fruit availability throughout the year. MRI, because of its versatility and non-destructive nature is an ideal tool for studying post-harvest fruit ripening. An interesting example involves the MRI study of post-harvest ripening of tomato that followed the changes observed on the same fruit over a period of three weeks.[36] Figure 9.3 presents a series of typical MRI images of a tomato fruit at three different

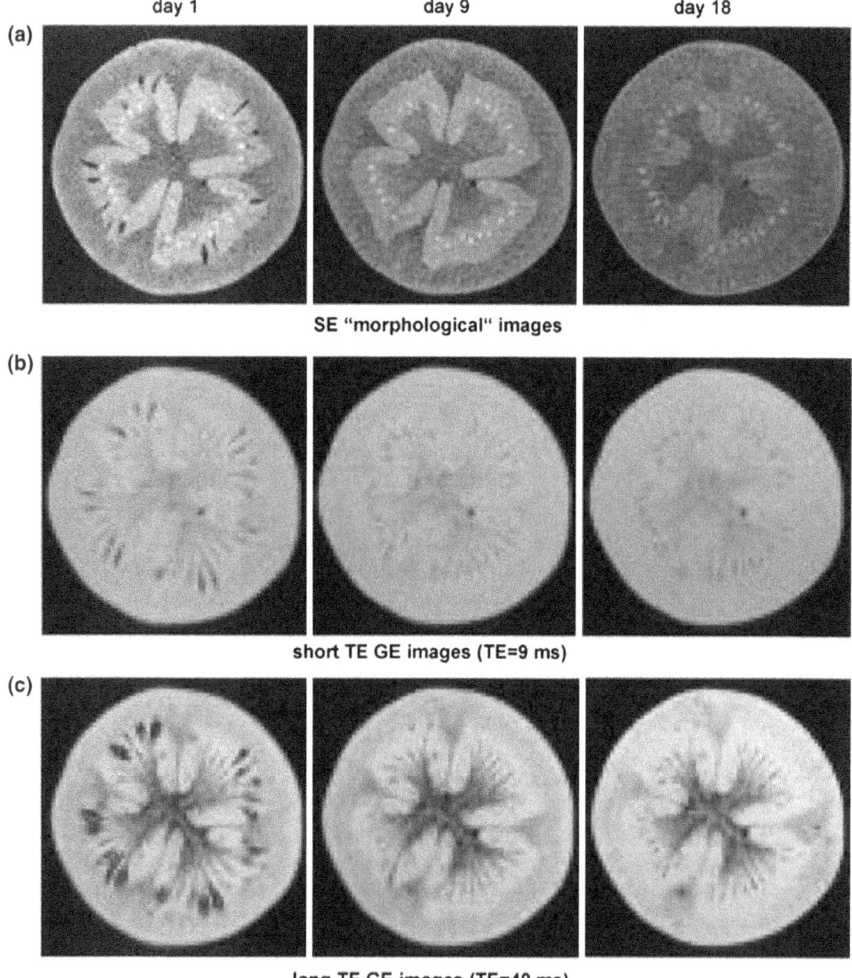

Figure 9.3   Typical 'morphological' SE image (a) along with short TE (b) and long TE (c) GE images of a tomato fruit at three different ripening stages. (Reprinted from ref. 36. Copyright (2009), with permission from Elsevier.)

ripening stages; obtained using spin-echo (SE, (a)), short TE (b) and long TE (c) gradient echo (GE) images, respectively.

The placenta, radial and outer pericarps and locular tissue, as well as the seeds, are obvious on the SE images acquired on days 1 and 9 and the 'long TE' GE images (c), while the contrast was poor on the 'short TE' GE images (b), in which different tissues could hardly be identified. The SE images change considerably as ripening progresses, while the contrast on the GE images remained almost unchanged. Small areas of signal void corresponding to air spaces can be observed close to seeds in the MRI images of day 1. The area of the air spaces, measured by pixel counting in the MRI images, was found to decrease continuously during ripening and disappeared entirely in most, but not all of the fruit samples studied. Another structural effect of post-harvest ripening reported in this study was the development of very small gas bubbles on the outer pericarp of the tomato fruit. Quantitative MRI images helped identify variations in the transverse (T_2) and longitudinal (T_1) relaxation times of tomato tissues. It was found that during post-harvest ripening T_2 decreased by about 25% and T_1 by about 25–30% from their initial values in the core, placenta, radial and outer pericarp, while some water redistribution within different tissues was also evident by the analysis of the relaxation time distribution. However, no obvious correlation between the chemical composition of tomato (sugars, water content) and the relaxation time changes observed during ripening was apparent.[36]

An elaborate attempt to characterise the distribution of sugars and lycopene during tomato post-harvest ripening by employing chemical shift MR imaging (CSI) analysis has been reported.[37] CSI allows the construction of images representing the spatial distribution of signal intensity corresponding to a particular chemical compound (in this case glucose, fructose, sucrose or lycopene). This should be contrasted with typical MRI where the signal providing image contrast is that of water. Figure 9.4 compares the pseudo-coloured images obtained for the same tomato in the mature green and red ripe states using 'classical' T_1-weighted GE (a, b), CSI fused sugar (c, d) and CSI fused lycopene (e, f) imaging.

The CSI sugar images (c, d) reveal non-uniform spatial distribution of sugar content in the tomato and (after averaging out 11 different samples of tomato) show no consistent trend in the variation of sugar signal intensity during ripening. On the other hand, the CSI results (e, f) indicate that the lycopene content significantly increased in the outer pericarp and columella of the tomatoes at the red ripe stage, as compared to the mature green stage. These observations regarding the sugar and lycopene distribution in various tissues were corroborated by HPLC analysis.[37]

Apart from tomato, structural changes due to post-harvest ripening and their relationship with T_1 and T_2 relaxation changes have been studied by NMR spectroscopy of extracts and MR imaging/relaxation for several other fruits.

In most cases, a straightforward relationship between relaxation and fruit compositional changes during ripening (such as water or sugar content) was not possible to be established. For example, in kiwi fruit (*Actinidia deliciosa var*

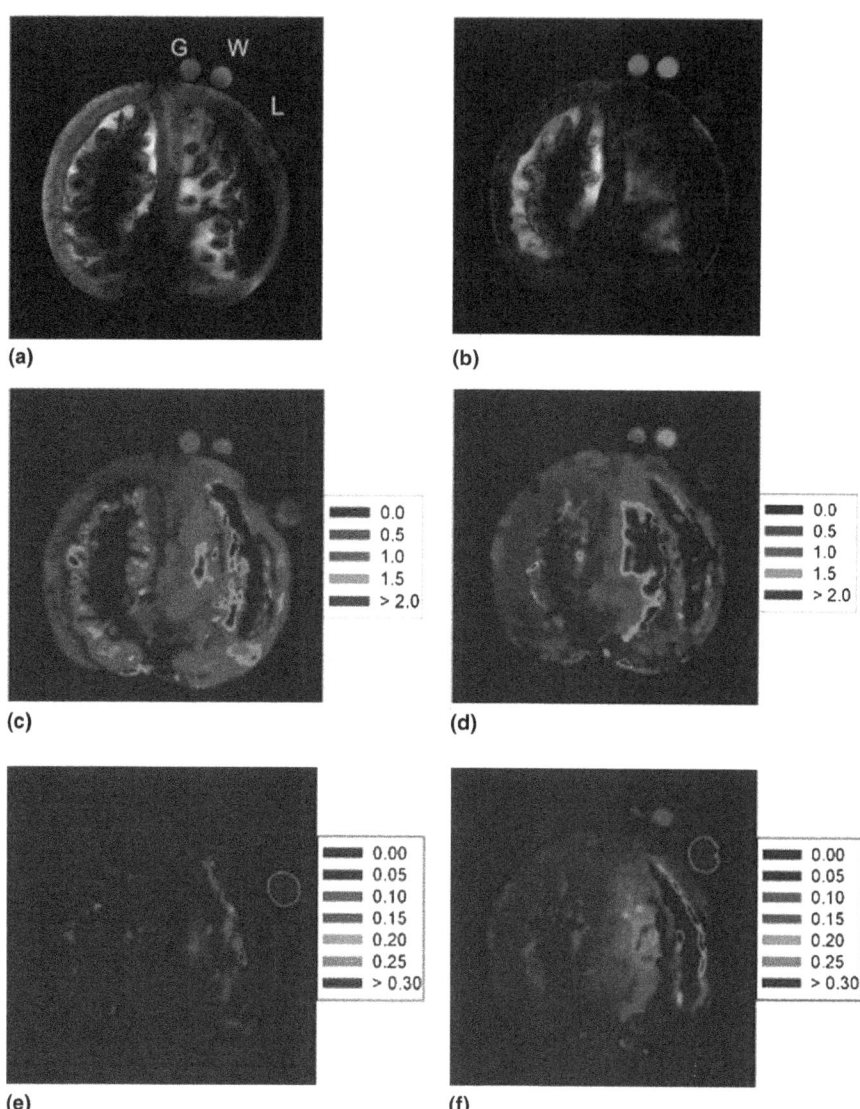

Figure 9.4 T_1-weighted GE image and CSI analyses of CSI fused sugar and lycopene images of a tomato in the mature green and red ripe stages. (a) T_1-weighted GE image of the tomato in the mature green stage; (b) T_1-weighted GE image of the tomato in the red ripe stage; (c) CSI fused sugar images of the tomato in the mature green stage; (d) CSI fused sugar images of the tomato in the red ripe stage. (e) CSI fused lycopene image of the tomato in the mature green stage; (f) CSI fused lycopene image of the tomato in the red ripe stage.
(Reprinted from ref. 37. Copyright (2011), with permission from Elsevier.)

deliciosa) one study found that a 200% increase in sugar content after one month of post-harvest ripening had no significant effect on the T_1 and T_2 relaxation of water protons,[38] while in another study only a small reduction in T_2 during ripening was observed, and associated with 'softening' of the fruit during storage.[39] Similarly, in persimmon fruit, spin–lattice relaxation times T_1 increased sigmoidally during growth but decreased abruptly during ripening in both vascular and mesocarp parenchyma tissue.[40] In freshly harvested cherimoya fruit mesocarp tissues T_1 increased during the first two days of post-harvest ripening, followed by a 33% decrease on the third day, associated with the increase of total soluble contents, mainly sugars.[41]

A comparison of the effect of natural and artificial (acetylene mediated) ripening on the molecular constituents of the harvested fruit of sweet lime (*C. limettioides*) was made with the help of volume-selective NMR spectroscopy.[42] Using this methodology, the ^{1}H NMR spectrum of a very small volume of a sample can be acquired selectively. Figure 9.5 presents representative GE images of sweet lime and the ^{1}H NMR spectrum of a small voxel indicated with a square in the first image indicating signals from glucose, sucrose, and citric acid.

Using this methodology, it was reported that during natural ripening the sugar-to-acid ratio and ethanol increased as a function of sweet lime ripening time. Acetylene mediated ripening was found to proceed faster than natural ripening, based on increased ethanol production in volume-selected spectra.[42]

Analysis of MRI images by mathematical modelling can be very useful in the characterisation of fruit ripening. Texture analysis (TA) is a generic name for a series of techniques used for the quantification of spatial variation of grey tones in images. As an example, MRI images of three apple varieties, Idared, Redspur and Topaz, obtained during post-harvest ripening and storage were analysed by TA to determine the correlation between TA parameters and variables related to ripening measured separately, such as firmness, soluble solids content (SSC) and titratable acids. A good correlation was demonstrated with several TA parameters showing that this technique can be used for characterising apple ripening.[43] Finally, water loss during post-harvest ripening has been studied by MRI and relaxometry in kiwi fruit[44] and pear.[45]

9.1.3 Quality Evaluation

There are several quality attributes of fruits that can be evaluated using magnetic resonance techniques, including mealiness, internal and external bruising, internal browning, dry regions, worm damage, stage of maturity, and presence of voids, seeds, and pits.[46] The development of on-line NMR sensors capable of recognising such attributes is a very active field of research with great implications for automated fruit evaluation in an industrial environment.[46–48]

9.1.3.1 *Morphological Analysis*

In the case of factors that affect the macroscopic shape of fruits, detection by MRI is facilitated since large voids or external bruising of fruits can be detected

(a)

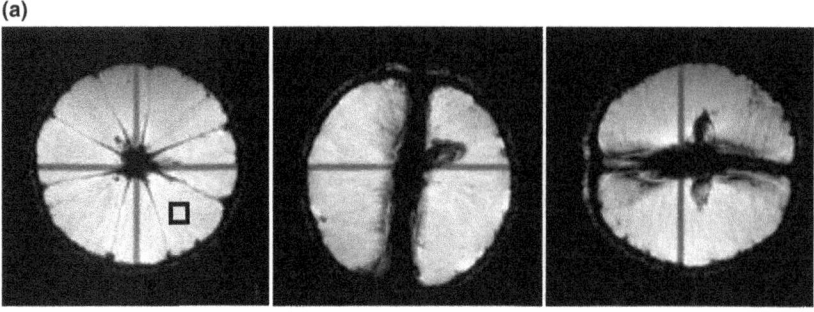

(b)

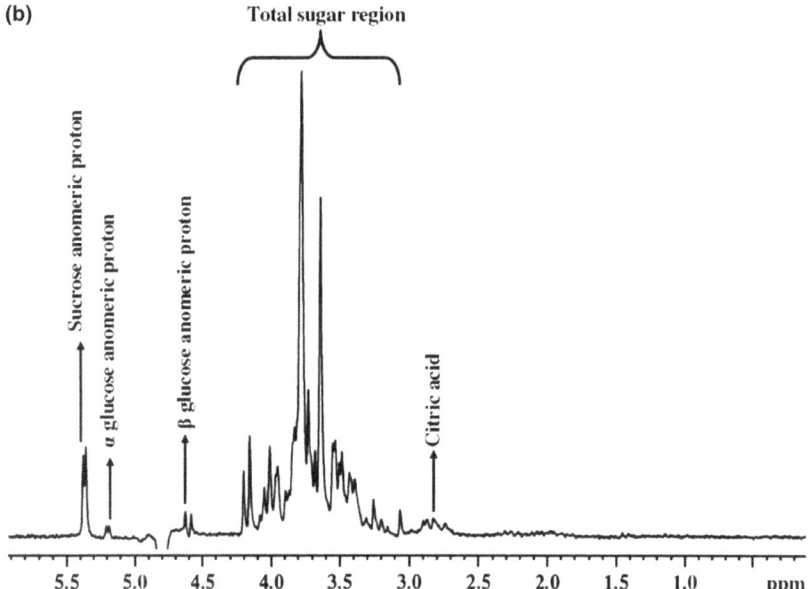

Figure 9.5   (a) GE images of sweet lime (*Citrus limettioides*) collected with a matrix size of 128 × 128: echo time, 6 ms; repetition time, 100 ms; slice thickness, 2 mm; FOV, 8 cm. (b) Spectrum obtained from a 4 mm voxel (marked with a square in the first image of (a)) of a sweet lime at the green-skinned stage. (Reprinted from ref. 42. Copyright (2009), with permission from American Chemical Society.)

by 1D profiling without the need to resort to spectroscopic means of image contrast enhancement, such as T_1 or T_2 differences. In fact, because of the time constraints imposed by the need to examine fruits very fast, usually on a conveyor line, 1D profiling is frequently the only feasible alternative. This for example is the case for detecting pits in olives[49] and cherries.[50] On the other hand, MRI has been used for the detection of voids in melon,[51] and bruising in pears[52] and peaches.[53] The detection of seeds in fruits that are supposed to be seedless is another interesting application for MRI. Seeds contain far less water

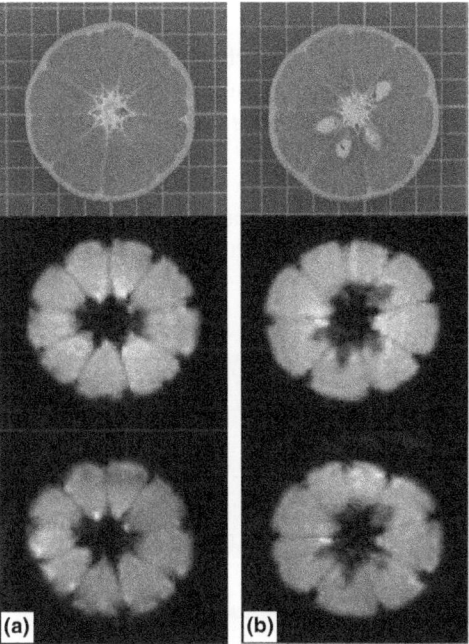

Figure 9.6 Examples of (a) seedless and (b) seed-containing mandarins; upper line corresponds to red green blue (RGB) images; middle line to static MR images, and bottom line to motion-corrected MR images.
(Reprinted from ref. 57. Copyright (2006), with permission from Elsevier.)

than the pulp, and thus provide relatively easy image contrast for identification. Efforts for fast MRI seed detection in mandarins have been attempted using both a high-field tomograph[54] and a low field permanent magnet[55] under static conditions. On-line seed detection in citrus[56] and mandarin[57] has been demonstrated using a fast low-angle shot sequence (FLASH) GE pulse sequence and subsequent image analysis. Figure 9.6 presents examples of (a) seedless and (b) seed-containing mandarin images, from top to bottom photographic, MR static and MR motion-corrected.

In this study, seedless mandarins were correctly sorted under motion conditions to 92.5%, but the classification performance decreased for fruits with seeds to 79.5%.[57]

Another example of morphological feature analysis in fruits using MRI is the determination of grape volume within a cluster of grapes, providing both volume measurements of single berries and their distribution within a cluster.[27]

MRI has been applied to study several structural aspects of the tomato fruit, such the analysis of differences in air-bubble content between tomato tissues towards their differentiation.[58] MRI has also been proposed as a new method for quantifying the apparent microporosity of tomatoes and apples,[59] with MRI results successfully validated using X-ray microtomography. However, MRI was not as successful as an indicator of tomato firmness during processing.[60]

9.1.3.2   Defects and Disorders

Mealiness is an internal quality defect characterised by a lack of juiciness and a soft, sandy texture when chewing and affects fruit such as apples, peaches and nectarines. Significant differences were found in the histograms of T_2 maps obtained from MRI images of mealy and non-mealy apples, as depicted in Figure 9.7.

Mealy apples showed a skew histogram combined with a 'tail' in the high T_2 extreme, a feature not present in the histograms of non-mealy apples.[61] In a

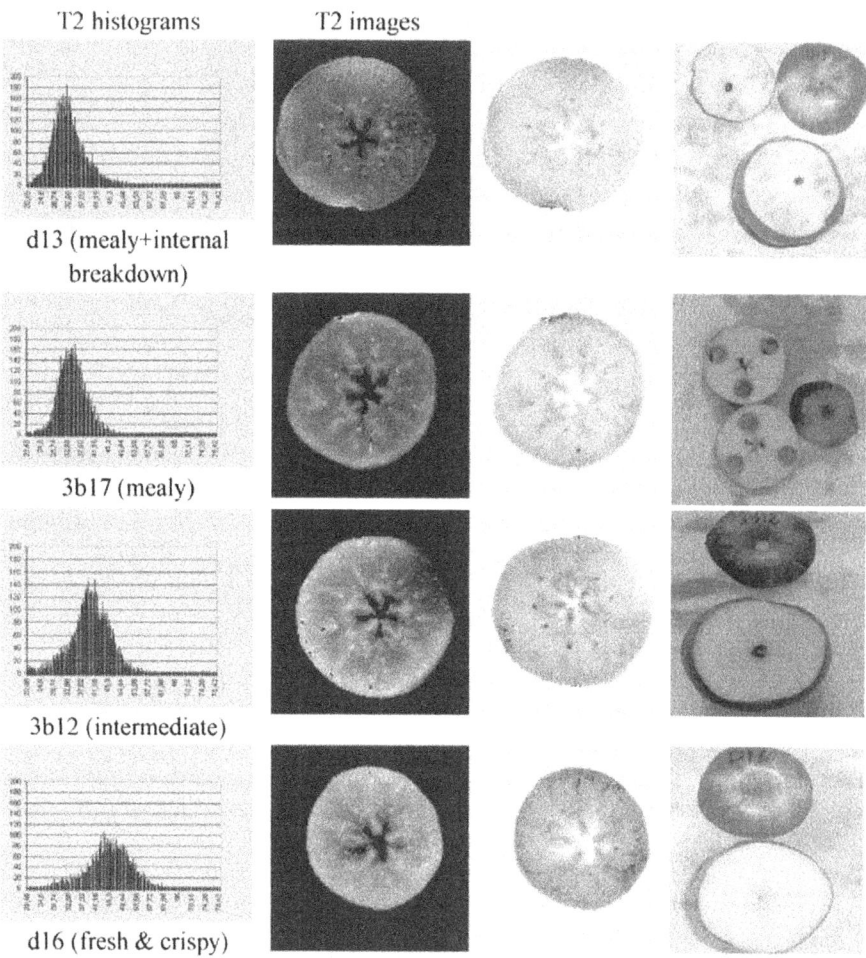

Figure 9.7 T_2 histograms and respective MRI images of two mealy apples (first and second line from top, the first one also showing internal breakdown) and intermediate and fresh fruits (third and fourth line, respectively). The third column shows the negative image of the T_2 map and the forth column a photograph of the halved fruits.
(Reprinted from ref. 61. Copyright (2000), with permission from Elsevier.)

previous study, a similarity was observed between histograms of an apple showing internal breakdown and a mealy one, indicating a possible association between internal breakdown and mealiness in apples,[62] and this is also confirmed in the top line histogram of Figure 9.7. Experiments on peaches showed that MRI was not as successful in identifying wooliness, the term used for mealiness in peaches.[61]

In a later study, it was reported that the main water T_2 peak measured in whole apple decreases with increasing mealiness index, and correlated well with mechanical and chemical methods of mealiness assessment, indicating that non-spatially resolved NMR relaxation measurements might be used for rapid on-line mealiness assessment in apples.[63] Mealiness in apples was also investigated with a comprehensive set of 2D NMR relaxation and diffusion experiments at several magnetic field strengths.[64] The T_1 relaxation time of the proton peak associated with the cell wall was found to be much longer in mealy apples than in fresh apples, and was suggested as a possible differentiating NMR parameter for on-line mealiness evaluation in combination with a second parameter, such as the NMR-derived °Bx determination.

The ability of MRI to unravel the internal structure of fruits non-destructively has been used for the detection and analysis of the effects of several physiological disorders of fruits. Watercore is a physiological disorder affecting apples, resulting in the filling of intercellular spaces with liquid. For some apple cultivars, watercore can be ameliorated during long-term storage without any effect on fruit quality. MRI was used successfully for the differentiation of normal and watercore tissues of *Red Delicious* apples (*Malus domestica* Borkh.),[65] and to follow the loss of watercore in *Fuji*[66] and *Braeburn*[67] apples during storage. In *Red Delicious* and other cultivars however, watercore may develop into severe internal browning, a condition that makes fruits unmarketable. Internal browning may also develop during prolonged storage of fruits under controlled atmosphere, especially elevated CO_2 levels. Several MRI and relaxometry studies have attempted to differentiate between normal and internal browning-affected apples[68–70] and pears.[71–73] Although differences in T_2 make the identification of the disorder possible in MR images, relaxation measurements on whole fruits using low field spectrometers are more promising for on-line detection of the disorder in conveyor belts, because of the rapidity of the analysis.

Other common fruit disorders have been studied by MRI, in efforts to ensure fruit quality for the consumer. Superficial scald is a physiological disorder that affects fruits such as apple and pear, characterised by damage to the surface of the fruit. [1]H MRI of scalded *Red Delicious* apples revealed an increase in the mobility of water protons or an increase in the ratio of free to bound water in the epidermal layer and peripheral cortical tissue.[74] MRI was also used to detect moisture distribution differences between sound and mouldy samples of 'Ichidagaki' dried persimmon fruit.[75]

Black heart, also know as 'heart rot', is a major pomegranate disease which although recognised as a post-harvest quality problem, fruit infection begins in the orchard. T_2-weighted fast SE images were acquired for healthy and pomegranates with black heart. Histogram features of images, including mean, median, mode, standard deviation, skewness, and kurtosis, were examined using PLS-DA. The PLS-DA model based on histogram features of MR images showed 92% accuracy in detecting the presence of black heart in pomegranate fruit.[76] Water-soaking, a physiological disorder characterised by a glassy texture of the flesh was studied using MRI in melons, demonstrating that water mobility increased in the diseased tissues.[77]

9.1.4 Processing

During processing, fruits face conditions that deviate from normal, such as increased heat, cold or freezing, high pressure, salinity, or interaction with various chemical substances used in processing. NMR spectroscopy and MRI have been used to understand the effect of such factors on fruit physiology and the quality of fruits as products after processing.

9.1.4.1 *Osmotic Dehydration*

Osmotic dehydration is a method used for the partial removal of water from fruits and vegetables by immersion in concentrated aqueous solutions of sugars and/or salts. It as also used in fruit processing as a pre-treatment prior to freezing, freeze-drying, vacuum drying and air-drying. The internal changes in the fruit and the kinetics of both moisture change and mobility during osmotic dehydration of strawberry,[78] apple[79,80] and kiwi[81,82] fruits has been studied using MRI and low-field NMR relaxometry in combination with other techniques.

1D MRI profiling was used to follow temporal and spatial changes in water mobility, water content, and structural shrinkage of strawberry slices during osmotic dehydration with aqueous sucrose over a 2 h interval.[78] After 2 h of osmotic dehydration, the T_2 values of water protons decreased, especially in the first few mm of the slice in contact with the sucrose solution, indicating reduced molecular mobility as water is lost to the sucrose solution, sucrose is added and the fruit cells collapse. The osmotic dehydration of apple in concentrated sucrose solutions was followed by LF-NMR T_1 and T_2 relaxation measurements.[79] A rapid decrease of both relaxation times was found between 0 and 3 h of dehydration, indicating a reduction of mobility of the water molecules in the cells due to greater sucrose concentration and interaction, along with shrinkage of the apple cells. Further experiments on apple osmotic dehydration using MRI showed that, macroscopically, water flows from the surface to the osmotic solution and from the core to the surface at the same time during dehydration, indicating that the osmotic treatment process cannot be explained by using only diffusion-based mechanisms.[80]

In kiwi fruits (*Actinidia deliciosa* and *Actinidia chinensis*), osmotic dehydration experiments were performed in 61.5 % (w/v) sucrose solutions at various temperatures.[81] The T_2 relaxation decay of water protons was deconvoluted using three exponential factors representing the cell vacuole, cytoplasm plus extracellular water and cell wall-bound water.[81,82] Figure 9.8 presents the evolution of the three different proton pools and their T_2 values during a 300 min osmotic dehydration treatment of an *A. chinensis* kiwi fruit at 25 °C.

The vacuole water T_2 component decreases with dehydration, while the extracellular water increases and the cell wall water remains approximately the same. Osmotic dehydration at higher temperatures intensified the leakage of water from the vacuole for both kiwi cultivars studied. The T_2 relaxation time of the vacuole water also decreases in Figure 9.8, as a result of the vacuole shrinkage during osmotic dehydration.[82]

9.1.4.2 Heat Stress

Heat disinfestation, either by vapour heat treatment or hot water dip, is a popular method employed for fruit disinfestation, but may lead to fruit injury. The possibility of using MRI to monitor hot water treatment-induced internal defects in mango fruit was examined.[83] Heat-injured areas could be identified already on the day of treatment by MRI, and were characterised by relatively low water levels (low signal intensity) corresponding to air filled cavities and 'islands' of starchy mesocarp. Physiological injury induced by vapour heat treatment has also been studied in papaya fruit,[84] where injured areas were also associated with low-intensity regions in the MRI images. The effect of heat stress has also been studied in tomatoes, where a change in water mobility during heating was observed by MR imaging.[85] When heating at 42 °C was continued for more than 12 h, a shortening of the T_2 relaxation time of the locular tissues was reported, and T_2 remained low even after one day. This result indicated that recovery of the tomato fruit from heat stress is not complete after one day, although ethylene production and respiration rate recovered almost immediately after heating.

9.1.4.3 High-Pressure Processing

High-pressure processing is a food processing method that uses elevated pressures (up to 6000 atm) and, sometimes, high temperature to achieve microbial inactivation. High-pressure can significantly alter the tissue of fruits and vegetables and induce changes in their structure and texture. ^{1}H NMR relaxometry has been used to investigate pressure-induced changes in the structure and water distribution of fresh strawberries (*Elsanta* variety), in conjunction with optical microscopy.[86] A low pressure of 35 MPa induced no visible tissue damage, but was detectable in the water relaxation behaviour suggesting subcellular water redistribution. Higher pressures of 100 MPa appeared to rupture membranes and cause major water redistribution, while cell wall damage was apparent at pressures of 300 MPa. Another study used MRI and T_1, T_2 and diffusion coefficient variations of water to study the effect of high-pressure processing on strawberries.[87] Figure 9.9 presents the values of

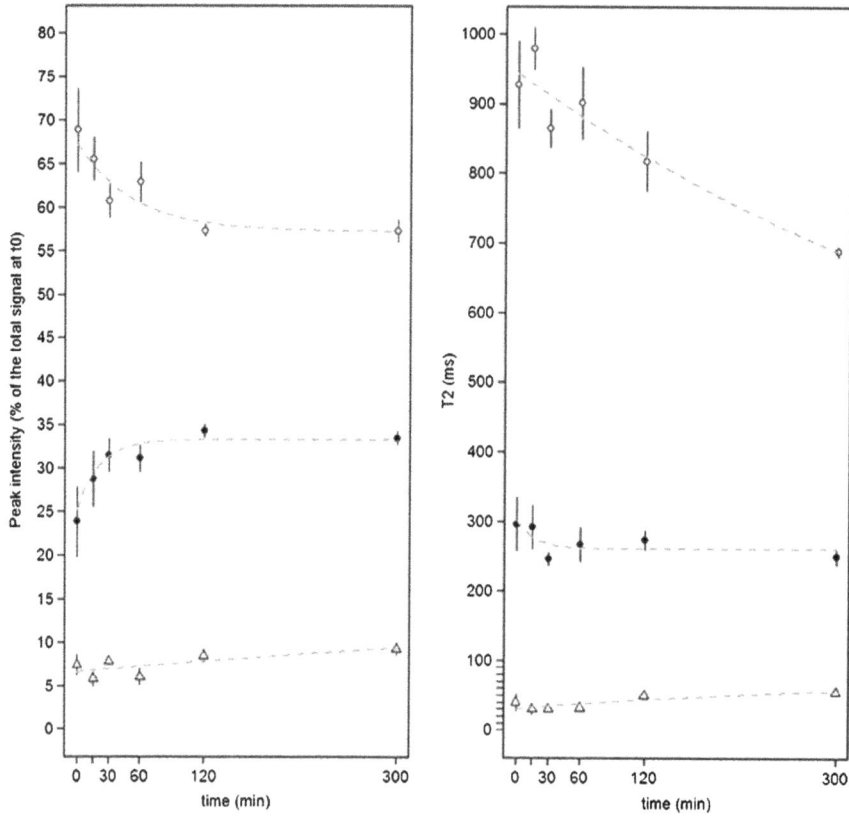

Figure 9.8   Intensities and T_2 of *A. chinensis* kiwi fruit proton pools during osmotic treatment at 25 °C: vacuole (*open circle*), cytoplasm and extracellular spaces (*filled circle*), cell walls (*open upright triangle*). To help in visualising the value trends, the points are fitted to monoexponential curves presented as dashed lines.
(Reprinted from ref. 81. Copyright (2011), with permission from Springer Science and Business Media.)

T_2, T_1 and apparent diffusion coefficient (ADC) extracted from the MR images of strawberries under different pressure treatments.

T_1 and the diffusion coefficient of water showed a clear dependence on the pressure applied, attributed to the destruction of biological barriers and the loss of cell compartments produced by pressure, inducing major water redistribution in the tissues. No significant compositional effects of high-pressure processing were reported by using HR-MAS ^{1}H NMR spectroscopy to analyse intact strawberry tissues.[87] An apparent decrease in the sucrose concentration of strawberry tissue as a function of pressure, accompanied by parallel increases in glucose and fructose, was attributed to increased enzymatic hydrolysis of sucrose during storage of the fruits, facilitated by pressure-induced cell de-compartmentalisation.

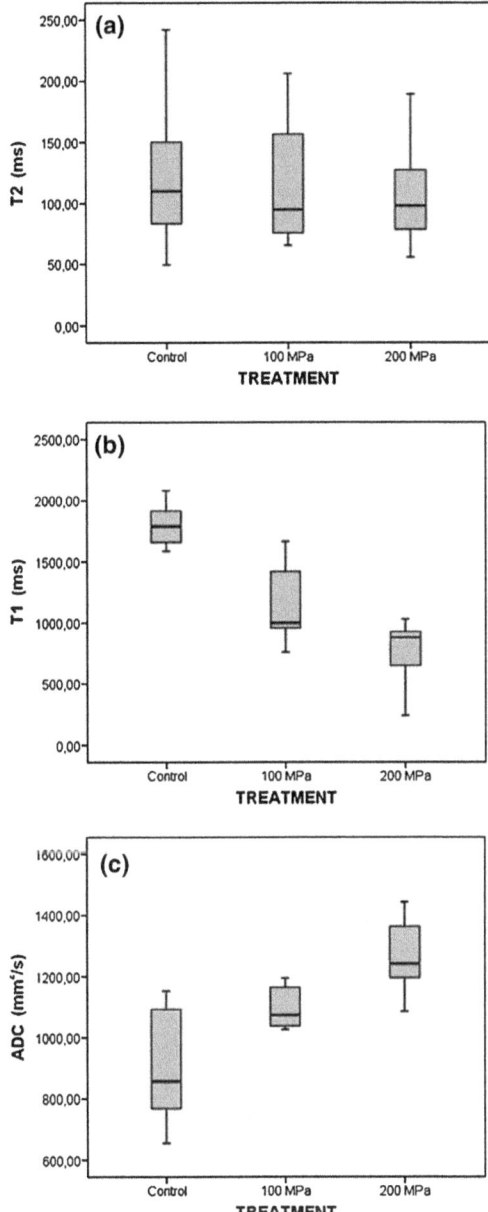

Figure 9.9 Box and whisker plots for MRI parameters in control and pressure treated strawberries: (a) spin–spin relaxation time (T_2); (b) spin–lattice relaxation time (T_1); and (c) apparent diffusion coefficient (ADC). Each box contains the middle 50% of the data and the end of the whiskers indicates the minimum and maximum data values. The dark line in the box represents the median.
(Reprinted from ref. 87. Copyright (2009), with permission from Elsevier.)

9.1.4.4   Storage and freezing

Freezing ($T<0\,^\circ$C) and chilling ($T>0\,^\circ$C) are important processing procedures aimed in ensuring the preservation of fruits during transportation and storage without (or with minimal) effect on their quality attributes. Freezing, which may also occur during fruit growth in the field (freeze damage), changes the water status and distribution in fruit tissues, since the ice crystals formed inside the fruit may cause migration of water out of the cells, or from one region to another. Exposure to low temperatures changes fruit physiology and may lead to fruit injuries upon thawing, creating quality issues that make fruits unsuitable for marketing. MRI and LF-NMR relaxometry can monitor changes in the water state and texture of fruits non-invasively, and are therefore, ideal tools for monitoring freezing and chilling injuries in fruits. Initial studies of MRI freezing and chilling injuries were focused on blueberry[88] and persimmon[89,90] fruits. The changes in blueberry tissue structure caused by freezing/thawing could be mapped by MRI both as a function of changes in water T_1 and T_2 relaxation.[88] The development of chilling injuries on persimmon fruit stored for several weeks at two different temperatures, 0 and 7 °C, was studied by MRI, with fruits either sealed in individual polyethylene bags under a protective atmosphere or under normal conditions. No clear sign of injury development during cold storage at 0 or 7 °C could be detected by MRI or relaxation time changes, although the chilling injury developed rapidly after fruits were returned to ambient temperature.[89,90]

Low-field relaxometry has been used successfully to monitor the changes in subcellular compartments in the parenchyma tissue of apple during drying, freezing and rehydration.[91] Figure 9.10 depicts the changes observed in the T_2 relaxation time distribution of water measured during the freezing of apple tissue from 22 °C to -25 °C (note the scale changes as the temperature decreases).

At -3 °C the peak corresponding to the vacuole at 652 ms has disappeared, indicating this is the first compartment to freeze (the T_2 of ice is only a few microseconds, and thus does not contribute to these experiments). Cytoplasm water (initially between 100–200 ms) vanishes at $ca.$ -10 °C and cell wall water (1–50 ms) below -25 °C, indicating the freezing order of these compartments.[91] MRI has been used to monitor the macroscopic freezing of whole kiwi fruit by visualising the progressive loss of signal starting from the epidermis and progressing towards the fruit core.[92] T_2 relaxation times were reported to be significantly shorter (and water diffusion greater) in frozen – thawed fruit compared with fresh fruit.

Freeze damage in oranges and mandarins is usually not visible on the surface of the fruit, and may affect fruits during cold weather conditions. Over time the freeze-damaged interior portions of the fruits become dry, causing the fruit to become unmarketable. MRI and relaxation measurements have been used as a means to evaluate orange and mandarin quality non-destructively using low field NMR spectrometers suitable for deployment in an industrial environment. The peel and flesh of oranges subjected to chilling (5 °C) and freezing (-7 °C)

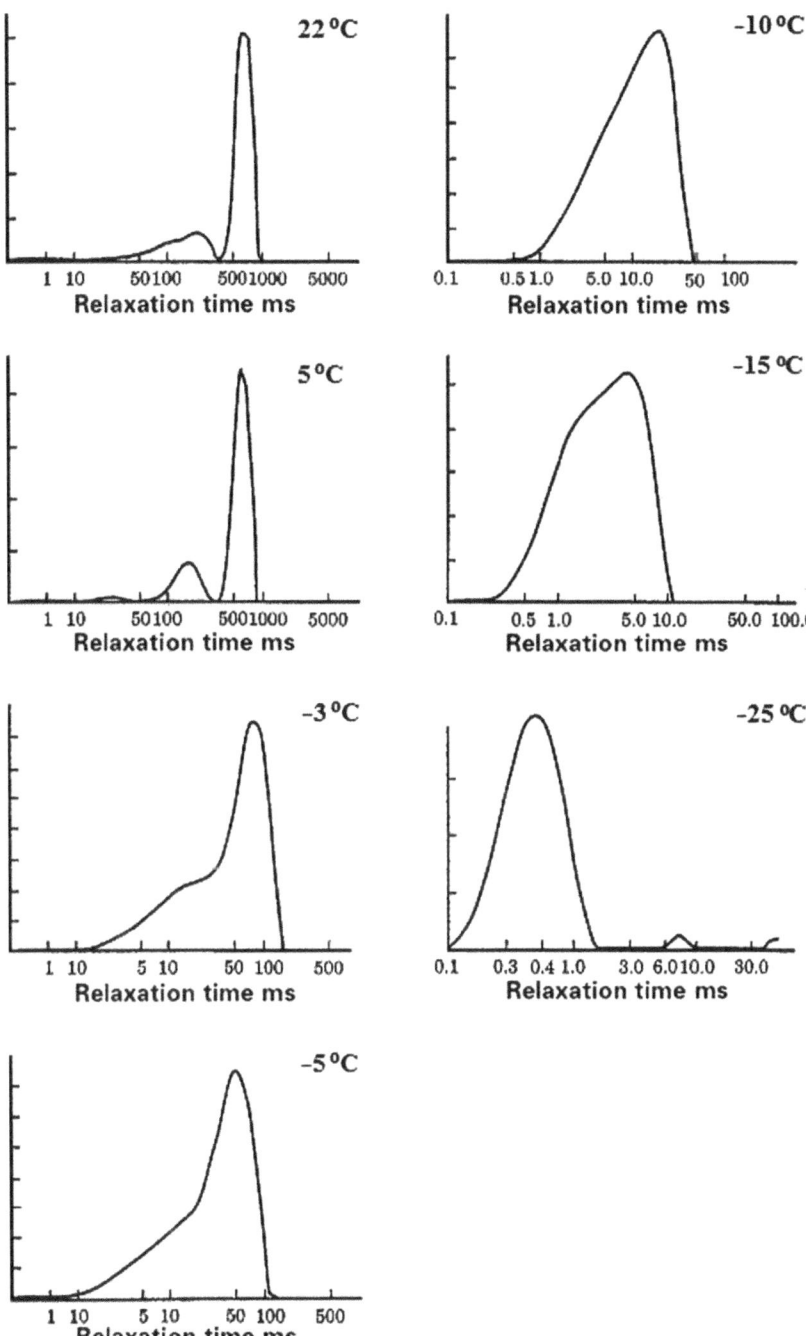

Figure 9.10 T_2 relaxation time distributions during the freezing of apple tissue from
− 25 to 22 °C.
(Reprinted from ref. 91. Copyright (1997), with permission from John
Wiley and Sons.)

temperatures for 20 h were studied by T_2 measurements.[93] Relaxation measurements showed that the orange peel did not freeze at these temperatures, but froze at $-20\,°C$. As in the case of kiwi fruit,[92] exposure to freezing caused an appreciable decrease ($\sim 15\%$) in the T_2 values of orange flesh, a feature that could provide adequate contrast enhancement in MRI experiments. Indeed, an on-line low-field MRI methodology examining fruits moving at fast speeds on a conveyor belt provided very encouraging results in detecting freeze damaged oranges using image acquisition times of less than 1s.[94] Under static conditions, a fast SE low-field MRI approach was also successful in detecting freeze-damaged mandarins.[55]

9.2   Vegetables

The term vegetable is used here in a very broad sense that encompasses all plants with edible tissue, such as leaf, stem, or roots, and also includes edible fungi, such as mushrooms and truffles. There is an enormous amount of published work in the literature dealing with the structural characterisation and biological or chemical properties of bioactive compounds in plants using NMR spectroscopy.[95] Plant NMR-based metabolomics have also been reviewed extensively.[96–98] We will focus our discussion on NMR studies that have a more or less direct aim at studying edible plants as food products or food components, and their properties.

9.2.1   Compositional Analysis and Metabolite Profiling

The chemical composition of vegetables can be very variable, depending on the type of plant tissue that is edible. Fleshy vegetables, for example, contain high amounts of water, while cereals and seeds contain significant amounts of carbohydrates and proteins, and nuts usually have a high lipid content. The liquid-state NMR analysis of the chemical composition of vegetables is usually performed using water or organic solvent extracts. The recent analysis of lettuce leaves is a solid example of the power of NMR spectroscopy in the compositional profiling of vegetables.[99] Figure 9.11 presents the ^1H–^{13}C HSQC 2D NMR spectrum of a water extract of lettuce leaves in buffered D_2O solution, indicating the signals of simple sugars and fructans (inulins), polyols, phenolic acids, organic acids and amino acids identified by 2D NMR spectroscopy.

Inulins are oligomers of the type $GpyF_n$ with $n = 2$–5, and their structure was elucidated with the help of DOSY experiments.[99] Glycerolipids, sulfolipids, phospholipids, sterols, pheophytins, carotenoids and fatty acids were also identified in the NMR spectra of the organic extracts of lettuce leaves. Other vegetables whose metabolite composition has been studied by 1D and 2D liquid state NMR spectroscopy include potato,[100] carrot,[101] brassica rapa,[102] maize,[103,104] wheat,[105–107] pine nuts,[108] and truffles.[109] There exists a large number of NMR-based metabolomics studies on vegetables, including maize,[110] lettuce[111,112] and potato,[113] dealing with attempts, mostly successful, to differentiate between genetically modified and conventional varieties.[114]

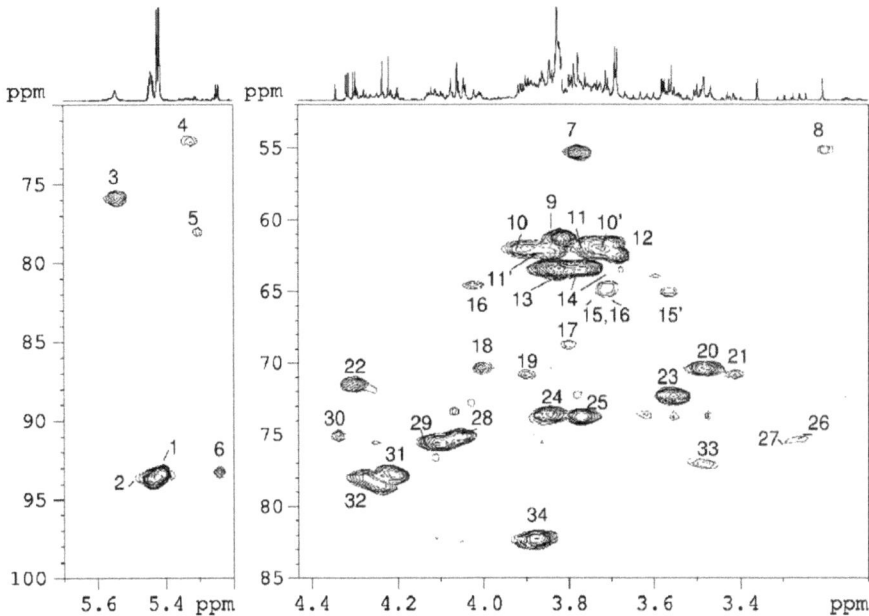

Figure 9.11 ^{1}H–^{13}C HSQC 2D NMR spectrum of lettuce water-soluble fraction, middle field and anomeric spectral regions. Key: α-Glc = D-α-glucose; β-Glc = D-β-glucose; α-FF = D-α-fructofuranose; β-FF = D-β-fructofuranose; β-FP = D-β-fructopyranose; Suc = sucrose; Inul = inulins. Assignments: **1**, CH-1, Suc (Glc); **2**, CH-1, Inul (Glc); **3**, *CH*(O)COOH, Chicoric acid; **4**, CH-3, Chlorogenic acid; **5**, CH(O)COOH, Monocaffeoyltartaric acid; **6**, CH-1 α-Glc; **7**, α-CH, Gln and Glu; **8**, N(CH$_3$)$_3$ choline; **9**, CH$_2$-6, Suc, Inul; **10** and **10'**, CH$_2$-6 β-Glc; **11**, and **11'**, CH$_2$-6 α-Glc; **12**, CH-10, Suc, Inul; **13**, CH$_2$-6', Suc, Inul; **14**, CH$_2$-6 β-FF; **15** and **15'**, CH$_2$-1 β-FP; **16** and **16'**, CH$_2$-6 β-FP; **17**, CH-3 β-FP; **18**, CH-5 β-FP; **19**, CH-4 β-FP; **20**, CH-4 Suc, Inul; **21**, CH-4 α- and β-Glc; **22**, α-CH malate; **23**, CH-2 α-Glc and Suc, Inul; **24**, CH-5 Suc, Inul; **25**, CH-3 Suc, Inul; **26**, CH-2 β-Glc; **27**, CH-4 myo-inositol; **28**, CH- 4' Suc, Inul; **29**, CH-4 β-FF, CH- 4' Inul; **30**, α-CH tartrate; **31**, CH- 3' Suc, Inul; **32**, CH- 3' Inul; **33**, CH-3 and CH-5 β-Glc; **34**, CH- 5' Suc, Inul.
(Reprinted from ref. 99.Copyright (2005), with permission from John Wiley and Sons.)

These, along with other studies dealing with cultivar, quality[115] and geographical origin classifications are summarised in Table 9.2.

As an example, Figure 9.12 presents the PCA analysis of metabolite signals obtained by NMR analysis of maize samples, showing clearly the differentiation in two distinct groups of samples, one group consisting of non-GM samples and, on the right side, a group of only GM samples.[104]

Due to the high water content of vegetables, HR-MAS NMR spectroscopy is an excellent tool for metabolite analysis of their soft tissue, and has been used for the identification of β-carotene in a raw vegetable matrix,[116] and the metabolic

Table 9.2   NMR-based chemometrics studies of vegetables.

Factor examined	Vegetable	NMR methodology	Multivariate analysis model	Ref.
Cultivar	Brassica rapa	HR	PCA	102
Pathogen infection	Brassica rapa	HR	PCA, PLS-DA	160
Adulteration	Spices	HR	PLS-DA	115
Geographical origin	Durum wheat	HR	LDA	106
Drought stress	Wheat	HR, HR-MAS	PFA	122
Cultivar	Wheat	HR	PCA	107
GM	Maize	HR	PCA, PLS-DA	110
GM, development	Maize	HR	OPLS-DA, PFA, PLS-DA	103
GM	Maize	HR	PCA	104
GM	Lettuce	HR	PCA	111
GM	Lettuce	HR	TCA, PCA	112
GM	Potato	HR	PCA	100
GM	Potato	HR	PCA	113
Drought and salt stress	Rice	HR, HR-MAS	PCA	121
Cultivar, geographical origin	Sweet pepper	HR-MAS	PLS-DA	117
Taste disturbance	Pine nut	HR	PCA, SIMCA	108

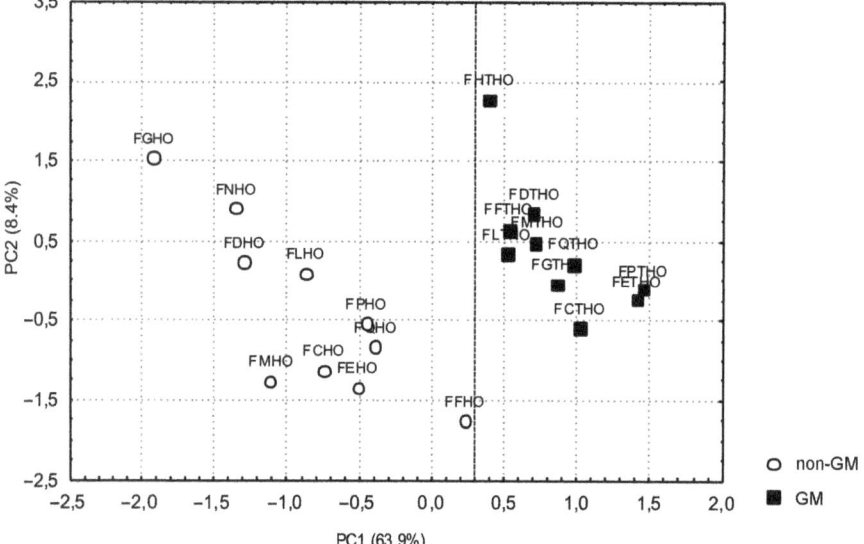

Figure 9.12   PCA analysis graph of the intensity analysis of 39 metabolites measured in GM and non-GM maize samples.
(Reprinted from ref. 104. Copyright (2009), with permission from American Chemical Society.)

profiling of sweet pepper.[117] HR-MAS NMR has also been applied for the study of nuts and grains, including almonds,[118] barley,[119] beans,[120] rice[121] and wheat.[122]

Vegetables contain large amounts of important biopolymers, such as cellulose, lignin, polysaccharides, starch (or pectins), and since they are mostly insoluble, solid-state ^{13}C CP-MAS NMR spectroscopy has been used extensively for their characterisation.[123]

The different crystalline alloforms of cellulose[124,125] and starch,[126,127] and the nature and relative proportions of amorphous, single- and double-helical components in starch granules can be determined by ^{13}C CP-MAS NMR spectra.[128] The cell wall biopolymer composition of several vegetables has been characterised in detail by ^{13}C CP-MAS NMR,[123] including wheat bran,[129] potato,[130] onion,[131] celery,[132] sugar beet,[133] and mung bean.[134] Figure 9.13 presents the ^{13}C CP-MAS NMR spectra of two different species of mushrooms *P. sapidus* (SMR 130) and *P. sajor-caju* (SMR 124).

The resonance due to protein components (170 ppm) is weaker in sample SMR 130 as compared to sample SMR 124, and the relative intensities of the resonances are markedly different in the two species, indicating that the protein/polysaccharide ratios may be used to differentiate between mushroom species.[135]

^{31}P NMR spectroscopy is a powerful tool for the analysis of phosphorus-containing compounds in vegetables, such as phospholipids in vegetable extracts.[99,109] Figure 9.14 presents the ^{1}H–^{31}P 2D HMBC NMR spectrum of an aqueous truffle extract, where the long range coupling between the ^{1}H and ^{31}P nuclei allowed the identification of several phosphodiesters

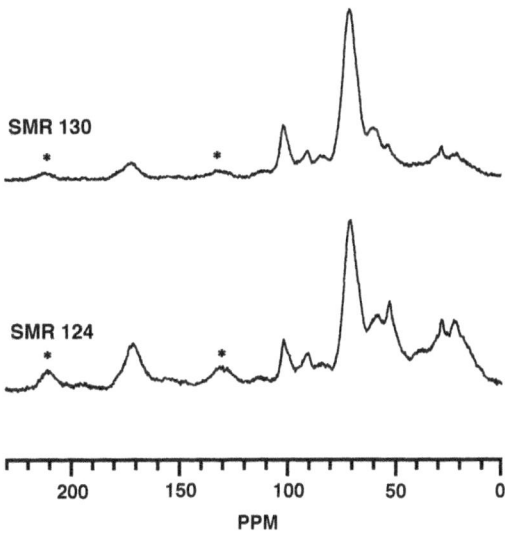

Figure 9.13 ^{13}C CP-MAS NMR spectra of mushroom samples of *P. sapidus* (SMR 130) and *P. sajor-caju* (SMR 124). Asterisks indicate the spinning sidebands.
(Reprinted from ref. 135. Copyright (2000), with permission from American Chemical Society.)

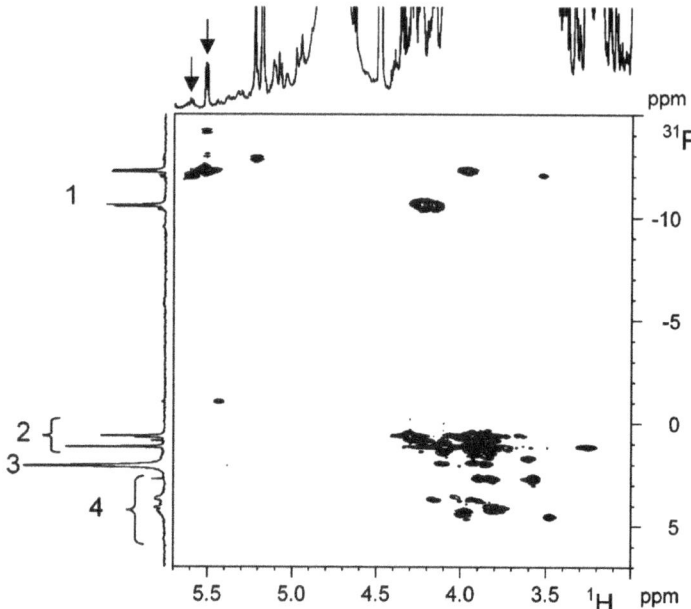

Figure 9.14   Long range coupling $^1H-^{31}P$ 2D HMBC NMR spectrum of an aqueous truffle extract; ^{31}P spectral assignment: **1**, uridine-5'-(diphospho-N-acetylglucosamine; **2**, phosphodiesters; **3**, inorganic phosphate; **4**, phosphomonoesters.
(Reprinted from ref. 109. Copyright (2004), with permission from American Chemical Society.)

(glycerophosphorylethanolamine, serine-ethanolamine phosphate, glycerophosphorylcholine, threonine-ethanolamine phosphate), phosphomonoesters, and uridine-5'-(diphospho-N-acetylglucosamine), an uridine diphosphate sugar.[109]

Starches contain both inorganic and organic phosphorus in the form of phospholipids and covalently bound phosphate groups (monoesters) at the O-6 or the O-3 position of the glucopyranosyl residues of starch.[136] ^{31}P NMR was used to study phosphorus speciation in root, tuber, cereal, waxy starches,[137] and modified starches,[138,139] while a quantitative ^{31}P NMR methodology has also been developed and validated for starch analysis.[140] ^{31}P NMR can also be used for the quantitative analysis of phytic acid (inositol hexaphosphate), a compound found in fruits, vegetables and grains,[141,142] and phospholipids in lecithins.[143,144] NMR-obtained phospholipid profiles have been used to study the source,[145] geographical origin and extraction conditions of lecithins.[146]

9.2.2   Quality Evaluation

The applications of MRI and relaxometry in the quality evaluation of intact fruits and vegetables have been reviewed,[46] and this non-invasive technique has also been evaluated and compared to other non-destructive instrumental means of analysis.[147,148] Here, we will present some work representative of the abilities of MRI and relaxometry in evaluating the internal structure of vegetables, followed by studies relating to vegetable processing in section 9.2.3.

9.2.2.1 Morphological Analysis

An MRI study of mushrooms at three magnetic field strengths (9.4, 4.7, and 0.47 T) showed that low fields and short echo times was the preferred MRI approach when studying water distribution, minimising the effect of magnetic susceptibility inhomogeneities that increase with increasing magnetic field strength.[149] Electron microscopy revealed that the inhomogeneities were related to the presence of extracellular air spaces, a common occurrence in many vegetables. Similar results were obtained while studying the water distribution in cucumber fruit, where MR image quality was reported to be improved at 1 T compared to 7 T with respect to observing low mobility (and short T_2) water.[150] A study of four raw potato varieties reported good correlation between dry matter content and parameters determined by LF-NMR relaxation, but found no correlation with the T_1-weighted MR images obtained.[151] The internal morphology of truffles has also been studied by MRI.[152]

Finally, there are several examples of NMR applications used to study the growth and development of cereals, such as barley grains,[153] maize[154] and wheat.[155] NMR imaging was used for the non-invasive acquisition of 3D images of developing barley grains from anthesis to maturity at 40 d after anthesis. Chemical shift MR imaging was able to detect changes in the tissue distribution of water, soluble carbohydrate and lipids in this study.[153] MRI and LF-NMR relaxation measurements were applied to obtain detailed information on mobility behaviour of water molecules in developing wheat seeds and this way it was possible to differentiate between two soft and hard isolines from common wheat cv. Enesco at early stages of seed development.[155]

9.2.2.2 Defects and Disorders

NMR spectroscopy and LF-NMR relaxation can provide information on the presence of internal defects in vegetables and identify the presence and effects of pathogen infection. In sweet potatoes, ^{1}H NMR relaxation of whole tubers was used to study the degeneration effects of long term storage.[156] The T_1 and T_2 relaxation times of water in tubers increased during one-year storage at 15 °C, and the relaxation time temperature dependence was used to indicate tissue viability after storage. Other interesting applications in potatoes include the use of MRI to detect non-visible internal bruises and Spraing disease symptoms,[157] and infection by several fungal pathogens.[158]

In all cases, the internal defects could be identified clearly in MRI images. Figure 9.15 presents several MRI images obtained from a potato with Spraing disease, which is caused by a soil-borne virus infection.

The appearance of Spraing spots on the proton and the T_1-weighted image (Figure 9.15(a) and (b)) is similar, with a large darker area surrounding the

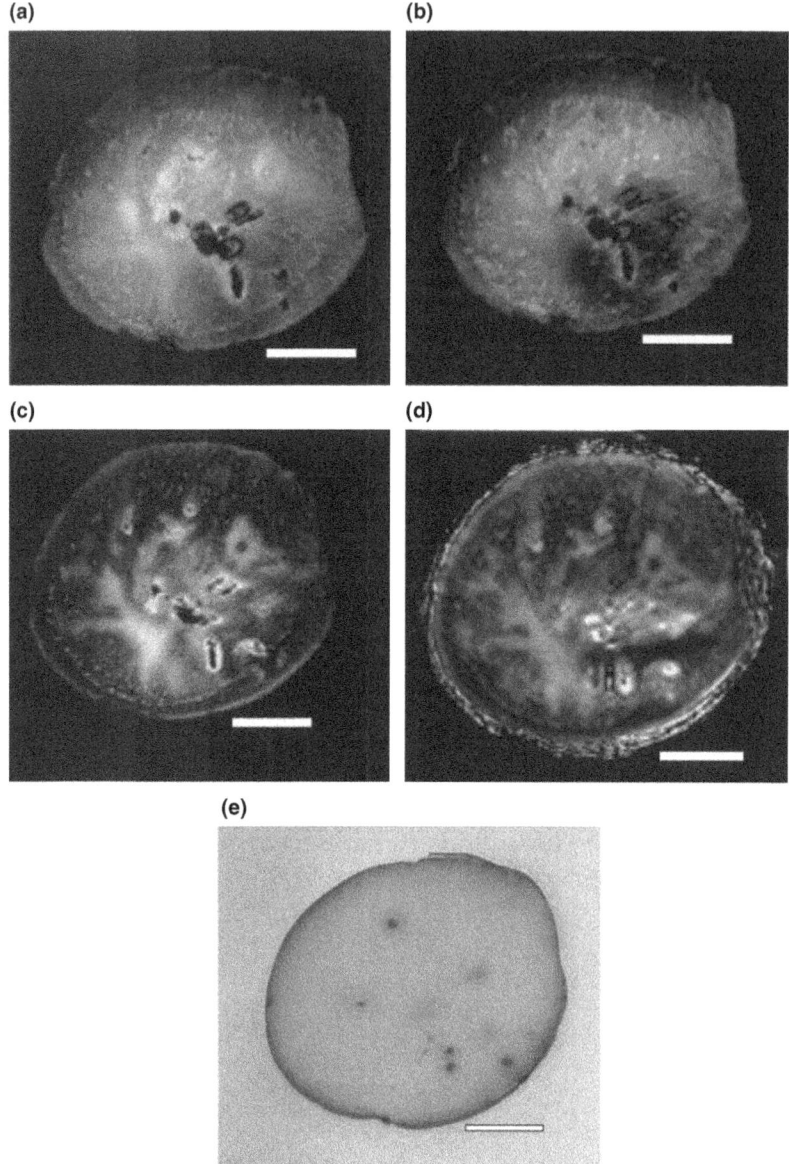

Figure 9.15   (a) Proton-, (b) T_1- and (c) T_2-weighted images, (d) a T_2 map and (e) digital photograph of a potato with Spraing disease (bar = 1 cm). (Reprinted from ref. 157. Copyright (2004), with permission from Elsevier.)

Spraing spots. In the T_2-weighted image and the T_2 map (Figure 9.15(c) and (d)), the rims of Spraing spots stand out more clearly, and the collective information from all the images indicates that these regions may be interpreted as containing relatively little water, but of high mobility.[157]

The successful detection and analysis of various types of internal defects was also reported non-destructively by MRI in dried normal and abnormal red ginseng samples,[159] while infection studies have been reported for truffle,[152] and brassica rapa.[160]

9.2.3 Processing

9.2.3.1 Freezing

MRI offers the ability to study freezing, an important means of preserving vegetables and foods in general, non-invasively and in real time, obtaining information on the rate and nature of freezing. MRI sees freezing as an abrupt decrease in the ^{1}H NMR signal of water, since ice has a T_2 value of only a few μs. MRI was used to monitor the formation of ice during freezing of potatoes, carrots, peas, and corn at temperatures of -30 to $-32\,°C$ using an air-blast freezer.[161] MR images of freezing potatoes (see Figure 9.16) showed a continuous advance of the ice interface, with ice forming first at the surface of the potato in direct contact with the cold air front and the freezing zone advancing towards the tuber interior.

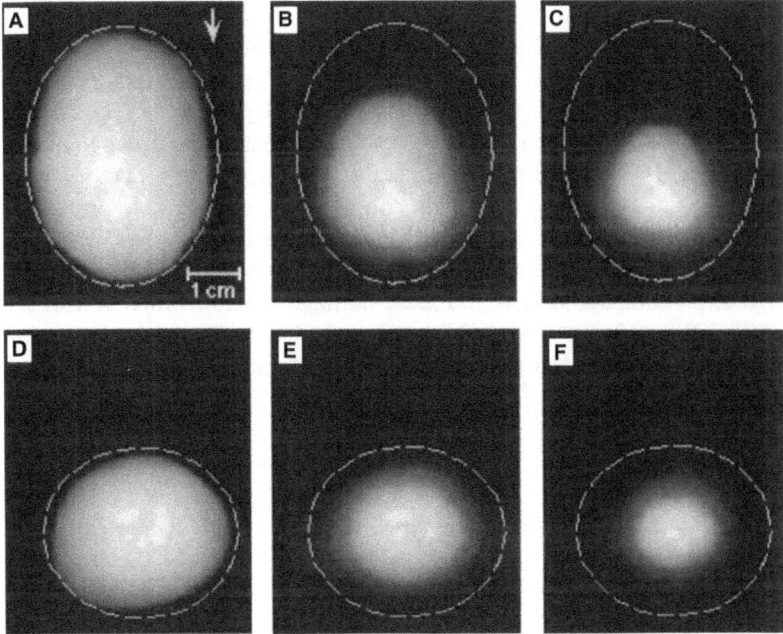

Figure 9.16 MRI images of a freezing potato showing orthogonal slices parallel (A, B, C) and perpendicular (D, E, F) to airflow (arrow). Freezer temperature $= -32\,°C$, air speed $= 5.1\,m\,s^{-1}$. A, D $= 0$ min; B, E $= 9$ min; C, F $= 15$ min. Dotted line shows boundary of initially unfrozen product. (Reprinted from ref. 161. Copyright (1998), with permission from Elsevier.)

The perpendicular MR images depicted in Figure 9.16(D–F) show that freezing was fairly symmetric about the central axis. Freezing dynamics were similar for all vegetables except whole corn, where kernels froze individually in a process possibly governed by ice nucleation.[161] A ^{1}H MRI study of freezing of single corn kernels has also been reported.[162] The freezing of carrot has also been studied by LF-NMR relaxation, using a numerical model to analyse the relaxation data in terms of tissue subcellular morphology changes.[163] It was reported that freezing causes initial ice crystal formation in the vacuolar compartment, and that even at $-25\,°C$ substantial quantities of non-freezing water are present in the carrot associated with the cell walls and dissolved biopolymers.

Thawing of frozen vegetables has also been studied by a combination of ^{1}H NMR, MRI and relaxation measurements using a small, 1.0 T static magnet.[164] ^{1}H NMR was used to study the thawing dynamics of green soybeans, broad beans, okra, asparagus and taro that were boiled and stored at $-20\,°C$ for an unknown amount of time. Differences were reported between vegetables with high water contents (okra, asparagus) and green leguminous seeds of lower water content.[164] MRI of fresh and freeze/thawed courgettes suggested that freezing ruptures the cell wall and alters tissue morphology, leading to MR images of deteriorated contrast compared to the fresh vegetable.[165] In a different approach, fresh and freeze/thawed onions were studied using ^{1}H pulsed field gradient NMR diffusometry.[166] Water diffusion was found to be less restricted after freeze–thawing, indicating a permanent increase in the permeability of the cell membrane is caused by freezing. Thus, it was suggested that freeze–thawing damaged the cell membrane rather than the cell wall. PFG-NMR was also used to study the protective effect of osmotic dehydro-freezing, a process in which a food is osmotically dehydrated before freezing (see section 9.1.4.1), on carrot texture softening after freeze–thawing.[167] NMR relaxation measurements have also been used to assess freeze damage in pickling cucumber.[168] Freeze-damaged regions in cucumber frozen at $-18\,°C$ for 150 min were associated with lower T_2 values than control samples in MR images, but no correlation was observed between T_2 values and firmness of the fruit.[168]

9.2.3.2   Drying

Prior to storage under low temperatures, the water content of vegetables and grains may be decreased by drying, in an effort to reduce the deteriorating effects on texture observed during the formation of ice. LF-NMR relaxation was used to study the effect of drying of carrot, indicating that drying removes water primarily from the vacuolar compartment, causing cell shrinkage and concentration of dissolved solutes, while little change was observed in intercellular air spaces.[163] MRI was used to measure moisture profiles and diffusivity during corn[169,170] and rice[171] drying, aiding in efforts to determine an accurate model of the drying procedure. A significant problem encountered during drying is the development of drying-induced stress cracks in grains and beans. MRI has been used to study analytically the formation of cracks in corn[172] and soya bean kernels.[173] In corn it was reported that cracks propagated from near the surface to the kernel centre, and that higher drying temperatures caused more severe cracks than lower ones.[172] In soya bean seed coat cracking, a positive correlation with drying temperature was also observed, and it was reported that moisture gradients present in the kernel enhanced stress and caused cracking during drying.[173]

MRI has also been used to study the tempering process on rice, the procedure of allowing adequate time for the redistribution of moisture in rice kernels after a drying step, in order to avoid moisture gradients that could lead to undesirable fissuring. MRI monitoring of the water redistribution in the rice kernel showed that after drying the moisture content of the interior starchy endosperm was higher than that of the exterior region, but this moisture gradient decreased gradually during the tempering process.[174]

9.2.3.3 Thermal Processing and Cooking

^{1}H-NMR relaxometry is an important tool for the study of the effects of thermal, high pressure and electric field processing on membrane permeability and cell compartmentalisation, two factors that determine vegetable tissue texture and affect their acceptability as food products.[175] The CPMG relaxation decay curves of water in onions can be deconvoluted to several components with different T_2 relaxation times representing distinct states of water molecules in the onion tissue. Figure 9.17 depicts the effect of thermal (top) and high pressure (bottom) processing on the values of the T_2 of different components.

The 40 and 50 °C thermal treatments resemble the raw control, while the 60, 70, and 90 °C show a lower T_2 of the water in the main component similar to that observed in the frozen–thawed tissue, an indication of loss of cell integrity and increased exchange of water between the different cell compartments.[176] High-pressure processing on the other hand has a milder effect on cell integrity, showing only a small T_2 decrease at the higher pressure used, 200 MPa. The efficiency of the above LF-NMR methodology has also been demonstrated in a study of electric field processing of onions, showing great promise as an in-line control tool.[177] Multi-dimensional relaxation studies combining T_1 and T_2 relaxation measurements have also been used for the characterisation of high pressure processing of potatoes[178] and thermal and high-pressure processing in carrot tissues.[179]

It was reported that raw potato cell walls are resistant to high pressure (500 MPa) damage, while potato starch was more resistant to pressure treatment than waxy maize starches.[178] In carrot, the ability of T_1-T_2 relaxation correlation spectroscopy to discriminate between the different components of phloem and xylem (extra- and intracellular water, protein, pectin, cellulose) was used to assess thermal and high pressure processing effects on the tissues and propose a treatment that could be applied commercially for tissue texture preservation.[179]

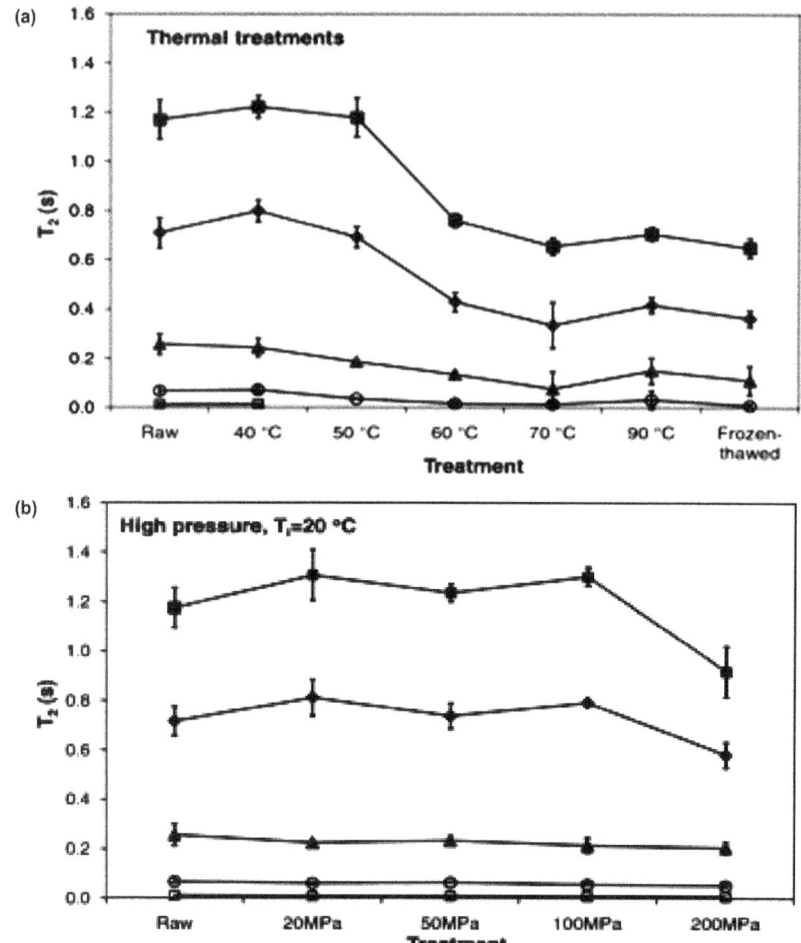

Figure 9.17 T_2 relaxation times of the different water signal components (component 1 [■], component 2 [◆], component 3 [▲], component 4 [○], component 5 [□]) of raw, thermally treated (40, 50, 60, 70, and 90 °C), and frozen–thawed onions (top) and raw and high pressure (at $T = 20$ °C) treated onions (bottom).
(Reprinted from ref. 176. Copyright (2010), with permission from John Wiley and Sons.)

The boiling of vegetables and grains is a common procedure before consumption, and a procedure also used during cooking. NMR spectroscopy and MRI have been used to study water migration and the effects of boiling and cooking on vegetable composition and texture. NMR spectroscopy can be used to study the extraction of sugars from fresh carrot boiled at different temperatures,[101] or the changes in the lipid profile of flax bolls during boiling.[180] MRI is an excellent tool to study water migration during cooking in peas,[181] wheat,[182] rice[183,184] and even food products obtained from grains, such as

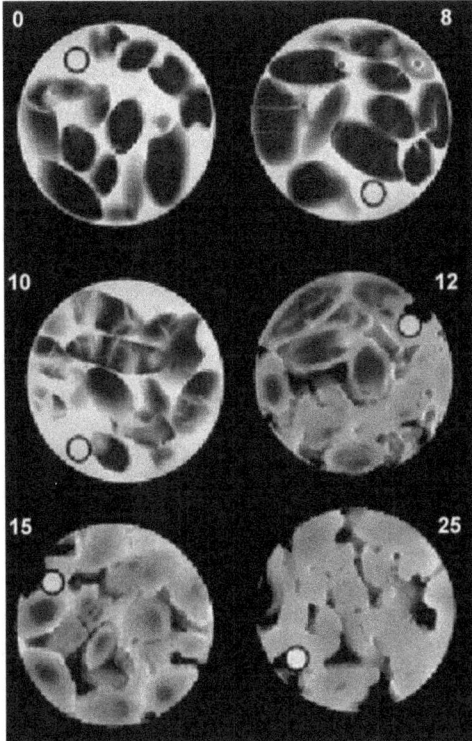

Figure 9.18   MRI images obtained at different stages during rice cooking. The cooking times in minutes are indicated. Each image is from a different sample. White indicates high water concentration and the small white circle in each image is from a reference water sample used for calibration purposes.
(Reprinted with permission from ref. 184. Copyright (2005), with permission from Elsevier.)

spaghetti,[185] noodles[186] and dough.[187] Using MRI, the continuing infusion of water inside the food during cooking can be monitored quantitatively, as depicted in Figure 9.18 for the case of rice.

Another interesting application involves the use of LF-NMR relaxation before cooking for the prediction of the sensory properties of potatoes after boiling.[188] LF-NMR performed better than traditional chemical descriptors of sensory texture quality due its sensitivity to the state of moisture in the sample, accessible through multi-component analysis of relaxation curves.

References

1. D. Capitani, L. Mannina, N. Proietti, A. P. Sobolev, A. Tomassini, A. Miccheli, M. E. Di Cocco, G. Capuani, R. De Salvador and M. Delfini, *Talanta*, 2010, **82**, 1826.

2. B. Biais, J. W. Allwood, C. Deborde, Y. Xu, M. Maucourt, B. Beauvoit, W. B. Dunn, D. Jacob, R. Goodacre, D. Rolin and A. Moing, *Anal. Chem.*, 2009, **81**, 2884.

3. M. Vermathen, M. Marzorati, D. Baumgartner, C. Good and P. Vermathen, *J. Agric. Food Chem.*, 2011, **59**, 12784.

4. S. K. Cho, S. O. Yang, S. H. Kim, H. Kim, J. S. Ko, K. Z. Riu, H. Y. Lee and H. K. Choi, *J. Pharm. Biomed. Anal.*, 2009, **49**, 567.

5. J. Schripsema, M. D. Vianna, P. A. B. Rodrigues, J. G. De Oliveira and R. W. A. Franco, *Acta Hortic.*, 2010, **851**, 505.

6. S. Moco, J. Forshed, R. De Vos, R. Bino and J. Vervoort, *Metabolomics*, 2008, **4**, 202.

7. C. Deborde, M. Maucourt, P. Baldet, S. Bernillon, B. Biais, G. Talon, C. Ferrand, D. Jacob, H. Ferry-Dumazet, A. de Daruvar, D. Rolin and A. Moing, *Metabolomics*, 2009, **5**, 183.

8. A. Malmendal, C. Amoresano, R. Trotta, I. Lauri, S. De Tito, E. Novellino and A. Randazzo, *J. Agric. Food Chem.*, 2011, **59**, 10831.

9. Q. W. Ni and T. M. Eads, *J. Agric. Food Chem.*, 1992, **40**, 1507.

10. Q. W. Ni and T. M. Eads, *J. Agric. Food Chem.*, 1993, **41**, 1026.

11. Q. X. Ni and T. M. Eads, *J. Agric. Food Chem.*, 1993, **41**, 1035.

12. A. M. Gil, I. F. Duarte, I. Delgadillo, I. J. Colquhoun, F. Casuscelli, E. Humpfer and M. Spraul, *J. Agric. Food Chem.*, 2000, **48**, 1524.

13. A. P. Deshmukh, A. J. Simpson and P. G. Hatcher, *Phytochemistry*, 2003, **64**, 1163.

14. E. M. Sanchez-Perez, M. J. Iglesias, F. L. Ortiz, I. S. Perez and M. M. Galera, *Food Chem.*, 2010, **122**, 877.

15. E. M. Sanchez-Perez, J. Garcia-Lopez, M. J. Iglesias, F. Lopez-Ortiz, F. Toresano and F. Camacho, *Food Res. Int.*, 2011, **44**, 3212.

16. I. Wawer, M. Wolniak and K. Paradowska, *Solid State Nucl. Magn. Reson.*, 2006, **30**, 106.

17. P. De Miranda Costa, M. I. B. Tavares, A. L. B. S. Bathista, E. O. Da Silva and J. S. Nogueira, *J. Appl. Polym. Sci.*, 2007, **105**, 973.

18. L. Zhang and M. J. McCarthy, *Postharvest Biol. Technol.*, 2012, **67**, 37.

19. P. Sequi, M. T. Dell'Abate and M. Valentini, *J. Sci. Food Agric.*, 2007, **87**, 127.

20. R. R. Milczarek, M. E. Saltveit, T. C. Garvey and M. J. McCarthy, *Postharvest Biol. Technol.*, 2009, **52**, 189.

21. R. R. Milczarek and M. J. McCarthy, *J. Food Process. Preserv.*, 2011, **35**, 631.

22. M. E. Saltveit, Jr, *Postharvest Biol. Technol.*, 1991, **1**, 153.

23. N. Ishida, M. Koizumi and H. Kano, *Sci. Hortic. (Amsterdam, Neth.)*, 1994, **57**, 335.

24. A. Ciampa, M. T. Dell'Abate, O. Masetti, M. Valentini and P. Sequi, *Food Chem.*, 2010, **122**, 1253.

25. C. Rondeau-Mouro, M. J. Crepeau and M. Lahaye, *Int. J. Biol. Macromol.*, 2003, **31**, 235.

26. N. Ishida, H. Ogawa, M. Koizumi and H. Kano, *Magn. Reson. Chem.*, 1997, **35**, S22.

27. J. E. Andaur, A. R. Guesalaga, E. E. Agosin, M. W. Guarini and P. Irarrázaval, *J. Agric. Food Chem.*, 2004, **52**, 165.
28. M. A. Brescia, T. Pugliese, E. Hardy and A. Sacco, *Food Chem.*, 2007, **105**, 400.
29. M. Gussoni, F. Greco, R. Consonni, H. Molinari, G. Zannoni, G. Bianchi and L. Zetta, *Magn. Reson. Imag.*, 1993, **11**, 259.
30. R. Chaughule, N. Ishida, S. Naito and H. Kano, *J. Food Sci. Technol.*, 2005, **42**, 162.
31. S. M. Shaarani, A. Cárdenas-Blanco, M. G. Amin, N. G. Soon and L. D. Hall, *Int. J. Agric. Biol.*, 2010, **12**, 101.
32. D. C. Joyce, P. D. Hockings, R. A. Mazucco and A. J. Shorter, *Funct. Plant Biol.*, 2002, **29**, 873.
33. A. Raffo, R. Gianferri, R. Barbieri and E. Brosio, *Food Chem.*, 2005, **89**, 149.
34. F. Z. Ribeiro, L. V. Marconcini, I. B. de Toledo, R. B. de Vasconcellos Azeredo, L. L. Barbosa and L. A. Colnago, *J. Sci. Food Agric.*, 2010, **90**, 2052.
35. B. A. Goodman, B. Williamson, E. J. Simpson, J. A. Chudek, G. Hunter and D. A. M. Prior, *Magn. Reson. Imag.*, 1996, **14**, 187.
36. M. Musse, S. Quellec, M. Cambert, M. F. Devaux, M. Lahaye and F. Mariette, *Postharvest Biol. Technol.*, 2009, **53**, 22.
37. Y. C. Cheng, T. T. Wang, J. H. Chen and T. T. Lin, *Postharvest Biol. Technol.*, 2011, **62**, 17.
38. C. J. Clark, L. N. Drummond and J. S. MacFall, *J. Sci. Food Agric.*, 1998, **78**, 349.
39. A. Taglienti, R. Massantini, R. Botondi, F. Mencarelli and M. Valentini, *Food Chem.*, 2009, **114**, 1583.
40. C. J. Clark and J. S. MacFall, *Magn. Reson. Imag.*, 2003, **21**, 679.
41. O. Goñi, M. Muñoz, J. Ruiz-Cabello, M. I. Escribano and C. Merodio, *Postharvest Biol. Technol.*, 2007, **45**, 147.
42. A. Banerjee, C. George, S. Bharathwaj and N. Chandrakumar, *J. Agric. Food Chem.*, 2009, **57**, 1183.
43. J. Létal, D. Jirák, L. Šuderlová and M. Hájek, *LWT – Food Sci. Technol.*, 2003, **36**, 719.
44. J. Burdon and C. Clark, *Postharvest Biol. Technol.*, 2001, **22**, 215.
45. T. A. Nguyen, T. Dresselaers, P. Verboven, G. D'Hallewin, N. Culeddu, P. Van Hecke and B. M. Nicolai, *J. Sci. Food Agric.*, 2006, **86**, 745.
46. B. P. Hills and C. J. Clark, *Annu. Rep. NMR Spectrosc.*, 2003, **50**, 75.
47. P. Chen, M. J. McCarthy, S. M. Kim and B. Zion, *Trans. Am. Soc. Agric. Eng.*, 1996, **39**, 2205.
48. M. McCarthy, S. Garcia, S. Kim and R. Milczarek, in *Emerging Technologies for Food Quality and Food Safety Evaluation*, ed. Y.-J. Cho, CRC Press, Florida, USA, 2011, pp. 149.
49. B. Zion, S. M. Kim, M. J. McCarthy and P. Chen, *J. Sci. Food Agric.*, 1997, **75**, 496.

50. S. M. Kim, P. Chen, M. J. McCarthy and B. Zion, *J. Agric. Eng. Res.*, 1999, **74**, 293.

51. K. Saito, T. Miki, S. Hayashi, H. Kajikawa, M. Shimada, Y. Kawate, T. Nishizawa, D. Ikegaya, N. Kimura, K. Takabatake, N. Sugiura and M. Suzuki, *Cryogenics*, 1996, **36**, 1027.

52. S. Zhou, D. Shang, Y. Ying and Y. Liao, *Trans. Chin. Soc. Agric. Mach.*, 2010, **41**, 107.

53. L. D. Hall, S. D. Evans and K. P. Nott, *Magn. Reson. Imag.*, 1998, **16**, 485.

54. P. Barreiro, C. Zheng, D. W. Sun, N. Hernández-Sánchez, J. M. Perez-Sáchez and J. Ruiz-Cabello, *Postharvest Biol. Technol.*, 2008, **47**, 189.

55. S. M. Kim, R. Milczarek and M. McCarthy, *Mod. Phys. Lett. B*, 2008, **22**, 941.

56. N. Hernández-Sánchez, P. Barreiro, M. Ruiz-Altisent, J. Ruiz-Cabello and M. Encarnación Fernández-Valle, *Concepts Magn. Reson., Part B*, 2005, **26**, 81.

57. N. Hernández-Sánchez, P. Barreiro and J. Ruiz-Cabello, *Biosyst. Eng.*, 2006, **95**, 529.

58. M. Musse, S. Quellec, M. F. Devaux, M. Cambert, M. Lahaye and F. Mariette, *Magn. Reson. Imag.*, 2009, **27**, 709.

59. M. Musse, F. De Guio, S. Quellec, M. Cambert, S. Challois and A. Davenel, *Magn. Reson. Imag.*, 2010, **28**, 1525.

60. S. S. Tu, Y. J. Choi, M. J. McCarthy and K. L. McCarthy, *Postharvest Biol. Technol.*, 2007, **44**, 157.

61. P. Barreiro, C. Ortiz, M. Ruiz-Altisent, J. Ruiz-Cabello, M. E. Fernádez-Valle, I. Recasens and M. Asensio, *Magn. Reson. Imag.*, 2000, **18**, 1175.

62. P. Barreiro, J. Ruiz-Cabello, M. E. Fernández-Valle, C. Ortiz and M. Ruiz-Altisent, *Magn. Reson. Imag.*, 1999, **17**, 275.

63. P. Barreiro, A. Moya, E. Correa, M. Ruiz-Altisent, M. Fernandez-Valle, A. Peirs, K. M. Wright and B. P. Hills, *Appl. Magn. Reson.*, 2002, **22**, 387.

64. N. Marigheto, L. Venturi and B. Hills, *Postharvest Biol. Technol.*, 2008, **48**, 331.

65. S. Y. Wang, P. C. Wang and M. Faust, *Sci. Hortic.*, 1988, **35**, 227.

66. C. J. Clark, J. S. MacFall and R. L. Bieleski, *Sci. Hortic.*, 1998, **73**, 213.

67. C. J. Clark and C. A. Richardson, *N. Z. J. Crop and Hortic. Sci.*, 1999, **27**, 47.

68. B. K. Cho, W. Chayaprasert and R. L. Stroshine, *Postharvest Biol. Technol.*, 2008, **47**, 81.

69. C. J. Clark and D. M. Burmeister, *HortScience*, 1999, **34**, 915.

70. J. J. Gonzalez, R. C. Valle, S. Bobroff, W. V. Biasi, E. J. Mitcham and M. J. McCarthy, *Postharvest Biol. Technol.*, 2001, **22**, 179.

71. N. Hernández-Sánchez, B. P. Hills, P. Barreiro and N. Marigheto, *Postharvest Biol. Technol.*, 2007, **44**, 260.

72. J. Lammertyn, T. Dresselaers, P. Van Hecke, P. Jancsók, M. Wevers and B. M. Nicolaï, *Magn. Reson. Imag*, 2003, **21**, 805.

73. J. Lammertyn, T. Dresselaers, P. Van Hecke, P. Jancsók, M. Wevers and B. M. Nicolaï, *Postharvest Biol. Technol.*, 2003, **29**, 19.

74. G. Paliyath, M. D. Whiting, M. A. Stasiak, D. P. Murr and B. S. Clegg, *Food Res. Int.*, 1997, **30**, 95.

75. F. Yoichi, T. Hirohiko and N. Shigehiro, *J. Jpn. Soc. Food Sci. Technol.*, 2011, **58**, 597.

76. L. Zhang and M. J. McCarthy, *Postharvest Biol. Technol.*, 2012, **67**, 96.

77. C. Du Chatenet, A. Latch, E. Olmos, B. Ranty, M. Charpenteau, R. Ranjeva, J. C. Pech and A. Graziana, *Physiol. Plant*, 2000, **110**, 248.

78. S. D. Evans, A. Brambilla, D. M. Lane, D. Torreggiani and L. D. Hall, *LWT – Food Sci. Technol.*, 2002, **35**, 177.

79. P. Cornillon, *LWT – Food Sci. Technol.*, 2000, **33**, 261.

80. A. Derossi, T. De Pilli, C. Severini and M. J. McCarthy, *J. Food Eng.*, 2008, **86**, 519.

81. U. Tylewicz, V. Panarese, L. Laghi, P. Rocculi, M. Nowacka, G. Placucci and M. D. Rosa, *Food Biophys.*, 2011, **6**, 327.

82. V. Panarese, L. Laghi, A. Pisi, U. Tylewicz, M. D. Rosa and P. Rocculi, *Food Chem.*, 2012, **132**, 1706.

83. D. C. Joyce, P. D. Hockings, R. A. Mazucco, A. J. Shorter and I. M. Brereton, *Postharvest Biol. Technol.*, 1993, **3**, 305.

84. K. Suzuki, A. Tajima, S. Takano, T. Asano and T. Hasegawa, *J. Food Sci.*, 1994, **59**, 855.

85. Y. Iwahashi, A. K. Horigane, K. Yoza, T. Nagata and H. Hosoda, *Magn. Reson. Imag.*, 1999, **17**, 767.

86. N. Marigheto, A. Vial, K. Wright and B. Hills, *Appl. Magn. Reson.*, 2004, **26**, 521.

87. L. Otero and G. Préstamo, *Innovative Food Sci. Emerging Technol.*, 2009, **10**, 434.

88. G. R. Gamble, *J. Food Sci.*, 1994, **59**, 571.

89. C. J. Clark and S. K. Forbes, *N. Z. J. Crop Hortic. Sci*, 1994, **22**, 209.

90. C. J. Clark and J. S. Macfall, *Postharvest Biol. Technol.*, 1996, **9**, 97.

91. B. P. Hills and B. Remigereau, *Int. J. Food Sci. Technol.*, 1997, **32**, 51.

92. W. L. Kerr, C. J. Clark, M. J. McCarthy and J. S. De Ropp, *Sci. Hortic.*, 1997, **69**, 169.

93. P. N. Gambhir, Y. J. Choi, D. C. Slaughter, J. F. Thompson and M. J. McCarthy, *J. Sci. Food Agric.*, 2005, **85**, 2482.

94. N. Hernández-Sánchez, P. Barreiro, M. Ruiz-Altisent, J. Ruiz-Cabello and M. Encarnación Fernández-Valle, *Appl. Magn. Reson.*, 2004, **26**, 431.

95. R. G. Ratcliffe, A. Roscher and Y. Shachar-Hill, *Prog. Nucl. Magn. Reson. Spectrosc.*, 2001, **39**, 267.

96. H. K. Kim, Y. H. Choi and R. Verpoorte, *Nat. Protoc.*, 2010, **5**, 536.

97. I. J. Colquhoun, *J. Pestic. Sci. (Tokyo, Jpn.)*, 2007, **32**, 200.

98. E. Holmes, H. Tang, Y. Wang and C. Seger, *Planta Med.*, 2006, **72**, 771.

99. A. P. Sobolev, E. Brosio, R. Gianferri and A. L. Segre, *Magn. Reson. Chem.*, 2005, **43**, 625.

100. M. Defernez, Y. M. Gunning, A. J. Parr, L. V. T. Shepherd, H. V. Davies and I. J. Colquhoun, *J. Agric. Food Chem.*, 2004, **52**, 6075.
101. A. Cazor, C. Deborde, A. Moing, D. Rolin and H. This, *J. Agric. Food Chem.*, 2006, **54**, 4681.
102. I. B. Abdel-Farid, K. K. Hye, H. C. Young and R. Verpoorte, *J. Agric. Food Chem.*, 2007, **55**, 7936.
103. C. Castro, M. Motto, V. Rossi and C. Manetti, *J. Exp. Bot.*, 2008, **59**, 3913.
104. F. Piccioni, D. Capitam, L. Zolla and L. Mannina, *J. Agric. Food Chem.*, 2009, **57**, 6041.
105. J. M. Baker, N. D. Hawkins, J. L. Ward, A. Lovegrove, J. A. Napier, P. R. Shewry and M. H. Beale, *Plant Biotechnol. J.*, 2006, **4**, 381.
106. R. Lamanna, L. Cattivelli, M. L. Miglietta and A. Troccoli, *Magn. Reson. Chem.*, 2011, **49**, 1.
107. S. Graham, E. Amigues, M. Migaud and R. Browne, *Metabolomics*, 2009, **5**, 302.
108. H. Kobler, Y. B. Monakhova, T. Kuballa, C. Tschiersch, J. Vancutsem, G. Thielert, A. Mohring and D. W. Lachenmeier, *J. Agric. Food Chem.*, 2011, **59**, 6877.
109. L. Mannina, M. Cristinzio, A. P. Sobolev, P. Ragni and A. Segre, *J. Agric. Food Chem.*, 2004, **52**, 7988.
110. C. Manetti, C. Bianchetti, M. Bizzarri, L. Casciani, C. Castro, G. D'Ascenzo, M. Delfini, M. E. Di Cocco, A. Lagani, A. Miccheli, M. Motto and F. Conti, *Phytochemistry*, 2004, **65**, 3187.
111. A. P. Sobolev, A. L. Segre, D. Giannino, D. Mariotti, C. Nicolodi, E. Brosio and M. E. Amato, *J. Agric. Food Chem.*, 2007, **55**, 10827.
112. A. P. Sobolev, G. Testone, F. Santoro, C. Nicolodi, M. A. Iannelli, M. E. Amato, A. Ianniello, E. Brosio, D. Giannino and L. Mannina, *J. Agric. Food Chem.*, 2010, **58**, 6928.
113. H. S. Kim, S. W. Kim, Y. S. Park, S. Y. Kwon, J. R. Liu, H. Joung and J. H. Jeon, *Biotechnol. Bioprocess. Eng.*, 2009, **14**, 738.
114. A. P. Sobolev, D. Capitani, D. Giannino, C. Nicolodi, G. Testone, F. Santoro, G. Frugis, M. A. Iannelli, A. K. Mattoo, E. Brosio, R. Gianferri, I. D'Amico and L. Mannina, *Nutrients*, 2010, **2**, 1.
115. C. V. Di Anibal, I. Ruisáchez and M. P. Callao, *Food Chem.*, 2011, **124**, 1139.
116. M. L. Miglietta and R. Lamanna, *Magn. Reson. Chem.*, 2006, **44**, 675.
117. M. Ritota, F. Marini, P. Sequi and M. Valentini, *J. Agric. Food Chem.*, 2010, **58**, 9675.
118. J. M. Ribó, J. Crusats, Z. El-Hachemi, M. Feliz, P. Sanchez-Bel and F. Romojaro, *J. Am. Oil Chem. Soc.*, 2004, **81**, 1029.
119. H. F. Seefeldt, F. H. Larsen, N. Viereck, B. Wollenweber and S. B. Engelsen, *Cereal Chem.*, 2008, **85**, 571.
120. L. M. Lião, R. Choze, P. P. A. Cavalcante, S. d. C. Santos, P. H. Ferri and A. G. Ferreira, *Quím. Nova*, 2010, **33**, 634.
121. E. Fumagalli, E. Baldoni, P. Abbruscato, P. Piffanelli, A. Genga, R. Lamanna and R. Consonni, *J. Agron. Crop Sci.*, 2009, **195**, 77.

122. H. Winning, N. Viereck, B. Wollenweber, F. H. Larsen, S. Jacobsen, I. Sondergaard and S. B. Engelsen, *J. Exp. Bot.*, 2009, **60**, 291.
123. F. Bertocchi and M. Paci, *J. Agric. Food Chem.*, 2008, **56**, 9317.
124. R. H. Atalla, J. C. Gast, D. W. Sindorf, V. J. Bartuska and G. E. Maciel, *J. Am. Chem. Soc.*, 1980, **102**, 3249.
125. H. Kono, S. Yunoki, T. Shikano, M. Fujiwara, T. Erata and M. Takai, *J. Am. Chem. Soc.*, 2002, **124**, 7506.
126. M. J. Gidley and S. M. Bociek, *J. Am. Chem. Soc.*, 1988, **110**, 3820.
127. C. Rondeau-Mouro, G. Veronese and A. Buleon, *Biomacromolecules*, 2006, **7**, 2455.
128. I. Tan, B. M. Flanagan, P. J. Halley, A. K. Whittaker and M. J. Gidley, *Biomacromolecules*, 2007, **8**, 885.
129. M. A. Ha, W. G. Jardine and M. C. Jarvis, *J. Agric. Food Chem.*, 1997, **45**, 117.
130. J. R. Garbow, L. M. Ferrantello and R. E. Stark, *Plant Physiol.*, 1989, **90**, 783.
131. M. A. Ha, D. C. Apperley and M. C. Jarvis, *Plant Physiol.*, 1997, **115**, 593.
132. K. M. Fenwick, M. C. Jarvis and D. C. Apperley, *Plant Physiol.*, 1997, **115**, 587.
133. C. M. G. C. Renard and M. C. Jarvis, *Plant Physiol.*, 1999, **119**, 1315.
134. T. J. Bootten, P. J. Harris, L. D. Melton and R. H. Newman, *J. Exp. Bot.*, 2004, **55**, 571.
135. L. Pizzoferrato, P. Manzi, F. Bertocchi, C. Fanelli, G. Rotilio and M. Paci, *J. Agric. Food Chem.*, 2000, **48**, 5484.
136. P. Muhrbeck and C. Tellier, *Starch/Staerke*, 1991, **43**, 25.
137. S. T. Lim, T. Kasemsuwan and J. L. Jane, *Cereal Chem.*, 1994, **71**, 488.
138. T. Kasemsuwan and J. Jane, *Cereal Chem.*, 1994, **71**, 282.
139. P. Y. Lin and Z. Czuchajowska, *Cereal Chem.*, 1998, **75**, 705.
140. T. Kasemsuwan and J. L. Jane, *Cereal Chem.*, 1996, **73**, 702.
141. E. P. Mazzola, B. Q. Phillippy, B. F. Harland, T. H. Miller, J. M. Potemra and E. W. Katsimpiris, *J. Agric. Food Chem.*, 1986, **34**, 60.
142. I. K. O'Neill, M. Sargent and M. L. Trimble, *Anal. Chem.*, 1980, **52**, 1288.
143. B. W. K. Diehl and W. Ockels, in *Phospholipids: Characterisation, Metabolism, and Novel Biological Applications*, ed. F. P. G. Cevc, AOCS Press, Champaign, USA, 1995, pp. 29.
144. G. Helmerich and P. Koehler, *J. Agric. Food Chem.*, 2003, **51**, 6645.
145. T. Glonek, *J. Am. Oil Chem. Soc.*, 1998, **75**, 569.
146. M. A. Cremonini, L. Laghi and G. Placucci, *J. Sci. Food Agric.*, 2004, **84**, 786.
147. P. Butz, C. Hofmann and B. Tauscher, *J. Food Sci.*, 2005, **70**, R131.
148. S. N. Jha and T. Matsuoka, *Food Sci. Technol. Res.*, 2000, **6**, 248.
149. H. C. W. Donker, H. T. Van As, H. T. Edzes and A. W. H. Jans, *Magn. Reson. Imag.*, 1996, **14**, 1205.
150. M. Koizumi, S. Naito, H. Kano and T. Haishi, *J. Jpn. Soc. Food Sci. Technol.*, 2009, **56**, 146.
151. A. K. Thybo, H. J. Andersen, A. H. Karlsson, S. DΓEnstrup and H. Stødkilde-Jørgensen, *LWT–Food Sci. Technol.*, 2003, **36**, 315.

152. G. Pacioni, M. Leonardi, A. Taglienti, S. Cozzolino, M. Ritota, P. Sequi and M. Valentini, *Plant Biosystems*, 2010, **144**, 826.
153. S. M. Glidewell, *J. Cereal Sci.*, 2006, **43**, 70.
154. L. Van der Weerd, M. M. A. E. Claessens, T. Ruttink, F. J. Vergeldt, T. J. Schaafsma and H. Van As, *J. Exp. Bot.*, 2001, **52**, 2333.
155. C. Castro, L. Gazza, R. Ciccoritti, N. Pogna, C. O. Rossi and C. Manetti, *J. Cereal Sci.*, 2010, **52**, 303.
156. M. Iwaya-Inoue, R. Matsui, N. Sultana, K. Saitou, K. Sakaguchi and M. Fukuyama, *J. Agron. Crop Sci.*, 2004, **190**, 65.
157. A. K. Thybo, S. N. Jespersen, P. E. Laerke and H. J. Stødkilde-Jørgensen, *Magn. Reson. Imag.*, 2004, **22**, 1311.
158. A. J. Snijder, R. L. Wastie, S. M. Glidewell and B. A. Goodman, *Biochem. Soc. Trans.*, 1996, **24**, 442.
159. S. Kim and C. Kim, in *Key Engineering Materials*, Trans Tech Publications, Switzerland, 2004, Vol. 270–273, pp. 1044.
160. M. Jahangir, H. K. Kim, Y. H. Choi and R. Verpoorte, *Food Chem.*, 2008, **107**, 362.
161. W. L. Kerr, R. J. Kauten, M. J. McCarthy and D. S. Reid, *LWT – Food Sci. Technol.*, 1998, **31**, 215.
162. C. Borompichaichartkul, G. Moran, G. Srzednicki and W. S. Price, *J. Food Eng.*, 2005, **69**, 199.
163. B. P. Hills and K. P. Nott, *Appl. Magn. Reson.*, 1999, **17**, 521.
164. M. Koizumi, S. Naito, T. Haishi, S. Utsuzawa, N. Ishida and H. Kano, *Magn. Reson. Imag.*, 2006, **24**, 1111.
165. S. L. Duce, T. A. Carpenter and L. D. Hall, *J. Food Eng.*, 1992, **16**, 165.
166. H. Ando, M. Fukuoka, O. Miyawaki, M. Watanabe and T. Suzuki, *Biosci., Biotechnol., Biochem.*, 2009, **73**, 1257.
167. H. Ando, K. Kajiwara, S. Oshita and T. Suzuki, *J. Food Eng.*, 2012, **108**, 473.
168. N. Kotwaliwale, E. Curtis, S. Othman, G. K. Naganathan and J. Subbiah, *Postharvest Biol. Technol.*, 2012, **68**, 22.
169. H. Song and J. B. Litchfield, *Trans. Am. Soc. Agric. Eng.*, 1990, **33**, 1286.
170. H. P. Song, J. B. Litchfield and H. D. Morris, *J. Agric. Eng. Res.*, 1992, **53**, 51.
171. J. M. Frías, L. Foucat, J. J. Bimbenet and C. Bonazzi, *Chem. Eng. J. (Lausanne)*, 2002, **86**, 173.
172. H. P. Song and J. B. Litchfield, *J. Agric. Eng. Res.*, 1994, **57**, 109.
173. X. Suzy Zeng, R. Roger Ruan, R. G. Fulcher and P. Chen, *Drying Technol.*, 1996, **14**, 1595.
174. S.-S. Hwang, Y.-C. Cheng, C. Chang, H.-S. Lur and T.-T. Lin, *J. Cereal Sci.*, 2009, **50**, 36.
175. M. E. Gonzalez and D. M. Barrett, *J. Food Sci.*, 2010, **75**, R121.
176. M. E. Gonzalez, D. M. Barrett, M. J. McCarthy, F. J. Vergeldt, E. Gerkema, A. M. Matser and H. Van As, *J. Food Sci.*, 2010, **75**, E417.

177. S. Ersus, M. H. Oztop, M. J. McCarthy and D. M. Barrett, *J. Food Sci.*, 2010, **75**, E444.
178. B. Hills, A. Costa, N. Marigheto and K. Wright, *Appl. Magn. Reson.*, 2005, **28**, 13.
179. M. E. Furfaro, N. Marigheto, G. K. Moates, K. Cross, M. L. Parker, K. W. Waldron and B. P. Hills, *Appl. Magn. Reson.*, 2009, **35**, 537.
180. C. W. Kirby, J. L. McCallum and B. Fofana, *Can. J. Chem.*, 2011, **89**, 1138.
181. P. K. Ghosh, D. S. Jayas and M. L. H. Gruwel, American Society of Agricultural and Biological Engineers Annual International Meeting 2008, Providence, RI, 2008.
182. A. G. F. Stapley, T. M. Hyde, L. F. Gladden and P. J. Fryer, *Int. J. Food Sci. Technol.*, 1997, **32**, 355.
183. A. Mohorič, F. Vergeldt, E. Gerkema, G. v. Dalen, L. R. v. d. Doel, L. J. v. Vliet, H. V. As and J. v. Duynhoven, *Food Chem.*, 2009, **115**, 1491.
184. M. Kasai, A. Lewis, F. Marica, S. Ayabe, K. Hatae and C. A. Fyfe, *Food Res. Int.*, 2005, **38**, 403.
185. A. K. Horigane, S. Naito, M. Kurimoto, K. Irie, M. Yamada, H. Motoi and M. Yoshida, *Cereal Chem.*, 2006, **83**, 235.
186. H. M. Lai and S. C. Hwang, *Food Res. Int.*, 2004, **37**, 957.
187. M. Fukuoka, T. Mihori and H. Watanabe, *J. Food Sci.*, 2000, **65**, 1343.
188. L. G. Thygesen, A. K. Thybo and S. B. Engelsen, *LWT–Food Sci. Technol*, 2001, **34**, 469.

CHAPTER 10
Milk and Dairy Products

10.1 Milk

Milk is a complex nutritional animal product consumed worldwide, and also forms the basis of a series of economically important products, such as cheese, yogurt and ice cream. The authentication and quality of dairy products is a very active field of analytical research,[1,2] and NMR spectroscopy has contributed significantly in the analysis of dairy products.[3]

10.1.1 Composition and Metabolite Profiling

Milk is an emulsion of fats in water that also contains high amounts of polysaccharides, proteins, phospholipids and minerals. The analysis of milk fat and phospholipids by NMR spectroscopic techniques has already been discussed in section 7.2.1; therefore our focus here will be the other constituents of milk. The high water content and the emulsion character of milk due to presence of casein micelles has precluded the use of high-resolution ^{1}H NMR spectroscopy of untreated milk until very recently, with most of the early NMR work on milk focusing on ^{13}C and ^{31}P NMR spectroscopy of milk samples that were pretreated (see below). Figure 10.1 presents the water-suppressed high-resolution ^{1}H NMR spectrum of whole untreated milk, indicating the major compound categories identified.[4]

Because of the fat globules in milk, which affect the magnetic field homogeneity of the sample, the resolution of the spectrum is quite low compared to that of beverages and juices (see chapter 8). Nevertheless, several key components of milk, namely fat triglycerides, D-lactose, citric acid, creatine and lecithin, can be identified, and have been assigned with the help of heteronuclear 2D NMR spectroscopy (^{1}H–^{13}C HSQC, [1]H–[13]C, ^{1}H–^{31}P and ^{1}H–^{15}N HMBC). In a follow-up study, ^{1}H 1D and ^{1}H–^{13}C HSQC 2D NMR spectra were used to quantify simultaneously and without any

RSC Food Analysis Monographs No. 10
NMR Spectroscopy in Food Analysis
By Apostolos Spyros and Photis Dais
© Apostolos Spyros and Photis Dais 2013
Published by the Royal Society of Chemistry, www.rsc.org

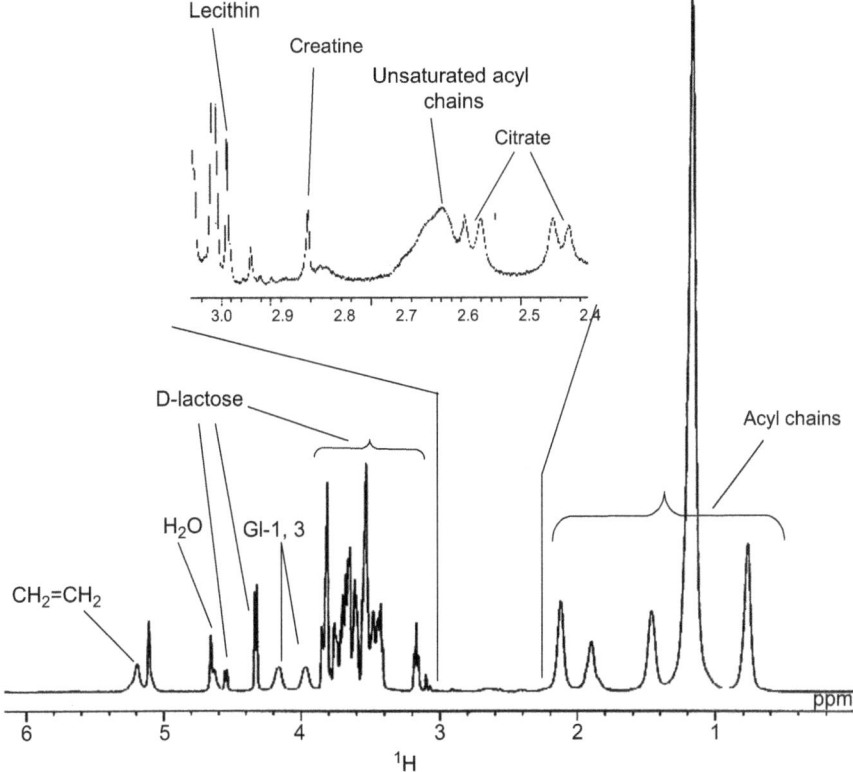

Figure 10.1 ¹H NMR spectrum of whole milk at a field of 500 MHz, high-field region. (Reprinted from ref. 4. Copyright (2004), with permission from American Chemical Society.)

pre-treatment the fat and lactose content of commercial whole milk, along with the concentrations of citrate, *N*-acetylcarbohydrates, trimethylamine, butyric acid, total monounsaturated fatty acids, and total polyunsaturated fatty acids.[5]

Apart from phospholipids, high-resolution liquid-state ³¹P NMR spectroscopy has been used for the analysis of other phosphorus-containing compounds in milk. Milk contains inorganic phosphate and phosphorus bonded to caseins as phosphoserine, and both of these can be readily quantified by ³¹P NMR.[6] The casein content of milk can be obtained through the ³¹P NMR quantification of casein phosphoserine residues, and provides an excellent correlation with the traditional Kjeldahl method,[7] while sphingolipids have also been quantified by ³¹P NMR in bovine milk.[8]

Solid-state MAS ³¹P NMR has been used extensively for the characterisation of native casein micelles isolated from milk.[9–11] Figure 10.2 depicts the solid-state MAS ³¹P NMR spectra of native casein micelles under various experimental NMR protocols.

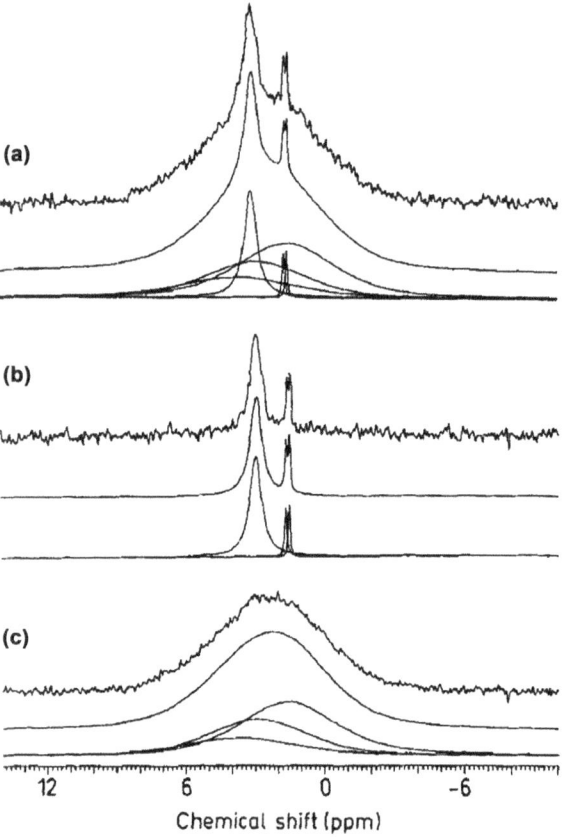

Figure 10.2   MAS ^{31}P-NMR spectra (121.5 MHz) of native casein micelles. Experimental (top) and computer deconvoluted (bottom) spectra are shown for (a) single pulse, (b) spin-echo and (c) cross-polarisation experiments.
(Reprinted from ref. 11. Copyright (1995), with permission from John Wiley and Sons.)

The two narrow peaks at δ 1.2–1.4 ppm are assigned to phosphoserine in κ-casein, the broader one at δ 3.1 ppm to colloidal calcium phosphate, while the broad resonance centred between δ 2–3 ppm results from immobile phosphorus compounds.[11] When casein micelles from ovine, caprine and bovine milk were compared, small differences in composition and mobility were noted.[10]

NMR-based metabolomic studies on milk are not too common, but are gaining attention recently, and will no doubt increase in forthcoming years. Some successful applications include the differentiation between different types of milk, cow/buffalo,[12] cow/sheep,[13] different breeds[14] and geographic origin.[15] Figure 10.3 depicts the successful differentiation of cow and sheep milk samples by the PCA analysis of the ^{1}H NMR spectra of their aqueous extracts.[13]

Finally, the NMR metabolic profiling of milk has proved to be very useful in unveiling potential prognostic biomarkers for cow ketosis risk,[16] and characterising the cow lactation state.[17]

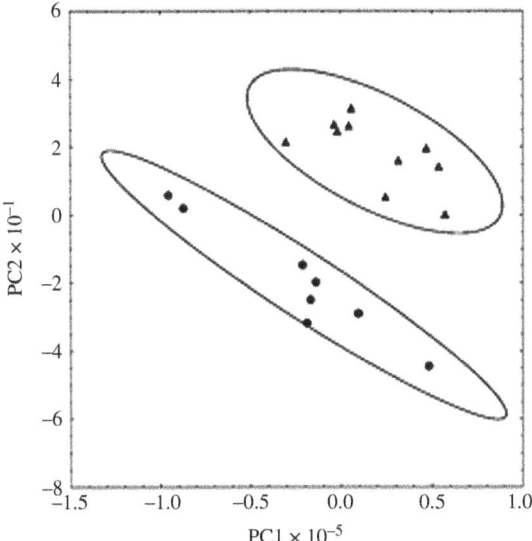

Figure 10.3   Score plot of the first two principal components of milk ^{1}H NMR spectra
of different animal origin (circles = cow; triangles = sheep). The variance
explained by PC1 and PC2 is, respectively, 69% and 23%. The ellipses
represent the 95% probability region for bivariate normal distribution
estimated from the score data.
(Reprinted from ref. 13. Copyright (2011), with permission from John
Wiley and Sons.)

10.1.2   Quality and Processing

Both high-resolution NMR and low-field NMR studies have addressed aspects
of quality evaluation of milk and its constituents in either whole milk or milk
powders.

In view of recent food crises involving melamine adulteration of infant milk
formulations, it is important that ^{1}H NMR was demonstrated to be suitable for
targeted melamine identification, and more importantly for routine non-targeted evaluation of milk preparations.[18] The ability of LF-NMR, in conjunction with PCA analysis for milk adulteration, has also been recently examined,
showing that milk samples adulterated with water, salt, urea or soya bean milk,
reconstituted milk and pure milk can be differentiated.[19] High-resolution ^{31}P
NMR spectroscopy has also contributed in studies dealing with the effect of
various processing procedures on milk composition. This methodology is
capable of quantifying additives such as polyphosphates used as stabilisers in
UHT milk.[6] It was also used to show that during storage, the phospholipid
hydrolysis in UHT milk depended on the microbial quality of the raw milk
utilised. High levels of the hydrolysis product, α-glucerophosphate, were
associated with UHT milks of poor original microbial quality, or expired
pasteurised milks, and were ascribed to increased phosphodiesterase activity of

bacteria that are able to survive UHT processing.[20] In another study, pasteurisation and UHT treatments did not alter significantly the casein content of milk as measured by ^{31}P NMR spectroscopy.[7]

Milk powders are obtained by dehydration and are important dairy products used in the manufacture of infant formula, confectionery and as milk replacement in pastry. ^{31}P HR-MAS NMR was used to study three types of skim milk powder having undergone different thermal treatments.[21] A study of the effect of high pressure treatment on skim milk powder dispersions by ^{31}P NMR reported an increase in the free phosphate concentration proportional to the magnitude of the pressure applied, but this effect was fully reversible at decompression to ambient pressure.[22] LF-NMR and relaxation measurements have been used to study the effect of composition on the hydration properties and the reconstitution of milk powders[23,24] and artificial milk protein dispersions containing lactose and calcium.[25,26] A kinetic NMR relaxometry method was used to study the water-holding capacity and follow the reconstitution process of milk powders, determining both the rate of solution and the delay for a complete reconstitution of milk powders.[24] Figure 10.4 presents the evolution of parameters A_p and R_{2s} during the reconstitution of native phosphocaseinate suspension powders of different animal origin.

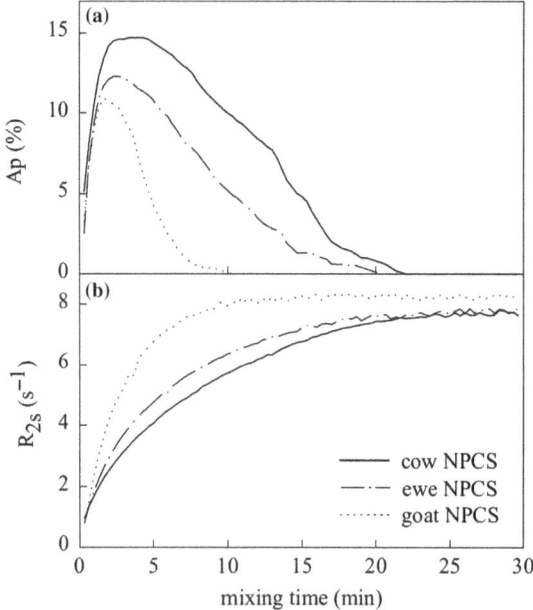

Figure 10.4   Effect of animal species origin on (a) the reconstitution period and (b) final relaxation rate R_{2s} of pure native phosphocaseinate suspension powders during reconstitution with water.
(Reprinted from ref. 24. Copyright (2002), with permission from EDP Sciences.)

A_p corresponds to the amount of protons in fast exchange with non-dissolved powder particles, and increases initially as the powder starts to absorb water, and after reaching a maximum decreases to zero because of total solubilisation of the powder. R_{2s} is the relaxation rate of slow relaxing protons associated with the amount of dissolved material, and reaches similar values for all three samples. However, the powder reconstitution periods differed significantly between the three animal species, being approx. 10, 20 and 22 for goat, ewe and cow, respectively, as observed in Figure 10.4(a).

10.2 Cheese

10.2.1 Composition and Metabolite Profiling

High-resolution liquid-state ^{1}H NMR spectroscopic studies of cheese can be performed in either chloroform extracts,[27] to characterise the lipid cheese fraction, or in aqueous extracts,[28-31] where low molecular weight polar compounds such as amino acids, organic acids and sugars can be characterised. In some studies a modification of the Bligh and Dyer method has been employed,[32,33] to minimise interference from lipid components in the aqueous phase studied. Figure 10.5 presents the full ^{1}H NMR spectrum of a *Parmigiano Reggiano* aqueous extract, indicating the major compounds identified with the help of 2D NMR spectroscopy.[29]

The ^{1}H NMR spectrum of the apolar extracts of cheese are dominated by lipid signals, while some minor components such as conjugated linoleic acid and 1-pentene were also assigned for *Asiago d'Allevo* cheese.[27] An exhaustive analysis of the ^{1}H and ^{13}C NMR spectra of the lipid fraction of *Pecorino Sardo* cheese, identified a large number of different fatty acyl chains, along with diacylglycerols and free fatty acids, and helped establish the very small levels of lipid oxidation in this cheese.[34]

^{1}H NMR has been recently validated as a fast quantitative methodology for the determination of histamine in different types of soft and hard cheeses.[35] The analytical NMR protocol suggested uses a perchloric acid solution for the quantitative extraction of histamine from cheese samples, and could be easily extended to other food systems. In another recent application, the *in vitro* digestibility of *Parmigiano Reggiano* cheese as a function of aging was studied by ^{1}H NMR metabolic profiling.[36]

The potential of solid-state NMR in cheese analysis was recognised very early.[37] Recently, ^{31}P single-pulse excitation and cross-polarisation MAS experiments (with dipolar proton decoupling) allowed the identification and quantification of soluble and insoluble phosphates, respectively, in semi-hard cow cheeses.[38] The single and double-quantum ^{23}Na NMR spectroscopy analysis of the same cheese samples provided information on the total and 'bound' Na^+ ions present, and the interactions between sodium, calcium and phosphates in this complex food matrix. HR-MAS ^{1}H NMR spectroscopy of cheese makes possible the analysis of both soluble amino acids and organic acids, and insoluble fatty acids in the same spectrum, without requiring any separation

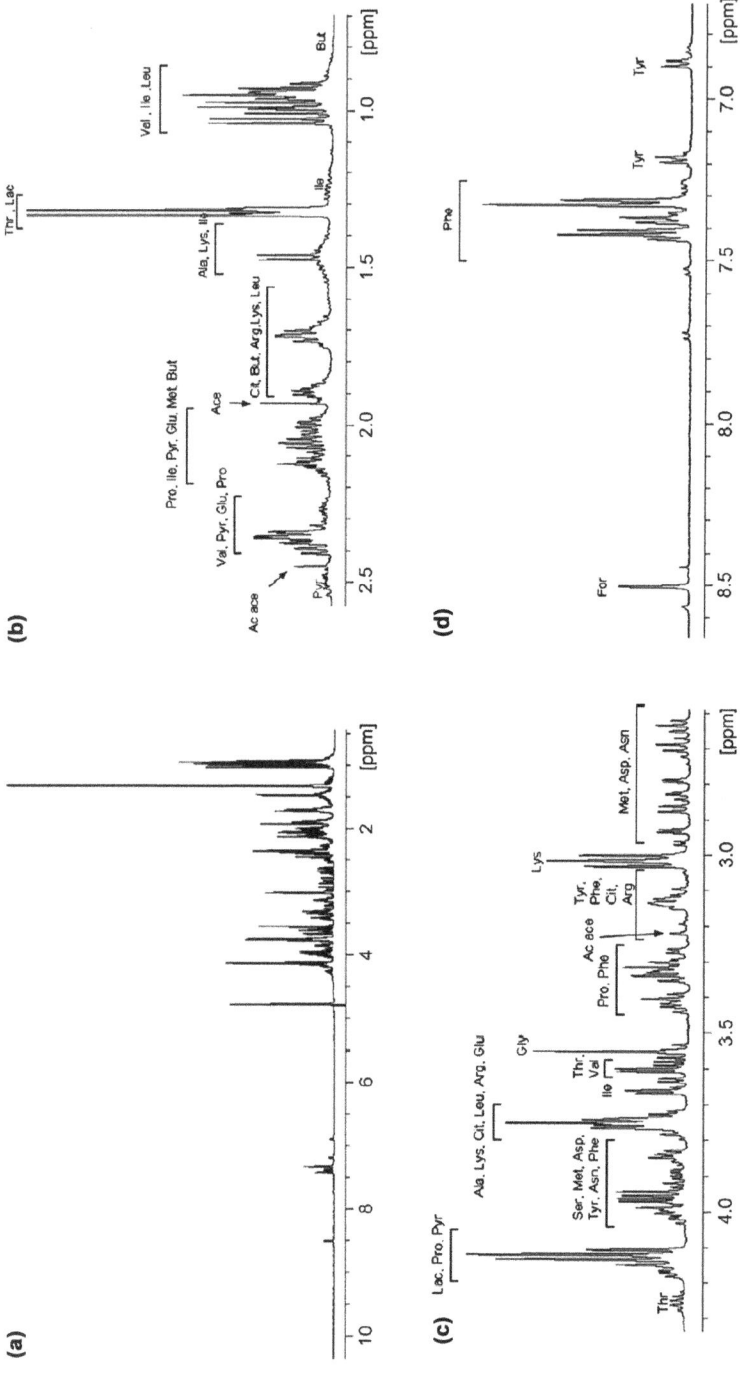

Figure 10.5 ¹H NMR spectrum of *Parmigiano Reggiano* aqueous extract. Complete spectrum (a) and expanded regions (b, c, d) with principal spin system assignments indicated. (Reprinted from ref. 29. Copyright (2008), with permission from Elsevier.)

Table 10.1 NMR-based chemometrics studies of dairy products.

Factor examined	Product	NMR methodology	Multivariate analysis model	Ref.
Type (cow/buffalo)	Milk	HR	PCA, HCA, DA	12
Type (cow/sheep)	Milk	HR	PCA, DA, ANN	13
Geographical origin	Milk	HR	PCA	15
Cow disease	Milk	HR	–	16
Cow lactation	Milk	HR	–	17
Breed	Milk	HR	PCA	14
Geographical origin	Cheese (*mozzarella*)	HR	PCA, HCA	28
Ripening	Cheese (*Parmigiano reggiano*)	HR-MAS	PCA, DA	40
Geographical origin	Cheese (*emmentaler*)	HR-MAS	PCA, DA	41
Production site	Cheese (*Asiago d'Allevo*)	HR	PCA, PLS-DA	27
Packaging	Cheese (*robiola*)	HR	PCA	30
Geographical origin, ripening	Cheese (*Parmigiano reggiano*)	HR	PCA, OPLS, PLS-DA	29
Strain, type, ripening	Cheese (probiotic, symbiotic)	HR	PCA, OPLS	31

step,[39–41] therefore providing a fast and efficient route towards the quality evaluation of cheese. Both liquid- and solid-state ^{1}H NMR methodologies have been utilised in metabolic profiling studies of different types of cheese, in efforts to study the effect of factors such as geographical origin, ripening and storage conditions (packaging). These studies are summarised in Table 10.1.

10.2.2 Quality and Processing

Early work uncovered the potential of MRI in the non-destructive analysis of the morphology and the study of various quality attributes of dairy products.[42] It was demonstrated that holes, cracks, or seams of mould in cheeses could be identified in 2D images through slice selection, while 3D imaging could be used to determine the shape and distribution of voids in cheese. Figure 10.6 presents an example of cheese morphological analysis, the MRI images obtained from a whole *Emmentaler* rind after 24 h of ripening.[43]

The presence of fat inclusions and air pockets can be discerned, and image analysis can further provide important information regarding factors that determine cheese quality during the cheese-making procedure. A morphological MRI analysis study of soft-blue veined cheeses during ripening has also been reported.[44]

Another interesting MRI application concerns the distribution of water and fat in cheese. With the use of suitable imaging pulse sequences, chemical shift selective (fat or water) MRI images can be used to quantify the fat and moisture content, and revealed that the moisture distribution was less uniform than fat distribution in cheese.[45] The moisture and heat transfer upon cooling of cheese

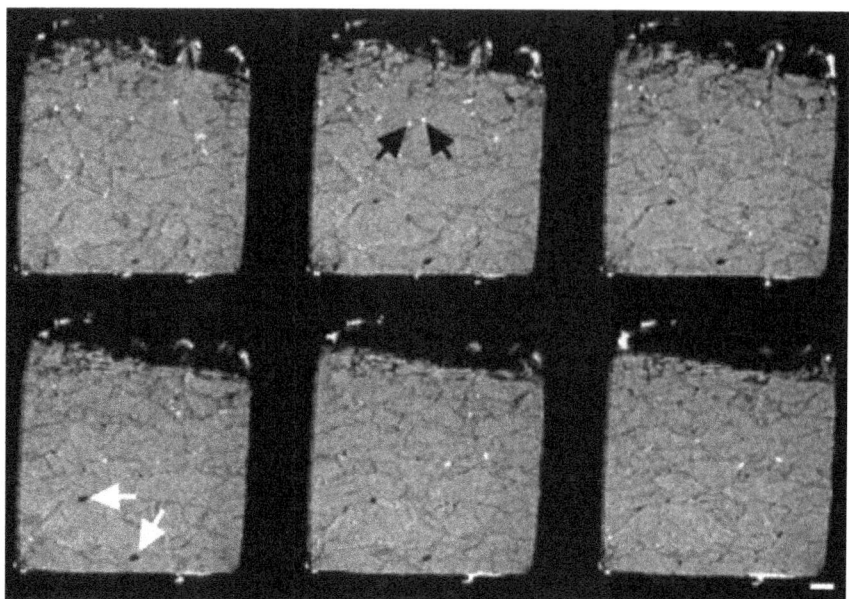

Figure 10.6 SE MR images of a whole *Emmentaler* rind after ripening for 24 h. The black arrows indicate small fat heaps, white arrows the presence of air pockets at the curd granule boundary. Bar = 1 mm.
(Reprinted from ref. 43. Copyright (2003), with permission from John Wiley and Sons.)

has also been studied by MRI.[46,47] Furthermore, the loss of signal from water protons upon freezing can be used to study the rate of ice formation during cheese freezing. In a study of *pasta filata* and *non-pasta filata mozzarella* it was reported that freezing begins symmetrically from the outer regions of the cheese sample and progresses toward the interior by a continuously advancing ice–water interface.[48]

Another cheese making process that can be followed by MRI is brining, as exemplified in a study of the changes occurring in feta cheese,[49] for which brining is very important for its organoleptic properties. MRI images of feta samples brined with a NaCl solution showed decreased image intensity during brining, as a result of the diffusion of water outside the cheese to the brining solution. This is a result of the osmotic pressure difference between cheese and brine solution caused by salt migration inside the cheese. The water loss was greater for more concentrated brines, as depicted in Figure 10.7, and was accompanied also by a slight shrinkage of the feta cheese.[49]

Low-field NMR and relaxometry have also found great application in cheese analysis, since they generally require instrumentation that is more affordable. The diffusion of fat and water in cheese was first studied by pulsed field gradient NMR almost thirty years ago.[50] Recently, an improved method based on a T_1-weighted SE sequence that takes advantage of the different T_1 properties of

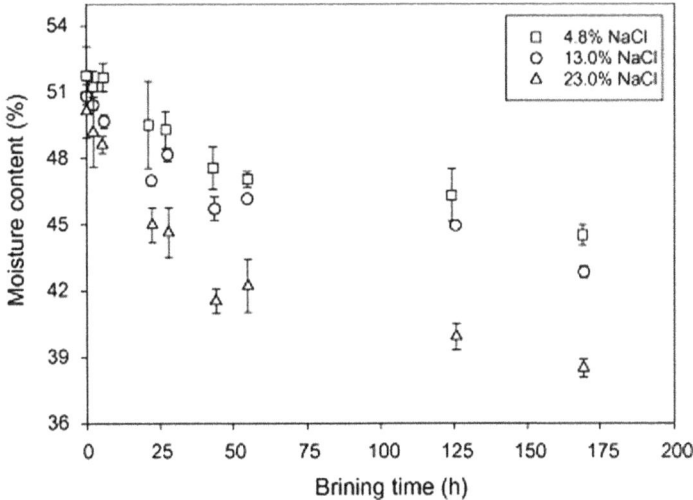

Figure 10.7 Effect of different brine concentration on moisture content ($\pm$ standard deviation) (%) of feta cheese during brining, obtained from MRI. (Reprinted from ref. 49. Copyright (2011), with permission from Elsevier.)

water and fat phases was proposed for the accurate determination of the water self-diffusion coefficient in cheese and food products in general.[51] T_2 spin–spin relaxation in cheese usually is deconvoluted in two components, representing water and fat,[52] although careful analysis of soft and hard cheeses has shown that two more components of reduced mobility can be identified, representing solid fat and protein protons (or protons in fast exchange with protein exchangeable protons).[53] It should be noted that other researchers have used a slightly different assignment of the four components lacking the solid fat contribution.[33]

T_2 relaxometry can be used to determine the moisture content of cheese[54] and study the effect of moisture content on the rheological properties of cheese analogues.[52,55]

Changes in water mobility using NMR relaxometry have been used to monitor the hydration of rennet casein and fat emulsification during the manufacture of imitation cheese,[56] and to characterise the water-holding capacity of differently treated fresh cheeses during manufacture.[57]

The proton T_1 and T_2 decay curves of *Mozzarella di Bufala Campana* samples were measured and analysed.[33,58] The T_2 values were also measured as a function of the aging time of mozzarella samples, and a decrease of free 'serum water' T_2 value with aging time was reported, a parameter possibly useful in monitoring this cheese after manufacture.[58] T_1 and T_2 relaxometry has also been employed to study cheese processing and storage effects, such as mozzarella storage[59] and freezing,[48] drying,[60] and feta cheese brining.[49] The T_1 and T_2 values of mozzarella cheeses generally increased during 10 d of storage, while the opposite behaviour was observed during feta cheese brining, especially at high salt concentrations. Finally cheese, along with other dairy products, has also been characterised by 2D $D–T_2$ and $T_1–T_2$ distribution functions in a static gradient.[61]

10.3 Yogurt

Yogurt is a dairy product of complex composition and structure, produced by the acidic fermentation of milk by bacterial cultures that transform lactose into lactic acid. Although the fat content of yogurt can be determined by NMR analysis in the same way as other dairy products,[62] high-resolution NMR studies dealing with yogurt compositional analysis have not been performed. Low-field NMR and relaxation measurements have been used to shed light on the complex phenomenon of gelation of milk during bacterial acidification to produce yogurt. The formation of yogurt by the acidification of heat-treated skim milk by lactic bacteria fermentation has been studied by ^{1}H NMR T_2 relaxation measurements at 30 and 42 °C.[63] During acidification, water T_2 relaxation times increase sigmoidally from an initial value of 150 ms, to reach a plateau value of 300 ms at pH 4.5 and lower, with small differences between the two temperatures.

The mobility of water in set yogurt during fermentation has also been studied using NMR relaxometry. T_2 relaxation was analysed using a biexponential model, with both components exhibiting rapid changes between 2 and 4 h of fermentation, coinciding with the drop in pH and a rise in lactic acid bacteria count. The water spin–lattice relaxation time T_1 also increased over the fermentation period. Both T_1 and T_2 provided evidence for increased water mobility upon gel formation during fermentation, attributed to water redistribution within the gel matrix due to casein aggregation and structure forming.[64]

The 2D $D–T_2$ and $T_1–T_2$ distribution functions of water in yogurt have also been reported. The increase in the T_2 relaxation time of water in yogurt, already mentioned above in the fermentation studies, is accompanied by a slight reduction of the diffusion coefficient of water in yogurt compared to milk, as expected from its semi-solid structure.[61]

NMR relaxometry in combination with other techniques has also been used to study yogurt rheology[65,66] and the water holding capacity of yogurt.[67]

10.4 Ice Cream

Ice cream is a complex dairy product containing water, fat, sugars, protein and emulsifiers. Since it is stored at subfreezing temperatures (< -13 °C) water and fat are present in both their liquid and solid state. As with yogurt, relaxometry has been the only NMR methodology that has been used consistently for ice cream analysis.[68–71]

Initial work focused on characterising the spin–spin (T_2) and spin–lattice (T_1) relaxations of water in frozen samples containing sucrose and/or casein at temperatures ranging from -13 to $20\,°C$. At the lower temperature the shorter T_2 component, belonging to ice crystals was used to quantify the ice content of the frozen solutions, is in good agreement with calorimetric data.[69] Then, in a series of papers, model ice cream formulations were prepared by varying the type of fat, sugar, protein and emulsifier, and their properties were studied by T_1 and T_2 relaxometry both before and after ice cream preparation. The effect of the composition of the formulations on the NMR relaxation properties of water/ice[70] and liquid/solid fat,[68] and the effect of cold storage were studied. It was reported that the factor most affecting the relaxation behaviour of water, ice crystals and liquid fat was the type of protein, while solid fat was affected mainly by the type of emulsifier. T_1 spin–lattice relaxation time measurements were used to quantify the amount of ice crystals and fat crystals in the model ice creams and to study mobility in the crystalline phases.[71]

References

1. R. Karoui and J. De Baerdemaeker, *Food Chem.*, 2007, **102**, 621.
2. I. S. Arvanitoyannis and N. E. Tzouros, *Crit. Rev. Food Sci. Nutr.*, 2005, **45**, 231.
3. J. Belloque and M. Ramos, *Trends Food Sci. Technol.*, 1999, **10**, 313.
4. F. Hu, K. Furihata, M. Ito-Ishida, S. Kaminogawa and M. Tanokura, *J. Agric. Food Chem.*, 2004, **52**, 4969.
5. F. Hu, K. Furihata, Y. Kato and M. Tanokura, *J. Agric. Food Chem.*, 2007, **55**, 4307.
6. J. Belloque, M. A. De La Fuente and M. Ramos, *J. Dairy Res.*, 2000, **67**, 529.
7. J. Belloque and M. Ramos, *J. Dairy Res.*, 2002, **69**, 411.
8. W. C. Byrdwell and R. H. Perry, *J. Chromatogr., A*, 2007, **1146**, 164.
9. M. Bak, L. K. Rasmussen, T. E. Petersen and N. C. Nielsen, *J. Dairy Sci.*, 2001, **84**, 1310.
10. L. K. Rasmussen, E. S. Sorensen, T. E. Petersen, N. C. Nielsen and J. K. Thomsen, *J. Dairy Sci.*, 1997, **80**, 607.
11. J. K. Thomsen, H. J. Jakobsen, N. C. Nielsen, T. E. Petersen and L. K. Rasmussen, *Eur. J. Biochem.*, 1995, **230**, 454.
12. M. A. Brescia, V. Mazzilli, A. Sgaramella, S. Ghelli, F. P. Fanizzi and A. Sacco, *J. Am. Oil Chem. Soc.*, 2004, **81**, 431.
13. R. Lamanna, A. Braca, E. Di Paolo and G. Imparato, *Magn. Reson. Chem.*, 2011, **49**, S22.
14. U. K. Sundekilde, P. D. Frederiksen, M. R. Clausen, L. B. Larsen and H. C. Bertram, *J. Agric. Food Chem.*, 2011, **59**, 7360.
15. D. Sacco, M. A. Brescia, A. Sgaramella, G. Casiello, A. Buccolieri, N. Ogrinc and A. Sacco, *Food Chem.*, 2009, **114**, 1559.

16. M. S. Klein, N. Buttchereit, S. P. Miemczyk, A. K. Immervoll, C. Louis, S. Wiedemann, W. Junge, G. Thaller, P. J. Oefner and W. Gronwald, *J. Proteome Res.*, 2012, **11**, 1373.

17. M. S. Klein, M. F. Almstetter, G. Schlamberger, N. Nürnberger, K. Dettmer, P. J. Oefner, H. H. D. Meyer, S. Wiedemann and W. Gronwald, *J. Dairy Sci.*, 2010, **93**, 1539.

18. D. W. Lachenmeier, H. Eberhard, F. Fang, S. Birk, D. Peter, S. Constanze and S. Manfred, *J. Agric. Food Chem.*, 2009, **57**, 7194.

19. C. Jiang, J. Han, J. Fan and S. Tian, *Trans. Chin. Soc. Agric. Eng.*, 2010, **26**, 340.

20. J. Belloque, A. V. Carrascosa and R. López-Fandiño, *J. Food Prot.*, 2001, **64**, 850.

21. T. Ishii, K. Hiramatsu, T. Tanaka, K. Sato and A. Tsutsumi, *Milchwissenschaft*, 2003, **58**, 178.

22. C. D. Hubbard, D. Caswell, H. D. Ldemann and M. Arnold, *J. Sci. Food Agric.*, 2002, **82**, 1107.

23. A. Davenel, P. Schuck and P. Marchal, *Milchwissenschaft*, 1997, **52**, 35.

24. A. Davenel, P. Schuck, F. Mariette and G. Brulé, *Lait*, 2002, **82**, 465.

25. A. Le Dean, F. Mariette, T. Lucas and M. Marin, *LWT – Food Sci. Technol.*, 2001, **34**, 299.

26. A. Le Dean, F. Mariette and M. Marin, *J. Agric. Food Chem.*, 2004, **52**, 5449.

27. E. Schievano, G. Pasini, G. Cozzi and S. Mammi, *J. Agric. Food Chem.*, 2008, **56**, 7208.

28. M. A. Brescia, M. Monfreda, A. Buccolieri and C. Carrino, *Food Chem.*, 2005, **89**, 139.

29. R. Consonni and L. R. Cagliani, *Talanta*, 2008, **76**, 200.

30. R. Lamanna, I. Piscioneri, V. Romanelli and N. Sharma, *Magn. Reson. Chem.*, 2008, **46**, 828.

31. D. Rodrigues, C. H. Santos, T. A. P. Rocha-Santos, A. M. Gomes, B. J. Goodfellow and A. C. Freitas, *J. Agric. Food Chem.*, 2011, **59**, 4955.

32. S. De Angelis Curtis, R. Curini, M. Delfini, E. Brosio, F. D'Ascenzo and B. Bocca, *Food Chem.*, 2000, **71**, 495.

33. R. Gianferri, M. Maioli, M. Delfini and E. Brosio, *Int. Dairy J.*, 2007, **17**, 167.

34. P. Scano, R. Anedda, M. P. Melis, M. A. Dessi, A. Lai and T. Roggio, *J. Am. Oil Chem. Soc.*, 2011, **88**, 1305.

35. E. Schievano, K. Guardini and S. Mammi, *J. Agric. Food Chem.*, 2009, **57**, 2647.

36. A. Bordoni, G. Picone, E. Babini, M. Vignali, F. Danesi, V. Valli, M. Di Nunzio, L. Laghi and F. Capozzi, *Magn. Reson. Chem.*, 2011, **49**, S61.

37. L. T. Kakalis, T. F. Kumosinski and H. M. Farrell, *J. Dairy Sci.*, 1994, **77**, 667.

38. M. Gobet, C. Rondeau-Mouro, S. Buchin, J. L. Le Quéré, E. Guichard, L. Foucat and C. Moreau, *Magn. Reson. Chem.*, 2010, **48**, 297.

39. L. Shintu, F. Ziarelli and S. Caldarelli, *Magn. Reson. Chem.*, 2004, **42**, 396.
40. L. Shintu and S. Caldarelli, *J. Agric. Food Chem.*, 2005, **53**, 4026.
41. L. Shintu and S. Caldarelli, *J. Agric. Food Chem.*, 2006, **54**, 4148.
42. S. L. Duce, M. H. G. Amin, M. A. Horsfield, M. Tyszka and L. D. Hall, *Int. Dairy J.*, 1995, **5**, 311.
43. R. Mahdjoub, J. Molegnana, M. J. Seurin and A. Briguet, *J. Food Sci.*, 2003, **68**, 1982.
44. A. Onea, G. Collewet, C. Fernandez, C. Vertan, N. Richard and F. Mariette, *SPIE-Int. Soc. Opt. Eng., Proc.*, 2003.
45. R. Ruan, K. Chang, P. L. Chen, R. G. Fulcher and E. D. Bastian, *J. Dairy Sci.*, 1998, **81**, 9.
46. R. R. Ruan, K. Chang, P. L. Chen and A. Ning, *Drying Technol.*, 1998, **16**, 1459.
47. K. Chang, R. Ruan, P. L. Chen and A. Ning, *Proceedings of the 1997 Annual International Meeting, American Society of Agricultural Engineers, Minneapolis, USA*, 1997, Vol. 3, pp.9.
48. M. I. Kuo, M. E. Anderson and S. Gunasekaran, *J. Dairy Sci.*, 2003, **86**, 2525.
49. A. Altan, M. H. Oztop, K. L. McCarthy and M. J. McCarthy, *J. Food Eng.*, 2011, **107**, 200.
50. P. T. Callaghan, K. W. Jolley and R. S. Humphrey, *J. Colloid Interface Sci.*, 1983, **93**, 521.
51. A. Metais and F. Mariette, *J. Magn. Reson.*, 2003, **165**, 265.
52. M. Budiman, R. L. Stroshine and O. H. Campanella, *J. Texture Stud.*, 2000, **31**, 477.
53. B. Chaland, F. Mariette, P. Marchal and J. De Certaines, *J. Dairy Res.*, 2000, **67**, 609.
54. M. Budiman, O. Campanella, S. Nielsen, R. Stroshine and P. Cornillon, *2000 ASAE Annual International Meeting, Technical Papers: Engineering Solutions for a New Century*, 2000.
55. M. Budiman, R. L. Stroshine and P. Cornillon, *J. Dairy Res.*, 2002, **69**, 619.
56. N. Noronha, E. Duggan, G. R. Ziegler, E. D. O'Riordan and M. O'Sullivan, *Int. Dairy J.*, 2008, **18**, 641.
57. R. Hinrichs, J. Gotz, M. Noll, A. Wolfschoon, H. Eibel and H. Weisser, *Food Res. Int.*, 2004, **37**, 667.
58. R. Gianferri, V. D'Aiuto, R. Curini, M. Delfini and E. Brosio, *Food Chem.*, 2007, **105**, 720.
59. M. I. Kuo, S. Gunasekaran, M. Johnson and C. Chen, *J. Dairy Sci.*, 2001, **84**, 1950.
60. A. Castell-Palou, C. Rosselló, A. Femenia, J. Bon and S. Simal, *J. Food Eng.*, 2011, **104**, 525.
61. M. D. Hürlimann, L. Burcaw and Y. Q. Song, *J. Colloid Interface Sci.*, 2006, **297**, 303.
62. G. Cartwright, B. H. McManus, T. P. Leffler and C. R. Moser, *J. AOAC Int*, 2005, **88**, 107.

63. A. Laligant, M. H. Famelart, D. Paquet and G. Brulé, *Lait*, 2003, **83**, 307.
64. C. Mok, J. Qi, P. Chen and R. Ruan, *Food Sci. Biotechnol.*, 2008, **17**, 895.
65. W. B. Yoon and K. L. McCarthy, *J. Texture Stud.*, 2002, **33**, 431.
66. A. Raudsepp, K. W. Feindel and Y. Hemar, *Rheol. Acta*, 2010, **49**, 371.
67. R. Hinrichs, J. Götz and H. Weisser, *Food Chem.*, 2003, **82**, 155.
68. T. Lucas, D. Le Ray, P. Barey and F. Mariette, *Int. Dairy J.*, 2005, **15**, 1225.
69. T. Lucas, F. o. Mariette, S. Dominiawsyk and D. Le Ray, *Food Chem.*, 2004, **84**, 77.
70. T. Lucas, M. Wagener, P. Barey and F. Mariette, *Int. Dairy J.*, 2005, **15**, 1064.
71. F. Mariette and T. Lucas, *J. Agric. Food Chem.*, 2005, **53**, 1317.

CHAPTER 11
Meat

In this chapter, we will discuss work pertaining to the analysis of meat and fish using high-resolution NMR in the liquid and solid state, MRI, and low-field NMR relaxometry. NMR spectroscopy studies concerning fish extracts and fish oils have been summarised in chapter 7.

11.1 Composition and Metabolite Profiling

Although the possibility to obtain high-resolution ^1H and ^{13}C NMR spectra of whole meat and meat extracts was demonstrated already more than 25 years ago,[1] this methodology has only recently started to be explored for the analysis of meat.[2–5] Figure 11.1 presents the high-resolution ^1H NMR spectrum of beef sirloin (or chuck) extracts obtained from samples originating in four different countries, indicating the major compounds assigned with the help of 2D NMR spectroscopy and spiking experiments.[2]

Twenty-five metabolites were identified, including amino acids and organic acids, and used in a metabolomics differentiation of geographic origin. Recently, a simple and rapid ^1H NMR methodology[6] has been proposed for the determination of conjugated linoleic acids in beef that could replace the more elaborate gas chromatographic techniques currently employed.

^1H HR-MAS solid-state NMR spectroscopy has also been proposed for the direct compositional analysis of meat,[7,8] and has been used successfully to follow lactate formation *post mortem* in rabbit muscle.[9] Meat may also be studied by using ^{13}C CP-MAS solid-state NMR analysis; signals from saturated and unsaturated carbons in fatty acids, carboxylic carbons, lactate and glycogen were identified in the solid-state CP-MAS NMR spectra of porcine muscle.[10,11] Lactate and glucogen can be determined, at least semi-quantitatively in both high-resolution[1] and solid-state[10] ^{13}C NMR spectra of meat. By far, ^{31}P is the most commonly employed nucleus for the NMR analysis of meat, and for good reason. ^{31}P has a natural abundance of 100%, and the NMR

RSC Food Analysis Monographs No. 10
NMR Spectroscopy in Food Analysis
By Apostolos Spyros and Photis Dais
© Apostolos Spyros and Photis Dais 2013
Published by the Royal Society of Chemistry, www.rsc.org

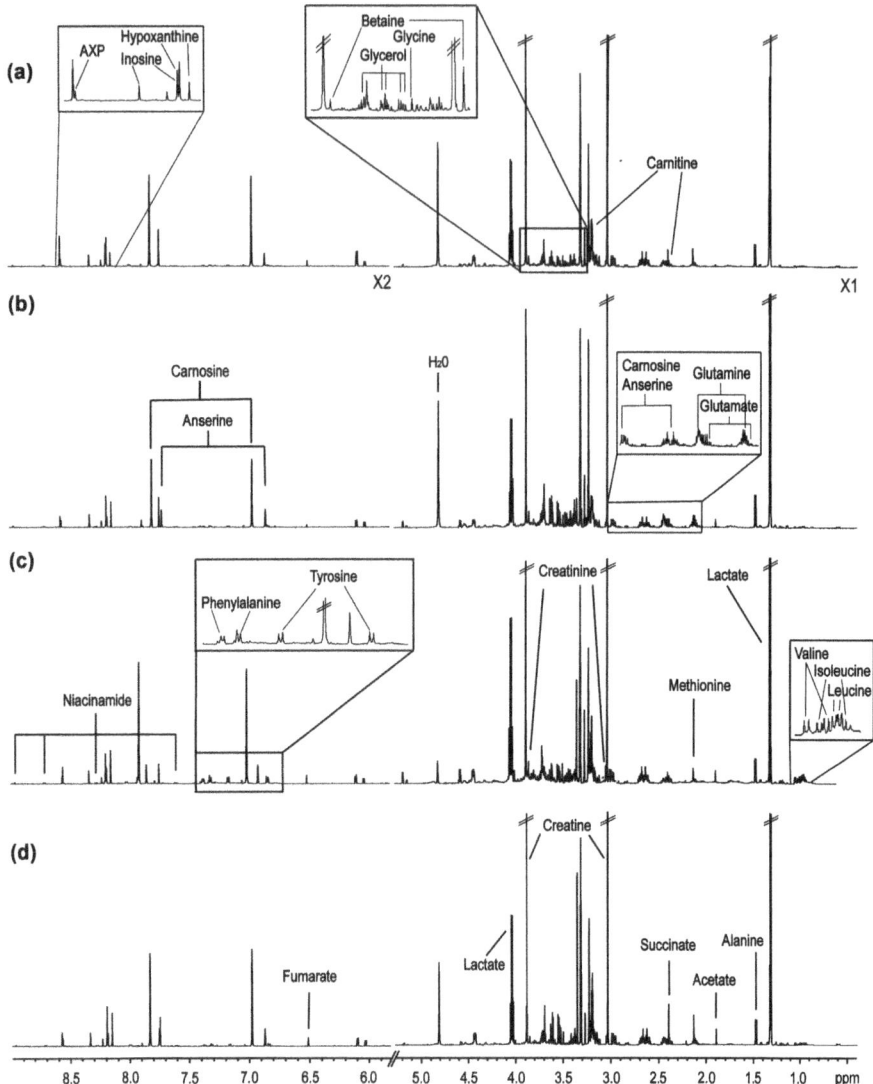

Figure 11.1 ¹H NMR spectra of beef sirloin (or chuck) extracts obtained from
Australia (a), Korea (b), New Zealand (c), and the United States (d). The
vertical scale of the aromatic region is doubled for better visibility. Some
signals are annotated.
(Reprinted from ref. 2. Copyright (2010), with permission from American Chemical Society.)

spectra of intact muscle can be obtained using high-resolution NMR spectrometers used for liquid analysis. Most importantly, meat contains several
phosphorus-containing metabolites involved in the *post mortem* metabolism of
muscle and the conversion to meat, procedures that play a crucial role in
defining several quality attributes of meat, including water holding capacity

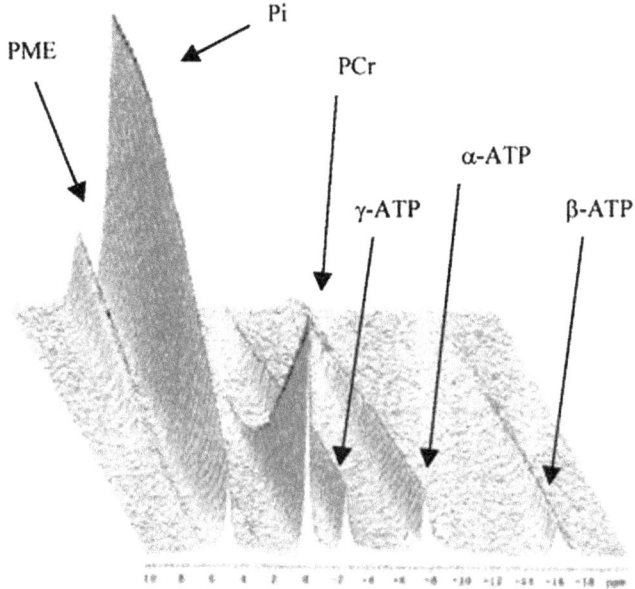

Figure 11.2 Stacked plot of ^{31}P NMR spectra recorded continuously from 20 min *post mortem* until 12 h *post mortem* on a muscle sample. Each spectrum is the average of 32 scans accumulated after 90° pulses at 8 s intervals. (Reprinted from ref. 12, Copyright (2001), with permission from Elsevier.)

(WHC) and tenderness. ^{31}P NMR provides a means to determine adenosine triphosphate (ATP), phosphocreatine (PCr), phosphomonoesters (PME), inorganic phosphate (Pi), and polyphosphates additives (PP) in meat quickly and efficiently. Figure 11.2 depicts the ^{31}P NMR spectra of pork muscle recorded continuously from 20 m until 12 h *post mortem*.[12] Due to the metabolic processes involved in the transformation of muscle to meat, signals due to ATP and PCr diminish, while Pi and PME signals increase. Another interesting aspect of ^{31}P NMR spectroscopy is that the ^{31}P NMR chemical shifts of ATP and Pi in meat can be used to measure the intracellular pH of the muscle.[13] Most studies use the chemical shift of the Pi peak for pH determination because of its favourable pK value. The width of the same signal serves as an indicator of pH heterogeneity in the muscle. In principle, ^{1}H NMR spectroscopy can be used for the determination of fat and water content of meat, using the lipid methylene resonance ($\sim$1.3 ppm) and the water resonance ($\sim$4.8 ppm), respectively, for quantification. Although this methodology can been applied both *in vitro* and *in vivo*,[14] the use of LF-NMR and relaxometry is far more common in meat analysis because of its rapidity and the fact that it requires less expensive instrumentation. Relaxation techniques have been used successfully for the determination of the fat and moisture content both in meat[15–19] and in fish.[20,21] Several studies utilising chemometrics and the multivariate analysis of

Table 11.1   NMR-based chemometrics studies of meat.

Factor examined	Type	NMR methodology	Multivariate analysis model	Ref.
Geographic origin	Beef	HR	PCA, OPLS-DA	2
Geographic origin	Dried beef	HR-MAS	PCA, DA	8
Pre-slaughter stress	Pork	HR	PCA, PLS-DA	22
Storage	Beef	HR	PCA	4
Sensory traits	Dry-cured ham	MRI	MLR[a]	68
Quality	Pork	HR	PCA	5
Sensory traits	Dry-cured pork	MRI	Texture Analysis	69

[a]MLR = Multiple Linear Regression.

NMR spectroscopic data obtained by HR, HR-MAS and MRI methodologies applied on meat have been published, examining factors such as geographic origin, sensory and quality traits, and animal stress[22] (see Table 11.1).

Figure 11.3 depicts the excellent discrimination between beef samples of different geographic origin based on ^{1}H NMR metabolic profiling.[2]

When compared to other foods such as beverages and fruit, NMR-based metabolomics on meat is still a field in its infancy, however it is expected to become the focus of much attention in the next few years.

11.2   Quality

The quality of meat depends on a large variety of factors that affect the *post mortem* metabolic reactions that are responsible for the conversion of muscle to meat. Some of these factors, such as genetic, animal feeding, and slaughter practices are already predetermined at the animal level. During the *post mortem* period, formation of lactate as a consequence of glycolysis under anaerobic conditions leads to an increase in pH, the muscle contracts and water channels between muscle fibres are formed that facilitate water loss. NMR spectroscopic techniques can provide information on meat pH, moisture and fat content and distribution and metabolite changes, and thus is an important tool in meat analysis.

11.2.1   Post Mortem Metabolism

NMR relaxometry can provide important information on the *post mortem* changes of water status and mobility in meat, mainly through T_2 relaxation time analysis, since T_1 appears to be monoexponential in most studies.[13] Water T_2 in meat is usually deconvoluted to two components, a major (80–95%) relatively fast-relaxing component with $T_2 \sim 35$–50 ms (within the protein-dense myofibrillar network) and a minor (5–15%) slow-relaxing component which represents mobile extra-myofibrillar water with $T_2 \sim 100$–250 ms. Sometimes a third, very slow T_2 component is observed, associated with water tightly associated with macromolecules ($T_2 < 10$ ms).[13] *Post mortem* LF-NMR

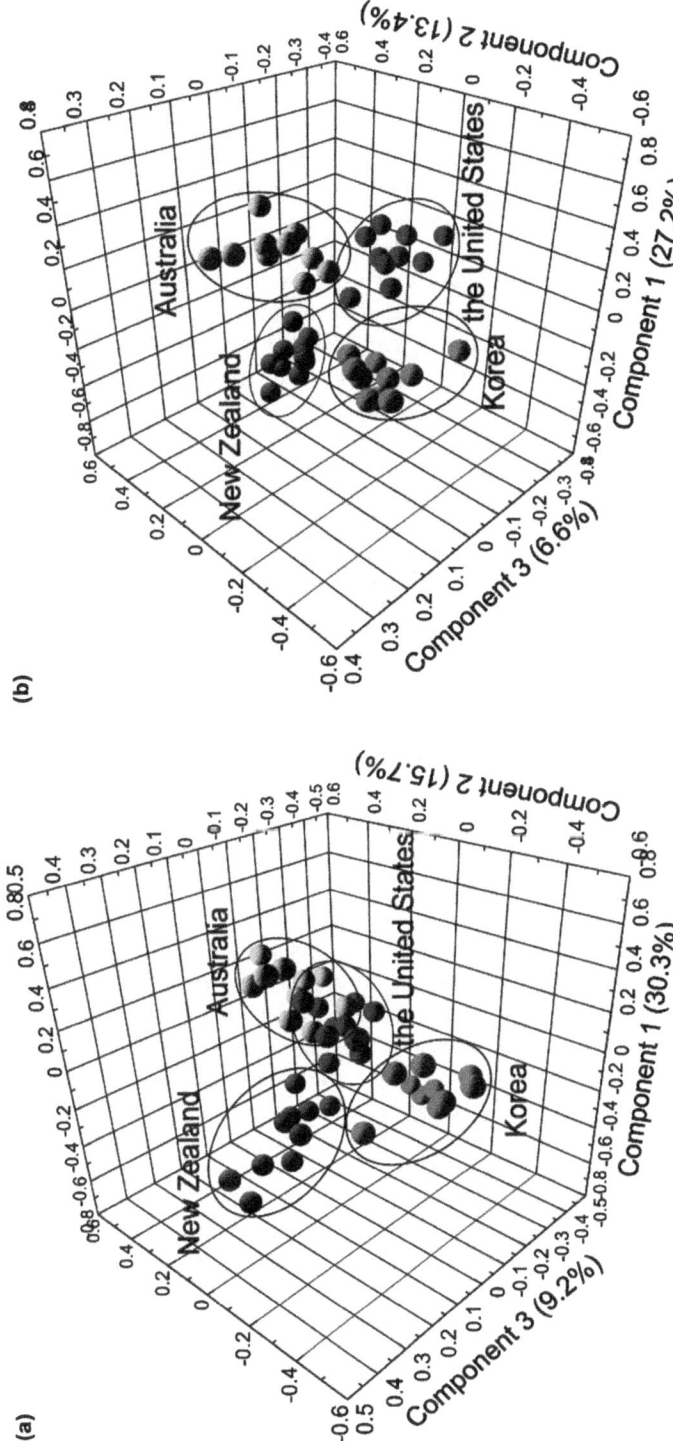

Figure 11.3 PCA (a) and OPLS-DA (b) 3D score plots derived from the ^{1}H NMR spectra of beef sirloin (or chuck) extracts obtained from Australia, Korea, New Zealand, and the United States. (Reprinted from ref. 2. Copyright (2010), with permission from American Chemical Society.)

relaxation measurements in porcine muscle showed that mobile extra-myofi-brillar water increased, suggesting that water is redistributed in the muscle during conversion to meat.[23] This redistribution is directly related to the formation of drip channels in muscle during its conversion to meat that facilitate water loss and are related to the WHC of meat. The formation of such drip channels in rabbit muscle has been successfully observed by using MRI.[24] Some representative examples of applications of LF-NMR relaxation in meat analysis include studies of pork,[25] processed pork,[26] sausage,[27] and turkey breast meat.[28] MRI has also contributed in meat quality analysis in a wide variety of applications, including the analysis of feeding background[29] and ripening[30] of fresh Iberian ham, carcass quality in lamb,[31] genotypic origin of bovine meat,[32] and the determination of lean meat percentage of pig carcass.[33]

In efforts to understand metabolic effects on meat quality, high-resolution ^{31}P NMR spectroscopy has been used to study the rate of post-mortem metabolism in muscles from several different species of interest to meat science, such as beef,[34,35] rabbit,[9,36] and pork.[35,37] The effect of fibre composition in muscle metabolism has been studied in rabbit,[36] and goat,[38] by ^{31}P NMR spectroscopy. The rate of ATP and PCr loss was found to be highest in oxi-dative and lowest in pure glycolytic muscles, but conflicting results were reported for the rates of pH reduction in different species, indicating that factors other than glycolysis might also play a role.[13] *Post mortem* metabolic studies in carp muscle,[39] oyster tissue[40] and crayfish[41] have also been performed.

Pork meat is usually classified in three different qualities: normal, pale soft exudative (PSE), possessing low WHC, and dark firm dry (DFD) displaying a high ultimate pH and high WHC. Figure 11.4 displays representative ^{31}P NMR spectra from the three types of pork meat obtained at 30 min *post mortem*.[37]

^{31}P NMR can be used to differentiate these pork qualities by their metabolic profile of phosphorus compounds, with high PME and low PCr concentrations indicative of DFD-prone meat, and high Pi combined with low PME signals associated with PSE-prone pork meat. ^{31}P NMR spectroscopy has been used extensively in the study of malignant hypothermia, a genetic syndrome in pigs, which is also known as halothane susceptibility. Malignant hypothermia affects muscle metabolism leading to pale soft exudative pork meat of inferior quality and low organoleptic properties. The determination of the PCr/Pi ratio of muscle samples by ^{31}P NMR spectroscopy allowed the discrimination between halothane-positive and halothane-negative pigs, with variations in pH, PCr and ATP being positively correlated with the rate of *post mortem* metabolism and meat quality traits of the animals after slaughter.[42]

Some important animal processing parameters that affect meat quality, such as feeding and slaughter conditions have also been studied by ^{31}P NMR spectroscopy through their effect on the metabolic processes during muscle conversion to meat.[13] Stunning before slaughter can be performed by a variety of methods, including CO_2, electrical, captive bolt pistol or anaesthesia. Apart from apparent ethical considerations, an optimised stunning strategy is important for minimising the effect of short-term stress immediately before

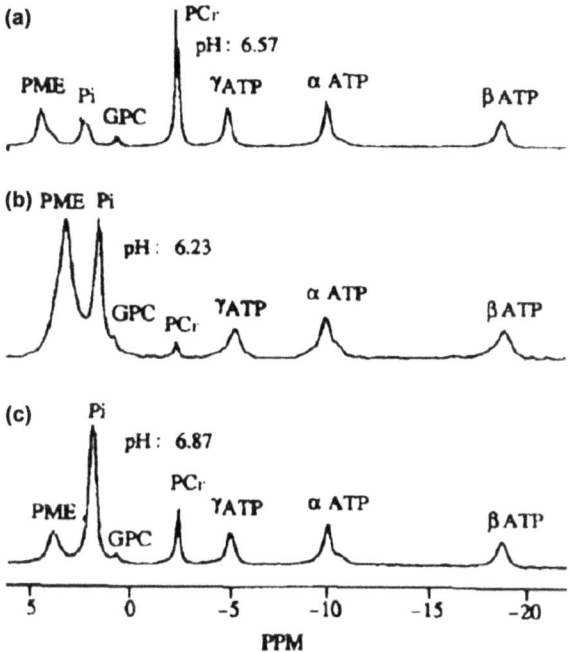

Figure 11.4 ^{31}P NMR spectra (161.9 MHz) of raw pig muscle obtained at 30 min *post mortem:* (a) normal muscle; (b) PSE-prone muscle; (c) DFD muscle. The labels denote: PME = phosphomonoesters; Pi = inorganic phosphate; GPC = glucerophosphorylcholine; PCr = phosphocreatine; α-, β- γ-ATP = adenosinetriphosphate.
(Reprinted from ref. 37. Copyright (1992), with permission from Elsevier.)

slaughter, a situation known to lead to pale soft exudative meat. The ^{31}P NMR spectroscopic determination of the levels of various phosphorus metabolites and the pH decrease in pig muscle *post mortem* from animals stunned with different methods showed that CO_2 stunning is more stressful than anaesthesia, although it is generally considered moderate compared to other stunning methods.[43]

Finally, ^{31}P NMR was used to study the effect of introducing dietary supplements in animal feeding such as magnesium[44] and vitamin E[45] on the *post mortem* metabolism of pig muscle.

11.2.2   Water Holding Capacity

The ability of fresh meat to retain moisture (WHC) is one of the most important quality characteristics of meat, and one of high economic importance to food industry, therefore methods to evaluate the WHC of meat and the parameters affecting it are necessary. Early work showed that WHC correlated with T_1 and T_2 relaxation parameters of water in pork meat,[46] and subsequent studies demonstrated a very good correlation ($r = 0.76$) between the slow relaxing T_2 population and the WHC, suggesting that WHC is mainly determined by the amount of loosely bound extra-myofibrillar water.[25] The decreased WHC characterising pork obtained from halothane gene carrier pigs was also correlated with increased extra-myofibrillar water as measured by T_2 relaxation. Good correlation was observed between T_2-determined WHC values and two traditional methods of analysis, indicating that LF-NMR is an efficient method for WHC determination.[47] The use of NMR relaxation measurements to characterise water populations in pork meat and investigate how the distribution and mobility of water changes *post mortem* has been summarised recently.[48]

11.3   Processing

11.3.1   Freezing

Both LF-NMR relaxation and MRI have been used for the study of the effect of freezing on the texture, sensorial and quality traits of meat and fish. T_2 relaxometry was used to study thawed beef and salmon samples that had been frozen for a period of up to two months at $-18\,^{\circ}$C. A decrease in the short T_2 component associated with myofibrillar water was observed for both beef and salmon upon frozen storage, indicating an increase in intercellular water, which was consistent with an increase in the squeezable drip also measured during storage.[49] In frozen cod, an extra relaxation component with a large T_2 appeared during frozen storage, possibly attributed to thaw exudate,[50] and similar observations were made in a cod mince relaxation study.[51] MRI investigations of freeze storage have been reported for both meat[52–54] and fish.[55–57] MRI was used to investigate the spatial distribution of the non-frozen water in beef during freeze storage in the temperature range between 20 and $-40\,^{\circ}$C.[54] The MRI analysis of T_1 and T_2 relaxation times, magnetisation transfer rates (MT) and apparent water diffusion coefficients (D) was used in an attempt to authenticate the effect of freezing/thawing in *longissimus dorsi* pig muscle.[53] MT refers to the magnetisation transfer rate between protons attached to slowly tumbling macromolecules and free water, and this was the only parameter that differed significantly for freeze/thawed samples, possibly attributed to a decrease in moisture content and denaturation of myofibrillar proteins. However, in a follow up study that utilised the simultaneous MRI analysis of a large number of beef, lamb, and pork samples, it was concluded that although T_1 and MT differences were present, inter-animal variability was too large to allow this methodology to be used for freeze/thawed authentication.[52] The same methodology was applied to trout,[56] cod and mackerel[55] frozen storage studies. In another study, trout samples subjected to different freezing methods were investigated using diffusion tensor imaging (DTI), taking advantage of the anisotropic orientation of fibres in the muscle.[57] Figure 11.5 presents DTI images of trout samples differing on the duration of freezing

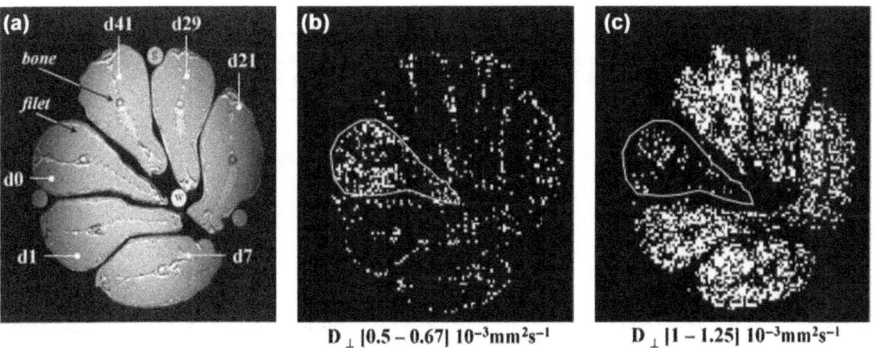

D⊥ |0.5 – 0.67| 10⁻³mm²s⁻¹   D⊥ |1 – 1.25| 10⁻³mm²s⁻¹

Figure 11.5 Morphological nuclear magnetic resonance images of whole trout: (a) anatomical and (b, c) diffusion parametric ($D_\perp$) images for the same set of six trout; (g) and (w) are reference tubes of gelatin gel and water, respectively.
(Reprinted from ref. 57. Copyright (2003), with permission from Elsevier.)

storage. Although frozen samples cannot be differentiated, the fresh trout sample (d0) is clearly differentiated by its lower radial diffusion coefficient.

11.3.2 Cooking

Cooking imparts strong morphological and compositional changes on meat, and these can be studied by NMR spectroscopy techniques. Using ^{1}H NMR spectroscopy it is possible to determine the decreasing water/fat ratio of hamburger, pork sausage, and various cuts of chicken during cooking by the integration of the respective NMR signals[58] (see Figure 11.6).

The difference in the NMR chemical shift of the same peaks can be used to measure the exact temperature of cooking within the food being cooked, with high accuracy for foods containing significant amounts of fat. The T_2 relaxation behaviour of water can also be used as a probe to obtain information on the changing structural conditions inside meat during cooking. During cooking the water is expelled from the food matrix, and thus the amount of relaxometrically 'free' water (with $T_2 > 1$ s), increases. Furthermore, the distribution of T_2 relaxation times is modified at a temperature between 40–50 °C during cooking, with the appearance of a 'new' transverse relaxation component with $T_2 \sim 15$–20 ms, originating from the initial ^{1}H pool of the myofibrillar slow relaxing water component.[13,59] The relaxation changes in this temperature region were found to depend on meat quality[60] and the processing history of pork and beef,[61] such as curing, aging and heat-pressure treatment. The structural changes between 40–50 °C during cooking were attributed to the thermal denaturation of meat proteins,[62] while further changes close to 56 °C were suggested to reflect the onset of collagen shrinkage.[60] Several studies have dealt with the effect of processing treatments by comparing the differences in the relaxation response of fresh and cooked meat samples.[63–67]

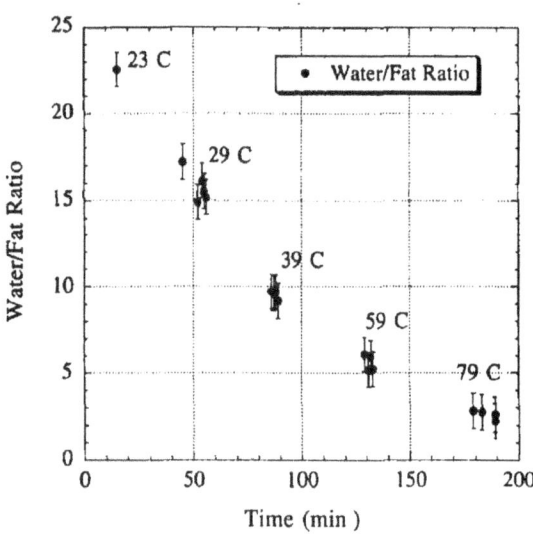

Figure 11.6   Water/fat ratio during cooking of chicken drum as obtained by ¹H NMR
spectroscopy.
(Reprinted from ref. 58. Copyright (1999), with permission from John
Wiley and Sons.)

11.3.3   Curing

Curing (brining) is a very important processing procedure for meat, since it is
well known that the addition of salts increases the WHC of meat and its ability
to take up additional water. Curing can be followed directly by ^{23}Na MRI of
the distribution of salt on meat, and by ^{23}Na NMR spectroscopy and relaxometry, although ^{1}H relaxometry and MRI have also contributed to curing
studies.[13,68,69]

The first application of ^{23}Na MRI involved following the effect of brine
injection in *pre rigor* rabbit muscles.[70] The distribution of sodium was found to
be inhomogeneous 6 h after injection and remained heterogeneous even after
24 h of brining, although tumbling was reported to improve somewhat the
homogeneity in sodium distribution, which was completely established upon
meat cooking. 1D ^{23}Na NMR imaging (MRI) has been used to follow the
ingress of sodium ions into *post rigor* porcine muscle during brining.[71] The
sodium diffusion coefficient was found to decrease exponentially with
increasing sodium content, probably reflecting the increased protein–ion
interactions with increasing NaCl concentration. In a similar study on pork
loin,[72] the salt concentration in pork loin was quantified successfully at salt
values above 0.9 g NaCl per 100 g sample in the meat, with the measurements
being calibrated against chemically determined chloride. ^{23}Na-MRI profiles
suggested that the diffusion of salt into whole meat cuts cannot be described by
simple ordinary Fickian diffusion but is affected by changes in NaCl concentration, meat swelling and degree of dehydration. ^{23}Na and ^{35}Cl NMR

spectroscopy have also been used to determine the mobility of sodium and chloride ions in intact snow crab (*Chionoecetes japonicus*) leg meat.[73] Heating treatments were reported to inhibit both sodium and chloride ion mobility as studied by T_1 relaxation time measurements. A combined ^{23}Na MRI and ^{23}Na NMR relaxometry study was applied for the study of meat curing.[74] A 20% decrease in the sodium diffusion coefficient was reported after 3 h of curing, suggesting curing imparts changes in the microscopic structure of the meat, while relaxometry revealed the presence of two sodium populations in cured meat. Figure 11.7 presents the [23]Na MR images obtained during the curing of a 1 kg meat sample, showing the progress in sodium interdiffusion.[74]

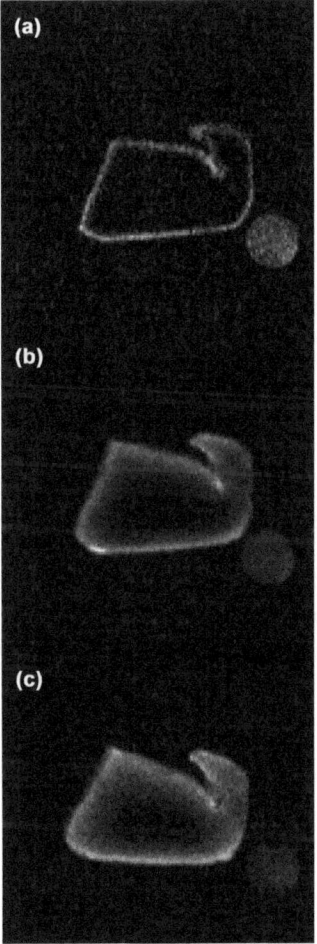

Figure 11.7 [23]Na images showing the progress in sodium inter-diffusion: (a) 3 h; (b) 52 h; (c) 100 h. The white circle is a phantom of 300 mM NaCl. (Reprinted from ref. 74. Copyright (2005), with permission from American Chemical Society.)

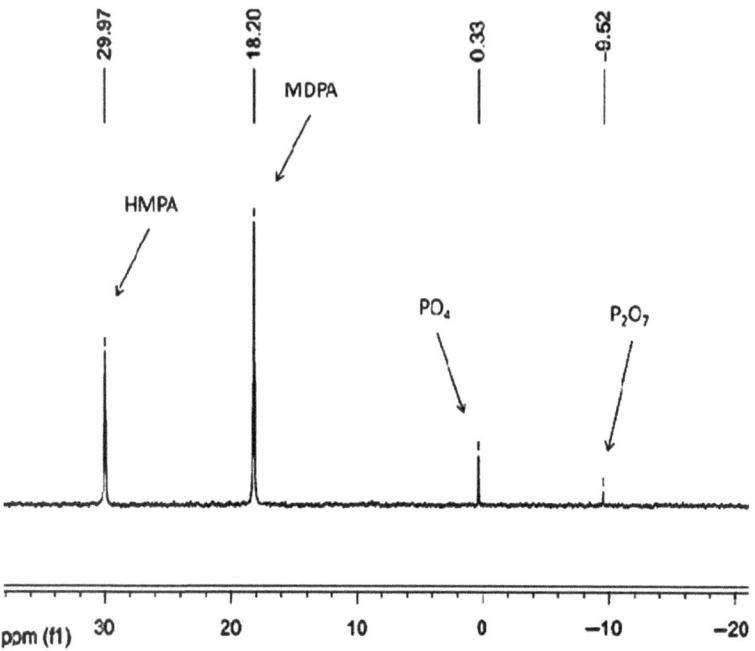

Figure 11.8   ^{31}P NMR spectrum of meat product extracts indicating phosphate signals. HMPA and MDPA are internal standards.
(Reprinted from ref. 77. Copyright (2011), with permission from Elsevier.)

Polyphosphates are used in combination with NaCl for the curing of processed meat products to increase their water-binding capacity. Before action, added phosphates must be hydrolysed either non-enzymatically or by muscle phosphatases, and this procedure has been studied extensively using ^{31}P NMR spectroscopy, reporting differences in the hydrolysis rate of polyphosphates.[75,76] ^{31}P NMR methodologies have also been validated for the analytical determination of polyphosphates in processed meat products.[77–79] Figure 11.8 presents the ^{31}P NMR spectrum of an extract of a processed meat product. Although triphosphate was declared as added by the manufacturer, only pyrophosphate and phosphate are observed in the spectrum, due to hydrolysis.[77]

References

1. P. Lundberg, H. J. Vogel and H. k. Rudarus, *Meat Sci.*, 1986, **18**, 133.
2. Y. Jung, J. Lee, J. Kwon, K. S. Lee, D. H. Ryu and G. S. Hwang, *J. Agric. Food Chem.*, 2010, **58**, 10458.

3. S. F. Graham, D. Farrell, T. Kennedy, A. Gordon, L. Farmer, C. Elliott and B. Moss, *Food Chem.*, 2012, **134**, 1633.
4. S. F. Graham, T. Kennedy, O. Chevallier, A. Gordon, L. Farmer, C. Elliott and B. Moss, *Metabolomics*, 2010, **6**, 395.
5. A. Reum Kim, S. H. Park, J. Nam, J. Kwon, M. H. Park, S. O. Kwon, E. J. Kwon, J. H. Jung, H. C. Park, B. Y. Park, G. S. Hwang, I. S. Jang, W. Y. Bang, C. W. Kim and J. S. Choi, *Afr. J. Biotechnol.*, 2011, **10**, 14209.
6. R. Manzano Maria, L. A. Colnago, L. Aparecida Forato and D. Bouchard, *J. Agric. Food Chem.*, 2010, **58**, 6562.
7. M. A. Brescia, A. C. Jambrenghi, V. Di Martino, D. Sacco, F. Giannico, G. Vonghia and A. Sacco, *Ital. J. Anim. Sci.*, 2002, **1**, 151.
8. L. Shintu, S. Caldarelli and B. M. Franke, *Meat Sci.*, 2007, **76**, 700.
9. H. C. Bertram, A. K. Whittaker, H. J. Andersen and A. H. Karlsson, *Int. J. Food Sci. Technol.*, 2004, **39**, 661.
10. H. C. Bertram, H. J. Jakobsen and H. J. Andersen, *J. Agric. Food Chem.*, 2004, **52**, 3159.
11. H. C. Bertram, H. J. Jakobsen, H. J. Andersen, A. H. Karlsson and S. B. Engelsen, *J. Agric. Food Chem.*, 2003, **51**, 2064.
12. H. C. Bertram, S. Donstrup, A. H. Karlsson, H. J. Andersen and H. Stodkilde-Jorgensen, *Magn. Reson. Imag.*, 2001, **19**, 993.
13. H. C. Bertram and H. J. Ersen, *Annu. Rep. NMR Spectrosc.*, 2004, **53**, 157.
14. A. D. Mitchell, T. H. Elsasser and P. C. Wang, *J. Sci. Food Agric.*, 1991, **56**, 265.
15. C. C. Corra, L. A. Forato and L. A. Colnago, *Anal. Bioanal. Chem.*, 2009, **393**, 1357.
16. T. P. Leffler, C. R. Moser, B. J. McManus, J. J. Urh, J. T. Keeton, A. Claflin, K. Adkins, A. Claflin, C. Davis, J. Elliot, P. Goin, C. Horn, J. Humphries, K. Kctteler, P. Perez and G. Steiner, *J. AOAC Int.*, 2008, **91**, 802.
17. E. Nagy and L. Kermendy, *Acta Aliment.*, 2003, **32**, 289.
18. J. T. Keeton, B. S. Hafley, S. M. Eddy, C. R. Moser, B. J. McManus and T. P. Leffler, *J. AOAC Int.*, 2003, **86**, 1193.
19. G. H. SΓErland, P. M. Larsen, F. Lundby, A. P. Rudi and T. Guiheneuf, *Meat Sci.*, 2004, **66**, 543.
20. C. A. Toussaint, F. Mcdale, A. Davenel, B. Fauconneau, P. Haffray and S. Akoka, *J. Sci. Food Agric.*, 2002, **82**, 173.
21. D. Nielsen, G. Hyldig, J. Nielsen and H. H. Nielsen, *LWT–Food Sci. Technol.*, 2005, **38**, 537.
22. H. C. Bertram, N. Oksbjerg and J. F. Young, *Meat Sci.*, 2010, **84**, 108.
23. H. C. Bertram, A. Schaefer, K. Rosenvold and H. J. Andersen, *Meat Sci.*, 2004, **66**, 915.
24. H. C. Bertram, A. K. Whittaker, H. J. Andersen and A. H. Karlsson, *Meat Sci.*, 2004, **68**, 667.
25. H. C. Bertram, P. P. Purslow and H. J. Andersen, *J. Agric. Food Chem.*, 2002, **50**, 824.

26. A. Hullberg and H. C. Bertram, *Meat Sci.*, 2005, **69**, 709.
27. S. M. Møller, A. Gunvig and H. C. Bertram, *Meat Sci.*, 2010, **86**, 462.
28. M. Bianchi, F. Capozzi, M. A. Cremonini, L. Laghi, M. Petracci, G. Placucci and C. Cavani, *J. Sci. Food Agric.*, 2004, **84**, 1535.
29. T. Pérez-Palacios, T. Antequera, M. L. Durin, A. Caro, P. G. Rodríguez and R. Palacios, *Food Chem.*, 2011, **126**, 1366.
30. T. Antequera, A. Caro, P. G. Rodriguez and T. Perez, *Meat Sci.*, 2007, **76**, 561.
31. S. V. Korn, U. Baulain, M. Arnold and W. Brade, *Zuchtungskunde*, 2005, **77**, 382.
32. D. Mahmoud-Ghoneim, J. M. Bonny, J. P. Renou and J. D. De Certaines, *J. Sci. Food Agric.*, 2005, **85**, 629.
33. G. Collewet, P. Bogner, P. Allen, H. Busk, A. Dobrowolski, E. Olsen and A. Davenel, *Meat Sci.*, 2005, **70**, 563.
34. H. J. Vogel, P. Lundberg, S. Fabiansson, H. Rudirus and E. Tornberg, *Meat Sci.*, 1985, **13**, 1.
35. P. Uhrin and T. Litpaj, *Gen. Physiol. Biophys.*, 1991, **10**, 83.
36. J. P. Renou, P. Canioni and P. Gatelier, *Biochimie*, 1986, **68**, 543.
37. A. Miri, A. Talmant, J. P. Renou and G. Monin, *Meat Sci.*, 1992, **31**, 165.
38. Y. Azuma, N. Manabe, F. Kawai, M. Kanamori and H. Miyamoto, *Anim. Sci. Technol.*, 1994, **65**, 416.
39. Y. Yokoyama, Y. Azuma, M. Sakaguchi, F. Kawai and M. Kanamori, *Fish. Sci.*, 1996, **62**, 267.
40. Y. Yokoyama, Y. Azuma, M. Sakaguchi, F. Kawai and M. Kanamori, *Fish. Sci.*, 1996, **62**, 416.
41. M. J. Gradwell, T. W. M. Fan and A. N. Lane, *Anal. Biochem.*, 1998, **263**, 139.
42. R. Lahucky, J. Mojto, J. Poltarsky, A. Miri, J. P. Renou, A. Talmant and G. Monin, *Meat Sci.*, 1993, **33**, 373.
43. H. C. Bertram, H. Stodkilde-Jorgensen, A. H. Karlsson and H. J. Andersen, *Meat Sci.*, 2002, **62**, 113.
44. B. Moesgaard, I. E. Larsen, B. Quistorff, I. Therkelsen, V. G. Christensen and P. F. Jørgensen, *Acta Vet. Scand.*, 1993, **34**, 397.
45. R. Lahucky, P. Krska, U. Kuchenmeister, K. Nurnberg, T. Liptaj, G. Nurnberg, I. Bahelka, P. Demo, G. Kuhn and K. Ender, *Arch. Tierz.*, 2000, **43**, 487.
46. J. P. Renou, G. Monin and P. Sellier, *Meat Sci.*, 1985, **15**, 225.
47. H. C. Bertram, H. J. Andersen and A. H. Karlsson, *Meat Sci.*, 2001, **57**, 125.
48. H. C. Bertram and H. J. Andersen, *J. Anim. Breed. Genet.*, 2007, **124**, 35.
49. S. Yano, M. Tanaka, N. Suzuki and Y. Kanzaki, *Food Sci. Technol. Res.*, 2002, **8**, 137.
50. P. Lambelet, F. Renevey, C. Kaabi and A. Raemy, *J. Agric. Food Chem.*, 1995, **43**, 1462.
51. C. Steen and P. Lambelet, *J. Sci. Food Agric.*, 1997, **75**, 268.

52. S. D. Evans, K. P. Nott, A. A. Kshirsagar and L. D. Hall, *Int. J. Food Sci. Technol.*, 1998, **33**, 317.

53. T. M. Guiheneuf, A. D. Parker, J. J. Tessier and L. D. Hall, *Magn. Reson. Chem.*, 1997, **35**, S112.

54. S. Lee, P. Cornillon and Y. R. Kim, *J. Food Sci.*, 2002, **67**, 2251.

55. K. P. Nott, S. D. Evans and L. D. Hall, *LWT – Food Sci. Technol.*, 1999, **32**, 261.

56. K. P. Nott, S. D. Evans and L. D. Hall, *Magn. Reson. Imag.*, 1999, **17**, 445.

57. J. P. Renou, L. Foucat and J. M. Bonny, *Food Chem.*, 2003, **82**, 35.

58. J. H. Walton and M. J. McCarthy, *J. Food Process Eng.*, 1999, **22**, 319.

59. E. Micklander, B. Peshlov, P. P. Purslow and S. B. Engelsen, *Trends Food Sci. Technol.*, 2002, **13**, 341.

60. H. C. Bertram, S. B. Engelsen, H. Busk, A. H. Karlsson and H. J. Andersen, *Meat Sci.*, 2004, **66**, 437.

61. H. C. Bertram, A. K. Whittaker, W. R. Shorthose, H. J. Andersen and A. H. Karlsson, *Meat Sci.*, 2004, **66**, 301.

62. H. C. Bertram, Z. Wu, F. van den Berg and H. J. Andersen, *Meat Sci.*, 2006, **74**, 684.

63. C. Li, D. Liu, G. Zhou, X. Xu, J. Qi, P. Shi and T. Xia, *Meat Sci.*, 2012.

64. S. M. Shaarani, K. P. Nott and L. D. Hall, *Meat Sci.*, 2006, **72**, 398.

65. I. K. Straadt, M. Rasmussen, H. J. Andersen and H. C. Bertram, *Meat Sci.*, 2007, **75**, 687.

66. Z. Wu, H. C. Bertram, A. Kohler, U. Bucker, R. Ofstad and H. J. Andersen, *J. Agric. Food Chem.*, 2006, **54**, 8589.

67. Z. Wu, H. C. Bertram, U. Bucker, R. Ofstad and A. Kohler, *J. Agric. Food Chem.*, 2007, **55**, 3990.

68. T. Pérez-Palacios, T. Antequera, R. Molano, P. G. Rodríguez and R. Palacios, *J. Food Eng.*, 2010, **101**, 152.

69. E. Cernadas, P. Carrioen, P. G. Rodriguez, E. Muriel and T. Antequera, *Comput. Vis. Image Und.*, 2005, **98**, 345.

70. J. P. Renou, S. Benderbous, G. Bielicki, L. Foucat and J. P. Donnat, *Magn. Reson. Imag.*, 1994, **12**, 131.

71. T. M. Guiheneuf, S. J. Gibbs and L. D. Hall, *J. Food Eng.*, 1997, **31**, 457.

72. C. Vestergaard, J. Risum and J. Adler-Nissen, *Meat Sci.*, 2005, **69**, 663.

73. T. Nagata, Y. Chuda, X. Yan, M. Suzuki and K. I. Kawasaki, *J. Sci. Food Agric.*, 2000, **80**, 1151.

74. H. C. Bertram, S. J. Holdsworth, A. K. Whittaker and H. J. Andersen, *J. Agric. Food Chem.*, 2005, **53**, 7814.

75. P. S. Belton, K. J. Packer and T. E. Southon, *J. Sci. Food Agric.*, 1987, **40**, 283.

76. R. Li, W. L. Kerr, R. T. Toledo and Q. Teng, *J. Sci. Food Agric.*, 2001, **81**, 576.

77. E. Szłyk and P. Hrynczyszyn, *Talanta*, 2011, **84**, 199.

78. P. Hrynczyszyn, A. Jastrzebska and E. Szłyk, *Anal. Chim. Acta*, 2010, **673**, 73.

79. A. Jastrzebska and E. Szłyk, *Chem. Pap.*, 2009, **63**, 414.

Subject Index